MECHANICS OF COMPOSITE MATERIALS

Mechanics of Composite Materials

R. M. CHRISTENSEN
Lawrence Livermore Laboratory,
University of California, Livermore

A WILEY-INTERSCIENCE PUBLICATION

JOHN WILEY & SONS, New York · Chichester · Brisbane · Toronto

Library of Congress Cataloging in Publication Data:

Christensen, Richard M
Mechanics of composite materials.

"A Wiley-Interscience publication."
Includes bibliographical references and index.
1. Composite Materials. 2. Continuum mechanics.
I. Title.

TA418.C6C5 620.1'1 79-14093
ISBN 0-471-05167-5

Printed in the United States of America

10 9 8 7 6 5 4 3 2 1

PREFACE

The field of composite materials is growing at a rapid rate, whether judged by measures of theoretical research, laboratory development, or product application. As with any rapidly expanding field, the contributions that provide clarification and enlightenment are diverse and sometimes fragmentary. Here we propose to assemble, interpret, and interrelate many of these contributions in order to provide a comprehensive account of the basic theory of the mechanical behavior of heterogeneous media.

The term composite materials is often used with a connotation of fiber reinforced solids. However, we take a much broader view. Our interest is in the behavior and special effects associated with a wide class of heterogeneous media types, only one of which is fiber reinforcement. In fact we do not even limit ourselves to solid media behavior; we also consider some types of fluid suspensions. Of course fluid suspensions constitute a large, self-contained topic in themselves. We explore this subject only far enough to reveal the common methodology and insights that can be gained from an interaction of solid and fluid oriented disciplines.

Our concern here is with the continuum mechanics aspects of behavior. However, even at this level, it would be unthinkable to analyze all effects occurring in typical composite systems which contain enormously large numbers of discrete phases. Accordingly we usually invoke idealized geometric models of the heterogeneous system, and proceed to obtain the theoretical predictions of the macroscopic properties in terms of the properties of the individual constituent materials. This theme will occur throughout our developments. It is through a rationale of this type that we can best understand the special mechanical performance capability of heterogeneous material combinations. It is also in this area of idealized geometric models that the greatest confusion exists, with attendant claims and counter claims. We intend to provide a detailed and complete clarification of that topic.

We cover a wide range of subjects, as a view of the contents will reveal. Included are the macroscopic stiffness properties previously mentioned, failure characterization, and wave propagation. Much of the work presumes familiarity with the theory of linear elasticity. However many important aspects of mechanical behavior are neither linear, nor elastic; thus we also treat types of behavior characterized by viscoelasticity and inviscid plasticity theories. Furthermore, we consider some problems having nonlinear kinematics. Because of the close relationship between mechanical and thermal effects, we also examine macroscopic, thermal properties of heterogeneous media. Finally we should mention that the initial chapter is used to summarize many basic results from continuum mechanics, results that are required in the subsequent work. Areas related to composite material performance that we do not cover are those of fatigue, fracture, and moisture diffusion in polymeric composites. These rather specialized topics are in a very active state of development, with work currently being performed by many investigators, including myself. However I believe that most researchers would agree that these important topics are not yet ready for definitive treatment by any source.

Although the emphasis herein is on the development of the theory, critical attention is also given to the practical assessment of the results and to the means of application. Comparisons between different approaches and comparisons with reliable experimental data are given at some main junctures. Also a brief exposure is given to important aspects of design. It is our contention that theory and design can be beneficially interactive, and we wish to so demonstrate in the present context.

As already implied, this book is meant to be of use not only for research, but also for advanced reference in applications. Further, it is designed for use as a graduate level textbook. In fact, in preparing the book, I benefited greatly from teaching a graduate course on composite materials at the University of California, Berkeley. I am grateful to colleagues at U. C. Berkeley, but especially to P. M. Naghdi and C.-L. Tien for that opportunity.

I am appreciative to many colleagues at Lawrence Livermore Laboratory. The Composites and Mechanics Project and the Organic Materials Division, in which I work, have active and stimulating environments. Much of the initial impetus for the LLL project in composites has been contributed by E. M. Wu and T. T. Chiao. The management in the Chemistry and Materials Science Department has been very helpful, and I explicitly wish to thank J. K. Lepper, H. F. Rizzo, L. W. Roberts, and C.

F. Bender. I am grateful to the Army Research Office and particularly to E. A. Saibel for contract support of my work on composites. Finally, thanks are due Loris C.-H. Donahue of U. C. Berkeley for typing the manuscript, and Kristine Christensen for helping with the proofreading.

RICHARD CHRISTENSEN

Danville, California
August 1979

CONTENTS

I SOME ELEMENTS OF MECHANICS 1

1.1 Elasticity Theory Results 2

Boundary Value Problem Formulation, 3
Strain Energy, 4
Symmetry Properties, 5
Minimum Theorems, 7
Relationship to Viscous Fluid Theory, 9
Notation, 10

1.2 Viscoelasticity Theory 11

Stress-Strain Relations, 11
Viscoelastic Fluid Behavior, 15
Elastic-Viscoelastic Correspondence Principle, 17
Wave Propagation, 17

1.3 Plasticity Theory 18

Constitutive Assumptions, 19
Yield Functions, 21
Flow Rule, 21
Work Hardening, 22

1.4 Eshelby's Formula 23

References 30

II EFFECTIVE MODULI: SPHERICAL INCLUSIONS 31

2.1 The Concept of Equivalent Homogeneity 32

Volumetric Averaging, 34
Dilute Suspensions, 37
Energy Methods, 38

2.2 Dilute Suspension, Spherical Inclusions 41

Shear Modulus, 41
Bulk Modulus, 47

2.3 Composite Spheres Model 47

Bulk Modulus, 48
Shear Modulus, 51

2.4 A Three Phase Model 52

2.5 The Self-Consistent Scheme 59

2.6 A Concentrated Suspension Model 61

2.7 Some General Observations 67

Problems 70

References 71

III EFFECTIVE MODULI: CYLINDRICAL AND LAMELLAR SYSTEMS 73

3.1 Transversely Isotropic Media 74

Modulus Properties, 75
Compliance Properties, 78
Bounds on Poisson's Ratios, 79

3.2 Composite Cylinders Model 80

Uniaxial Modulus, 81
Other Properties, 83

3.3 A Model for the Transverse Shear of a Fiber System 84

3.4 Finite Length Fiber Effects 89

General Method, 90
Shear Properties, 93
Uniaxial Modulus and Other Properties, 96

3.5 A Dilute Suspension of Randomly Oriented Platelets 100

Problems 104

References 105

IV BOUNDS ON EFFECTIVE MODULI AND FURTHER RESULTS 106

4.1 Bounds on Isotropic Effective Moduli 107

Upper Bound Development, 109
Strain Field, 112
Explicit Bounds, 115

4.2 Bounds on Transversely Isotropic Effective Moduli 118

4.3 Interpretation of Effective Moduli and Associated Bounds 119

4.4 Some Isotropic Properties of Fiber Systems 124

Three-Dimensional Case, 124
Two-Dimensional Case, 129
Asymptotic Predictions, 131

4.5 Some Isotropic Properties of Platelet Systems 137

Two-Dimensional Case, 137
Three-Dimensional Case, 139
$cE_p \ll E_m$ Case, 143
$cE_p \gg E_m, cE_p \gg k_m$ Case, 143
Incompressible Case, 144

4.6 Summary of Effective Stiffness Properties and Conclusions 146

Problems 150

References 151

V LAMINATES 152

5.1 Transformation Relations 153

Tensor Transformations, 156
Quasi-Isotropic Case, 158

5.2 Classical Theory of Laminated Plates 160

Governing Relations, 161
Special Cases, 165

5.3 Cylindrical Bending **167**

Exact Solution, 167
Classical Solution, 170
Example, 171

5.4 High-Order Theory **174**

5.5 Interlaminar Edge Effects **179**

Problems **186**

References **187**

VI ANALYSIS, STRENGTH, AND DESIGN **188**

6.1 A Boundary Layer Theory of Anisotropic Elasticity **189**

Two-Dimensional Theory, 190
Concentrated Force Example, 193
Crack Example, 196

6.2 Strength and Failure Criteria **199**

Generalized Mises Criterion, 200
Tensor Polynomial Criterion, 200
Other Criteria and Considerations, 205

6.3 Design Example I: Fiber Reinforced Pressure Vessels **208**

Cylindrical Vessel, 210
Spherical Vessel, 212
Hybrid Systems, 213

6.4 Design Example II: Fiber Reinforced Flywheels **213**

Analysis, 214
Evaluation of Results, 219
Design Considerations, 219

Problems **222**

References **223**

VII WAVE PROPAGATION **225**

7.1 Wave Character in Equivalent Homogeneous Media **226**

7.2 Transmission and Reflection in Layered Media **228**

Wave Speeds, 228
Wave Reflections, 230

7.3 Dispersion Relations **232**

Wave Direction Normal to Layers, 232
Wave Direction Along Layers, 236
Typical Dispersion Curves, 241

7.4 Transient Wave Propagation: Layered Media **243**

General Solution, 243
Initial Value Problems, 248

7.5 Transient Wave Propagation: Three-Dimensionally Periodic Media **253**

7.6 Attenuation Due to Random Inhomogeneities **258**

Wave Scattering Effects, 258
Mixture Theory Formulation, 262
Wave Dispersion and Attenuation, 264

7.7 A Mixture Theory Application: Dynamic Instability in Fluidized Columns **265**

Problems **271**

References **271**

VIII INELASTIC AND NONLINEAR EFFECTS 274

8.1 Plastic Deformation of Porous Media **275**

8.2 Plastic Deformation of Fiber Systems **280**

Yield Function, 281
Hardening Rule, 284
Flow Rule, 285

8.3 Viscoelastic Properties of Composites **287**

Correspondence Principle Application, 288
Viscoelastic Bounds and Other Results, 290
Wave Behavior Example, 292

8.4 A Viscoelastic Fluid Suspension Model **295**

Small Deformation Characterization, 295
Non-Newtonian Flow Behavior, 298

8.5 **Large Deformation of Fiber Systems** 301

General Theory, 302
Deformation of a Cantilever Plate, 306

Problems 308

References 309

IX **EFFECTIVE THERMAL PROPERTIES** 311

9.1 **Thermoelasticity Theory** 311

Helmholtz Free Energy Derivation, 312
Gibbs Free Energy Derivation, 315

9.2 **Thermal Conductivity** 316

Spherical Inclusion Model, 317
Anisotropic Results, 319

9.3 **Thermal Expansion Coefficient** 321

9.4 **Specific Heats** 325

General Formulation, 326
Macroscopically Isotropic Case, 329
Transversely Isotropic Case, 333

9.5 **Discussion of Results** 336

Problems 337

References 338

Author Index 341

Subject Index 345

MECHANICS OF COMPOSITE MATERIALS

CHAPTER I
SOME ELEMENTS OF MECHANICS

We are concerned here with many different aspects of the mechanical behavior of heterogeneous media. A subject of this type can be approached from several different directions, and at various levels. For example, we could gather the very extensive data obtained from testing, collate it, interpret it, and, ultimately, manipulate it into a form with master curves in terms of nondimensional variables. Or we could go one step further and seek empirical analytical expressions that seem to model the data. We do not, however, take these approaches. Even though such approaches can be useful and convenient, the results are limited to the conditions under which data were obtained. The method has no power to predict behavior outside the range of laboratory experience. We seek a more fundamental approach, one that will give us a predictive capability.

In pursuing a general approach to the mechanics of heterogeneous media, we place a premium on this goal of predictive power. But a goal of this type is not obtained quickly or cheaply. The price to be paid is that of the expenditure of the time and effort to develop a rigorous theory(ies) of behavior. Carefully derived theories of behavior have as their basis certain assumptions or hypotheses that give the boundaries of applicability of the results. Within these boundaries the theory has a full and complete capability to model actual behavior.

Our objective then is to develop the theoretical framework for the behavior of heterogeneous media. It would be possible to approach the problem at a level of utmost generality, with completely nonlinear kinematics and the most general possible constitutive assumptions. Although this approach can be completely rigorous, it is too general for our interests in specific applications. We therefore pursue a middle level approach, one in which we do not hesitate to make assumptions and hypothesis that are in accordance with physical reality. However, after making these assumptions, we seek a rigorous mathematical structuring of the theory(ies) and its application. In fact, much of our work is based on mechanical behavior described by the linear theory of elasticity, which itself is a highly developed and rigorous theory.

Of the various aspects of the mechanics of deformable media (continuum mechanics), probably linear elasticity has had the most far-reaching impact. The tremendous success of linear elasticity theory can be attributed to several factors. First and foremost, it provides a realistic model of behavior for a wide class of materials. Second, the topic is highly developed and sophisticated; there is a vast library of methods and results to draw on. Third, in many practical problems the results take simple, but general forms that are amenable to design evaluation and application. For these reasons we develop the subject of the behavior of heterogeneous media primarily from the point of view of linear elasticity theory. We do not, however, wish to convey the impression that linear elasticity answers all the problems; it most certainly does not. The two most common generalizations of elasticity theory are those of viscoelastic behavior and elastic-plastic behavior. We are also concerned with these types of inelastic behavior for heterogeneous media. Furthermore, we also study some aspects of problems with completely nonlinear kinematics of deformation, to show that despite the inherent complications of nonlinearity, practical problems for heterogeneous media can still be approached at that level.

With our intentions to employ elasticity, viscoelasticity, and plasticity theories, we find it useful to review the elements of these three theories that will be of later use to us in constructing theories of behavior for heterogeneous media. We review these elements in this chapter. In addition, we present a fundamental result, due to Eshelby, that is of great value to us in our applications to heterogeneous media in the next few chapters.

1.1 ELASTICITY THEORY RESULTS

In this section we summarize many of the results from the theory of linear elasticity under homogeneous material conditions. There are many references on the subject; we here follow Sokolnikoff [1.1] as much as any single reference.

Boundary Value Problem Formulation

The most general anisotropic form of linear elastic stress-strain relations is given by

$$\sigma_{ij} = C_{ijkl}\varepsilon_{kl} \qquad i,j=1,2,3 \tag{1.1}$$

where σ_{ij} and ε_{ij} are the respective linear stress and strain tensors and C_{ijkl} is the fourth order tensor of elastic moduli, the stiffness tensor. We are employing rectangular Cartesian coordinates, with the usual Cartesian tensor notation, involving summation on repeated indices. The stress and strain tensors are required to be symmetric. The stiffness tensor has 81 independent components, as a fourth order tensor. However, the symmetry of σ_{ij} and ε_{ij} reduces the number of independent components to 36. As we note later, the existence of a strain energy function reduces the number of independent components still further. In this section, assuming homogeneity, then C_{ijkl} are independent of position. Our main objective in the subsequent work is to relax the restriction to homogeneity.

The infinitesimal strain tensor is defined in terms of displacement components as

$$\varepsilon_{ij} = \tfrac{1}{2}(u_{i,j} + u_{j,i}) \tag{1.2}$$

where the comma denotes partial differentiation with respect to the coordinate of the index following the comma. The six independent strain components are derived from three independent displacement components; therefore all of the strain components cannot be independent. This gives rise to the compatibility conditions on the strain components. The form of the compatibility equations given by Sokolnikoff [1.1] is

$$\varepsilon_{ij,kl} + \varepsilon_{kl,ij} = \varepsilon_{ik,jl} + \varepsilon_{jl,ik} \tag{1.3}$$

Of course, most of the 81 equations in (1.3) are not independent. Typically, equations (1.3) are written as a group of six equations; however only three of the equations are independent. The compatibility equations can be written in terms of stresses rather than strains, using the stress-strain relations.

The governing balance of momentum relations are given by

$$\sigma_{ij,j}(x_k,t) + F_i(x_k,t) = \rho\frac{\partial^2 u_i(x_k,t)}{\partial t^2} \tag{1.4}$$

where ρ is the mass density and F_i are the body force components. In the case where the inertia terms in (1.4) can be neglected, the problem is of a static nature and (1.4) comprises the equations of equilibrium.

When adjoined by the proper specification of boundary conditions and initial conditions, relations (1.1)–(1.4) comprise the complete and governing set of relations to be solved to obtain the distribution of the field variables in particular boundary value problems. Uniqueness of solution can be proved by many different methods. Sokolnikoff [1.1] gives a proof based on the existence of a strain energy function. We turn next to the relationship between stress, strain, and stored energy of deformation.

Strain Energy

The most satisfactory way to show the relationship between stress, strain, and energy is through a proper thermodynamical treatment. At this point we do not wish to take an extensive excursion into thermodynamics. The subject is taken up in Chapter IX in connection with nonisothermal states. For present purposes we merely note from the thermodynamical treatment of Section 9.1 that the stress can be expressed as the derivative of the strain energy W with respect to strain as

$$\sigma_{ij} = \frac{\partial W}{\partial \varepsilon_{ij}} \tag{1.5}$$

where

$$W = \tfrac{1}{2} C_{ijkl} \varepsilon_{ij} \varepsilon_{kl} \tag{1.6}$$

For relations (1.5) and (1.6) to give the stress-strain relation (1.1) it is necessary that C_{ijkl} have the symmetry

$$C_{ijkl} = C_{klij} \tag{1.7}$$

With the restriction (1.7) the number of independent components of C_{ijkl} is reduced to 21. Any further reduction in the number of independent components can only be made through restrictions imposed by the symmetry properties of the material. We consider next these types of restrictions.

Symmetry Properties

To represent C_{ijkl} in compact form, it is convenient to introduce a contracted notation. Let

$$
\begin{aligned}
\sigma_1 &= \sigma_{11} \\
\sigma_2 &= \sigma_{22} \\
\sigma_3 &= \sigma_{33} \\
\sigma_4 &= \sigma_{23} \\
\sigma_5 &= \sigma_{13} \\
\sigma_6 &= \sigma_{12}
\end{aligned} \tag{1.8}
$$

with a similar convention for strain. With this notation the stress-strain relations (1.1) with 21 independent components for C_{ijkl} can be written as

$$
\begin{bmatrix} \sigma_1 \\ \sigma_2 \\ \sigma_3 \\ \sigma_4 \\ \sigma_5 \\ \sigma_6 \end{bmatrix} =
\begin{bmatrix}
C_{11} & C_{12} & C_{13} & C_{14} & C_{15} & C_{16} \\
 & C_{22} & C_{23} & C_{24} & C_{25} & C_{26} \\
 & & C_{33} & C_{34} & C_{35} & C_{36} \\
 & & & C_{44} & C_{45} & C_{46} \\
 & & & & C_{55} & C_{56} \\
 & & & & & C_{66}
\end{bmatrix}
\begin{bmatrix} \varepsilon_1 \\ \varepsilon_2 \\ \varepsilon_3 \\ \varepsilon_4 \\ \varepsilon_5 \\ \varepsilon_6 \end{bmatrix} \tag{1.9}
$$

where the C_{ij} matrix is symmetrical, and in most compact form we write

$$
\sigma_i = C_{ij}\varepsilon_{ij} \qquad i,j = 1 \ldots 6 \tag{1.10}
$$

Following Green and Zerna [1.2], for symmetry with respect to a plane, C_{ij} has 13 independent components as

$$
C_{ij} = \begin{bmatrix}
C_{11} & C_{12} & C_{13} & 0 & 0 & C_{16} \\
 & C_{22} & C_{23} & 0 & 0 & C_{26} \\
 & & C_{33} & 0 & 0 & C_{36} \\
 & & & C_{44} & C_{45} & 0 \\
 & & & & C_{55} & 0 \\
 & & & & & C_{66}
\end{bmatrix} \tag{1.11}
$$

where coordinate x_3 is normal to the plane of symmetry.

Next we consider the case of symmetry with respect to three mutually orthogonal planes. This class is known as *orthotropy*, and there remain nine independent components of C_{ij}, as

$$C_{ij}=\begin{bmatrix} C_{11} & C_{12} & C_{13} & 0 & 0 & 0 \\ & C_{22} & C_{23} & 0 & 0 & 0 \\ & & C_{33} & 0 & 0 & 0 \\ & & & C_{44} & 0 & 0 \\ & & & & C_{55} & 0 \\ & & & & & C_{66} \end{bmatrix} \tag{1.12}$$

For a transversely isotropic material one of the planes for the orthotropic case is taken to be a plane of isotropy. Letting x_1 be normal to the plane of isotropy, we then have five independent components as

$$C_{ij}=\begin{bmatrix} C_{11} & C_{12} & C_{12} & 0 & 0 & 0 \\ & C_{22} & C_{23} & 0 & 0 & 0 \\ & & C_{22} & 0 & 0 & 0 \\ & & & \frac{1}{2}(C_{22}-C_{23}) & 0 & 0 \\ & & & & C_{66} & 0 \\ & & & & & C_{66} \end{bmatrix} \tag{1.13}$$

Finally, in the case of complete isotropy there are only two independent components of C_{ij}, and we have

$$C_{ij}=\begin{bmatrix} C_{11} & C_{12} & C_{12} & 0 & 0 & 0 \\ & C_{11} & C_{12} & 0 & 0 & 0 \\ & & C_{11} & 0 & 0 & 0 \\ & & & \frac{1}{2}(C_{11}-C_{12}) & 0 & 0 \\ & & & & \frac{1}{2}(C_{11}-C_{12}) & 0 \\ & & & & & \frac{1}{2}(C_{11}-C_{12}) \end{bmatrix} \tag{1.14}$$

The stress-strain relations (1.10) can be easily inverted to express strains in terms of stresses in general, or in any of the symmetry classes mentioned here. This result is given explicitly in Section 3.1 for the case of transversely isotropic media, which is of special relevance for fiber reinforced materials.

In the case of isotropy the stress-strain relations can be written as

$$\sigma_{ij} = \lambda \varepsilon_{kk} \delta_{ij} + 2\mu \varepsilon_{ij} \tag{1.15}$$

where λ and μ are the Lamé elastic constants and δ_{ij} is the Kronecker symbol. Alternatively, the stress-strain relations can be written compactly in terms of deviatoric and dilatational components. Let s_{ij} and e_{ij} be the deviatoric components of stress and strain, defined as

$$s_{ij} = \sigma_{ij} - \tfrac{1}{3}\delta_{ij}\sigma_{kk}$$

$$e_{ij} = \varepsilon_{ij} - \tfrac{1}{3}\delta_{ij}\varepsilon_{kk} \tag{1.16}$$

With (1.16) the stress-strain relations (1.15) take the form

$$s_{ij} = 2\mu e_{ij}$$

and

$$\sigma_{kk} = 3k\varepsilon_{kk} \tag{1.17}$$

where now we see constant μ as the shear modulus and constant k as the bulk modulus, which governs volumetric changes. Of course, only two of the three properties, λ, μ, and k, are independent, the relationship between these properties as well as those of other stress states, are given in Sokolnikoff [1.1]. We merely note the commonly used uniaxial modulus E and Poisson's ratio ν are related to μ and k through

$$E = \frac{9k\mu}{3k+\mu}$$

and

$$\nu = \frac{3k-2\mu}{2(3k+\mu)}$$

Minimum Theorems

There are two fundamental energy principles or theorems in linear elasticity theory. These are the minimum energy theorems, and they are of indispensable usefulness to us in our work with heterogeneous media. In physical terms, these energy theorems state that certain energy type functionals have a minimum value for the unique values of the field variables that comprise the solution of the boundary value problem, compared with the value of the functionals for other "admissible" values of the field variables. We now proceed to state these two minimum energy theorems: the theorem of minimum potential energy and the theorem of minimum complementary energy.

Consider a problem of static elasticity with body forces $F_i(x_k)$ and boundary conditions

$$\sigma_{ij} n_j = f_i \qquad \text{on } S_\sigma \tag{1.18}$$

$$u_i = U_i \qquad \text{on } S_u \tag{1.19}$$

where S_σ and S_u are complementary portions of the surface of the body of volume V, and n_j are the components of the unit outward normal to the surface. Next define the potential energy functional

$$U_\varepsilon = \int_V \left[W(\varepsilon_{ij}) - F_i u_i \right] dv - \int_{S_\sigma} f_i u_i \, ds \tag{1.20}$$

where $W(\varepsilon_{ij})$ is given by (1.6).

We define as admissible displacement fields, $\mathring{u}_i(x_j)$, any continuous displacement field that satisfies the displacement boundary condition (1.19), but is otherwise arbitrarily chosen (except for the usual regularity requirements on derivatives). The theorem of minimum potential energy can now be stated as:

Of all the admissible displacement fields, the one that satisfies the equations of equilibrium makes the potential energy functional (1.20) an absolute minimum.

Mathematically, this result is stated as

$$\mathring{U}_\varepsilon - U_\varepsilon \geqslant 0$$

where $\mathring{U}_\varepsilon$ is the functional (1.20) evaluated for any admissible displacement field, $\mathring{u}(x_i)$. The proof of this theorem, as given in many sources, is based on the positive definite character of the strain energy (1.6), as

$$W(\varepsilon_{ij}) \geqslant 0$$

The theorem of minimum complementary energy is the analog of that just given. Define the complementary energy functional

$$U_\sigma = \int_V W(\sigma_{ij})\,dv - \int_{S_u} \sigma_i U_i\,ds \tag{1.21}$$

where the strain energy (1.6) is expressed in terms of stresses, as

$$W = \tfrac{1}{2} S_{ijkl}\sigma_{ij}\sigma_{kl} \tag{1.22}$$

where S_{ijkl} is the tensor of elastic compliances. Define admissible stress states $\mathring{\sigma}_{ij}(x_i)$ as those stress states that satisfy the equilibrium equations and the stress boundary conditions (1.18) but are otherwise arbitrary. Again we assume continuity of the stresses and their derivatives sufficient to meet the needs of the proof of the theorem.

The theorem of minimum complementary energy can now be stated as:

Of all the admissible stress fields, the one that satisfies the compatibility equations makes the complementary energy functional (1.21) an absolute minimum.

The mathematical statement of this result is

$$\mathring{U}_\sigma - U_\sigma \geqslant 0$$

where $\mathring{U}_\sigma$ is the functional (1.21) evaluated for any admissible stress field. The key step in the proof is the positive definite character for (1.22).

Relationship to Viscous Fluid Theory

Until this point we have restricted attention to linear elastic behavior. There is a noteworthy connection between the governing equations for elasticity theory and those for the motion of viscous fluids. We note the relationship here, as it is of use to us in the study of certain fluid suspension problems. For this purpose we first state the equations of motion (1.4) in terms of displacements, using (1.1) and (1.2); it is found that

$$(\lambda + \mu) u_{k,ki} + \mu u_{i,jj} + F_i = \rho \frac{\partial^2 u_i}{\partial t^2} \tag{1.23}$$

In the case of an incompressible material we have the requirement

$$u_{k,k}=0 \tag{1.24}$$

and the incompressibility condition implies that the modulus $\lambda \to \infty$. The first term in (1.23) is indeterminate, and, accordingly, is written in terms of the reactive hydrostatic pressure p; thus

$$-p_{,i}+\mu u_{i,jj}+F_i=\rho\frac{\partial^2 u_i}{\partial t^2} \tag{1.25}$$

Now, for comparative purposes, we state the governing Navier-Stokes equations of motion for an incompressible Newtonian viscous fluid. From Batchelor [1.3] these equations are given by

$$-p_{,i}+\eta v_{i,jj}+\rho F_i=\rho\left(\frac{\partial^2 v_i}{\partial t^2}+v_j v_{i,j}\right) \tag{1.26}$$

where v_i is the velocity vector, η is the coefficient of viscosity, the body force F_i is expressed per unit mass, and again p is the reactive pressure. The equation of continuity takes the form

$$v_{k,k}=0 \tag{1.27}$$

We note that the equations (1.24) and (1.25) governing the displacement vector in the elastic solid have exactly the same form as the respective equations (1.27) and (1.26) governing the velocity vector in the viscous fluid. Not only do the equations have the same form, they have term-by-term equivalence with one exception. The material derivative in (1.26) introduces a nonlinear term $v_j v_{i,j}$ that does not have a counterpart in (1.25). However, under so-called creeping flow conditions, where the velocities in the fluid are very small relative to some norm, the nonlinear term, $v_j v_{i,j}$ in (1.26), is of higher order than the linear terms, and can be neglected. Thus under creeping flow conditions there is a direct analogy between the solutions in viscous flow problems and solutions in elasticity problems, simply by identifying viscous fluid velocity with elastic solid displacement. This analogy is of use to us in some of the future developments.

Notation

Finally, we finish this section with an observation on notation. For the most part we use the Cartesian tensor notation already employed. In some derivations, however, we use other forms of notation, particularly, a direct notation such that the stress-strain relations of (1.1) have the form

$$\boldsymbol{\sigma} = \mathbf{C}\boldsymbol{\varepsilon}$$

In direct notation of this type the symbols have direct interpretation as vectors and tensors. In mathematical terminology the **C** tensor, is actually a linear transformation over the appropriate vector space.

1.2 VISCOELASTICITY THEORY

Many materials, particularly polymers, exhibit a time and rate dependence that is completely absent in the constitutive relations of elasticity theory. Although these types of materials have a capability to respond instantaneously, as in elasticity, they also exhibit a delayed response. In recognition of these effects the materials are said to have a capacity for memory. Another characteristic of these materials is that of the combined capacity of an elastic type material to store energy with the capacity of a viscous type material to dissipate energy. Accordingly, such materials are said to be *viscoelastic*. The theory of viscoelastic materials is highly developed, and amenable to broad application. Gross [1.4] has given the most general treatment of the various forms viscoelastic stress-strain relations can take. Further aspects of the general theory of viscoelasticity have been treated by Christensen [1.5] and Pipkin [1.6].

Stress-Strain Relations

The most general form of the linear viscoelastic stress-strain relation is given by

$$\sigma_{ij}(t) = \int_{-\infty}^{t} C_{ijkl}(t-\tau)\frac{d\varepsilon_{kl}(\tau)}{d\tau}\,d\tau \tag{2.1}$$

where the tensor $C_{ijkl}(t)$ has components that are said to be the relaxation functions of the material. They are the basic properties of the material and as such are the counterparts of the elastic moduli. In fact, if the relaxation function tensor $C_{ijkl}(t)$ were independent of time, relations (2.1) could be integrated directly to give elasticity type relations. Relations (2.1) are written in direct notation for further purposes, as

$$\boldsymbol{\sigma}(t) = \int_{-\infty}^{t} \mathbf{C}(t-\tau)\dot{\boldsymbol{\varepsilon}}(\tau)\,d\tau \tag{2.2}$$

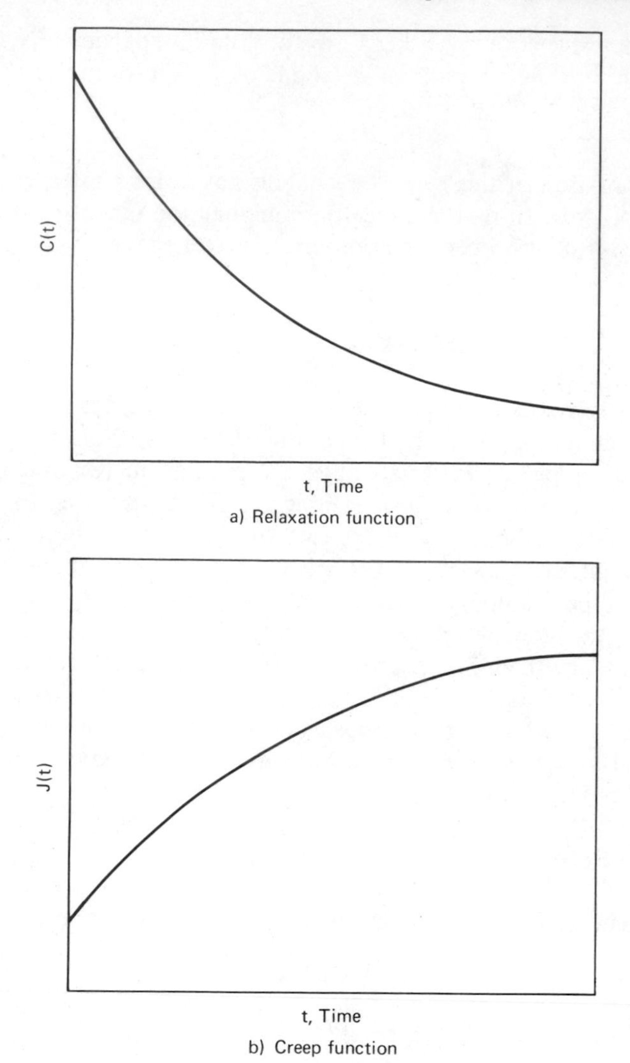

Fig. 1.1 Relaxation and creep functions.

Relation (2.2) can be written in an alternate form through integration by parts as

$$\boldsymbol{\sigma}(t)=\mathbf{C}(0)\boldsymbol{\varepsilon}(t)-\int_{-\infty}^{t}\frac{d\mathbf{C}(t-\tau)}{d\tau}\boldsymbol{\varepsilon}(\tau)d\tau \tag{2.3}$$

where $\boldsymbol{\varepsilon}(t)$ is taken as vanishing as $t\rightarrow-\infty$.

The relaxation functions $\mathbf{C}(t)$ are observed to be positive monotone decreasing functions of time, as in Fig. 1.1*a*. The restrictions imposed on $\mathbf{C}(t)$ by thermodynamics and a requirement of fading memory are given in [1.5].

An alternative form of the viscoelastic stress-strain relations can be obtained by expressing strain as a time functional of stress, through

$$\boldsymbol{\varepsilon}(t)=\int_{-\infty}^{t}\mathbf{J}(t-\tau)\dot{\boldsymbol{\sigma}}(\tau)\,d\tau \tag{2.4}$$

where $\mathbf{J}(t)$ is termed the tensor of creep functions. The creep functions are observed to be monotone increasing functions of time, Fig. 1.1*b*, which may or may not approach a time independent asymptote. These latter possibilities are discussed further. Obviously, the creep functions $\mathbf{J}(t)$ and the relaxation functions $\mathbf{C}(t)$ must have some type of reciprocal relationship. To establish this relationship it is convenient to employ the Laplace transform. Taking the Laplace transforms of (2.2) and (2.4), using the convolution theorem, gives

$$\bar{\boldsymbol{\sigma}}(s)=s\bar{\mathbf{C}}(s)\bar{\boldsymbol{\varepsilon}}(s)$$

and

$$\bar{\boldsymbol{\varepsilon}}(s)=s\bar{\mathbf{J}}(s)\bar{\boldsymbol{\sigma}}(s) \tag{2.5}$$

where s is the transform variable and $\bar{\boldsymbol{\varepsilon}}(s)$ denotes the transform of $\boldsymbol{\varepsilon}(t)$, and so on. From (2.5) it is found that

$$\bar{\mathbf{J}}(s)=\frac{1}{s^{2}\bar{\mathbf{C}}(s)} \tag{2.6}$$

Still a further form of the viscoelastic stress-strain relations is that involving the complex modulus. Specifically, let strain be a harmonic function of time, as in the real or imaginary parts of

$$\boldsymbol{\varepsilon}(t)=\boldsymbol{\varepsilon}_0 e^{i\omega t} \tag{2.7}$$

where ω is the frequency of oscillation. Substituting (2.7) into the stress-strain relation (2.2) gives

$$\boldsymbol{\sigma}(t)=\mathbf{C}^*(\omega)\boldsymbol{\varepsilon}_0 e^{i\omega t} \tag{2.8}$$

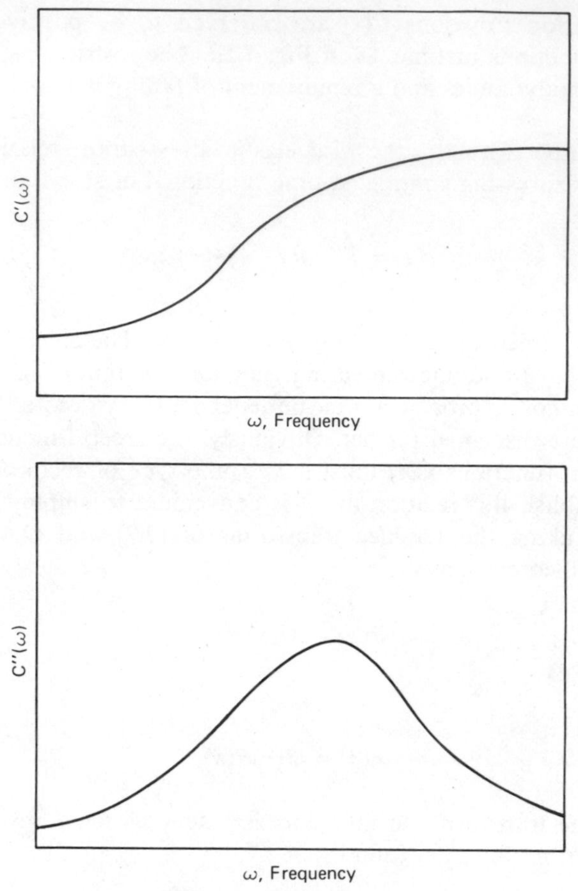

Fig. 1.2 Complex modulus.

where $\mathbf{C}^*(\omega)$ is the complex modulus:

$$\mathbf{C}^*(\omega)=\mathbf{C}'(\omega)+i\mathbf{C}''(\omega) \tag{2.9}$$

The real and imaginary parts of $\mathbf{C}^*(\omega)$ are given by

$$\mathbf{C}'(\omega)=\mathbf{C}_\infty+\omega\int_0^\infty \hat{\mathbf{C}}(s)\sin\omega s\,ds$$

$$\mathbf{C}''(\omega)=\omega\int_0^\infty \hat{\mathbf{C}}(s)\cos\omega s\,ds \tag{2.10}$$

where $\mathbf{C}(t)$ is decomposed as

$$\mathbf{C}(t)=\mathbf{C}_\infty+\hat{\mathbf{C}}(t)$$

such that $\mathbf{C}_\infty$ is the long time asymptote required by

$$\lim_{t\to\infty} \hat{\mathbf{C}}(t)=0$$

Typical behavior of the real and imaginary parts of the complex modulus are as shown in Fig. 1.2. The complex modulus can be used to write the stress-strain relations in Fourier transformed form, if desired. A complex compliance function $\mathbf{J}^*(\omega)$ can also be defined that is the tensor inverse of $\mathbf{C}^*(\omega)$.

The ratio of the imaginary and real parts in (2.9) is often noted as a property form. Designate a particular component of $\mathbf{C}(t)$ by $C(t)$, and correspondingly $\mathbf{C}^*(\omega)$ by $C^*(\omega)$. The loss tangent of $C^*(\omega)$ is defined by

$$\tan\phi=\frac{C''(\omega)}{C'(\omega)} \tag{2.11}$$

The angle ϕ has the interpretation as providing the phase angle by which the strain lags behind stress in steady state harmonic oscillation in viscoelastic materials.

Viscoelastic Fluid Behavior

At this point we give the characteristic of the viscoelastic properties that distinguishes solid behavior from fluid behavior. Since we are dealing with fluids, we restrict attention to isotropic materials and consider states of shear deformation with governing shear deformation properties

$\mu(t)$ shear relaxation function

$J(t)$ shear creep function

$\mu^*(\omega)$ shear complex modulus

A viscoelastic fluid has the capacity to undergo steady state shear flow. Referring to relation (2.4), the creep function is seen to give the strain response to a step function in stress. Under constant stress the strain increases in an unlimited manner, but the strain rate approaches a constant value, as the steady state condition is approached. Accordingly, the creep function in shear must have the form

$$J(t)=\hat{J}(t)+\frac{t}{\eta_{\text{eff}}} \tag{2.12}$$

where $\hat{J}(t)$ approaches an asymptote at $t\to\infty$. The term η_{eff} in (2.12) is the governing effective viscosity of the viscoelastic fluid under steady state conditions. Using the relaxation function form of the stress-strain relations (2.2), it can be shown that the viscosity η_{eff} is related to the relaxation function in shear by

$$\eta_{\text{eff}} = \int_0^\infty \mu(s)\,ds \tag{2.13}$$

The relations just obtained are meaningful only under conditions of a vanishing small rate of deformation. It must be emphasized that for a viscoelastic fluid there are full memory and rate effects. The material behaves as a viscous fluid with an effective viscosity given by (2.13) only in the case of steady state flow conditions.

There is one item of importance to us concerning fluids that is of use in studying the behavior of a fluid suspension. This matter concerns the form that the viscoelastic functions take in the limiting case when the material is a Newtonian viscous fluid. We have already seen how the viscoelastic stress-strain relations reduce very simply and directly to elastic material behavior. The comparable results for viscous fluids complete the delineation of the limiting case behavior possible with viscoelastic constitutive relations. For a Newtonian viscous fluid the stress-strain relations in shear are given by

$$s_{ij} = 2\eta d_{ij} \tag{2.14}$$

where s_{ij} is the deviatoric stress and the rate of deformation tensor d_{ij} is given in terms of velocity gradients by

$$d_{ij} = \tfrac{1}{2}(v_{i,j} + v_{j,i}) \tag{2.15}$$

Under infinitesimal strain rate conditions, (2.14) is equivalent to

$$s_{ij} = 2\eta \frac{d\varepsilon_{ij}}{dt} \tag{2.16}$$

From relation (2.2) we see that the relaxation function $\mu(t)$ must have the form

$$\mu(t) = \eta\,\delta(t) \tag{2.17}$$

for relation (2.2) to reduce to relation (2.16); the function $\delta(t)$ is the Dirac delta function. For a Newtonian viscous fluid the creep function in shear has the form of the last term in (2.12). Finally, the complex modulus

$\mu^*(\omega) = \mu'(\omega) + i\mu''(\omega)$ must have the form

$$\mu'(\omega) = 0$$
$$\mu''(\omega) = \omega\eta \tag{2.18}$$

These relations may be obtained by using the Newtonian viscous fluid relaxation function (2.17) in the forms (2.10).

Elastic-Viscoelastic Correspondence Principle

Another important item to be covered here is that of the elastic-viscoelastic correspondence principle. We note that the governing equations and conditions for elastic and for viscoelastic boundary value problems are of identical form except for the forms of the stress-strain relations in the two cases. However, even the viscoelastic stress-strain relations can be brought to the same form as the elastic forms, through the use of integral transforms, such as the Laplace or Fourier transform. Note that the Laplace transformed viscoelastic stress-strain relations (2.5) are of the same form as the corresponding elastic results if we identify $s\bar{\mathbf{C}}(s)$ with the elastic moduli $\mathbf{C}$. This identification has far-reaching consequences. If all the governing relations for an elastic boundary value problem are subjected to an integral transform, these relations are seen to be identical with those of the transformed viscoelastic problem if elastic moduli $\mathbf{C}$ are identified with $s\bar{\mathbf{C}}$. It now follows that solutions of static elastic problems can be converted to transformed solutions of the corresponding viscoelastic problems, simply by replacing elastic moduli $\mathbf{C}$ by $s\bar{\mathbf{C}}(s)$ and reinterpreting the elastic field variables in the solution as transformed viscoelastic field variables. The time domain viscoelastic solution follows, then, through a transform inversion. This process for obtaining viscoelastic solutions has great utility, and it has been widely applied. In the case of steady state harmonic conditions the process is even more simple. Static elastic solutions can be converted to steady state harmonic viscoelastic solutions simply by replacing elastic moduli $\mathbf{C}$ by the corresponding complex viscoelastic moduli $C^*(\omega)$, and reinterpreting elastic field variables as complex harmonic viscoelastic field variables. The correspondence principle obviously applies to heterogeneous as well as homogeneous material conditions.

Wave Propagation

As the final matter here we consider wave propagation in viscoelastic materials. These results aid us in later interpretations of wave behavior in heterogeneous media. A plane, time harmonic wave in an isotropic viscoelastic material has the displacement solution of the equation of motion

(1.4) given by

$$u(x,t) = Ae^{-\alpha x} \exp\left[i\omega\left(t - \frac{x}{c}\right)\right] \tag{2.19}$$

where ω is the given frequency, $c = c(\omega)$ is the wave speed, and $\alpha = \alpha(\omega)$ is the attenuation coefficient. The wave speed and attenuation coefficient are given by

$$c(\omega) = \frac{|C^*(\omega)|^{1/2}}{\rho^{1/2}} \sec \frac{\phi(\omega)}{2}$$

$$\alpha(\omega) = \frac{\omega}{c} \tan \frac{\phi(\omega)}{2} \tag{2.20}$$

where ϕ is the loss angle defined through the loss tangent (2.11). Constant A is the wave amplitude and C^* is given by $\lambda^* + 2\mu^*$ in the case of a longitudinal wave, μ^* in the case of a shear wave, and E^* in the case of propagation along a bar (under long wave length conditions). The distinguishing features of wave propagation in homogeneous viscoelastic media are the dispersion effects due to the dependence of phase velocity c on frequency ω and the wave attenuation effect, related to the loss tangent, which results from the conversion of mechanical energy to heat. Neither of these effects are present in homogeneous elastic media.

1.3 PLASTICITY THEORY

Plasticity theory models inelastic effects that are very different from the inelastic effects inherent in viscoelasticity. Viscoelasticity theory accounts for a specific rate dependence in material response, in addition to elastic type effects. Plasticity theory, on the other hand, models nonlinear material response beyond the elastic range, with no dependence given to the rates of loading. Plasticity theory (in the absence of inertia effects) then is static, with no time dependence.

The type of inviscid plasticity theory to be recalled here is often spoken of as being a *rate type theory*. This term may sound directly contradictory to the previously described characteristics of plasticity theory. Actually, it is not contradictory. The term *rate type theory* simply means that terms with the appearance of a rate are involved; however, any time dependence is explicitly suppressed. The entire scope of the three-dimensional theory of plasticity has as its objective the multidimensional generalization of the uniaxial form of the stress-strain relation, involving an elastic region,

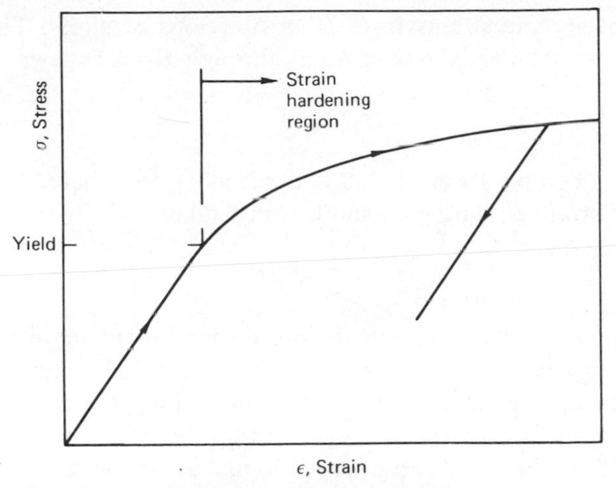

Fig. 1.3 Plasticity behavior.

limited by the yield point, outside of which there is a nonlinear range of behavior, but involving elastic unloading, as shown in Fig. 1.3. We now state the proper theoretical forms for this theory. References on the subject are the books by Prager and Hodge [1.7], Hill [1.8], Kachanov [1.9], and Mendelson [1.10]; and the review article by Naghdi [1.11].

Constitutive Assumptions

We first state the basic constitutive assumptions of the theory. There is an elastic range of behavior governed by a certain yield function, or loading function. We generally use the term *loading function*, reserving the term *yield function* to imply the initial form of the loading function, operative during the first cycle of loading from the underformed state. The general form of the loading function is taken as

$$f(\sigma_{kl}, \varepsilon''_{kl}) = \kappa \tag{3.1}$$

where strain is decomposed into elastic and plastic parts as

$$\varepsilon_{ij} = \varepsilon'_{ij} + \varepsilon''_{ij} \tag{3.2}$$

with ε''_{ij} being the plastic component of strain. The scalor function κ in (3.1) is a general function of the primary variables of the theory. The manner in which the loading function (3.1) is used to specify elastic or plastic deformation is shown shortly.

Next we specify the strains in (3.2) as functions of stress. The elastic component ε'_{ij} we take to be related to σ_{ij} through Hooke's law,

$$\sigma_{ij} = C_{ijkl}\varepsilon'_{kl}$$

Typically, the isotropic form of this is used, as is done here. The plastic component of strain is, during plastic loading, taken as

$$\dot{\varepsilon}''_{ij} = F_{ij}(\sigma_{kl}, \dot{\sigma}_{kl}, \varepsilon''_{kl}) \tag{3.3}$$

where time rates are employed to denote an increment of plastic strain obtained during an increment of time. Now for the actual, physical behavior to be independent of rates of deformation, it is necessary that F_{ij} ( ) in (3.3) be homogeneous of degree one in the stress rate $\dot{\sigma}_{ij}$. Rather than writing (3.3) in terms of rates, we could just as well write it in terms of increments $d\varepsilon''_{ij}$ and $d\sigma_{kl}$. We are now at the point at which we can state the constitutive assumptions of the theory of inviscid plasticity. We state these forms relative to three states of deformation: loading, unloading, and neutral loading.

The entire deformation is elastic until the yield surface has first been reached. Subsequent loading changes the size and shape of the loading function (3.1). Relative to an arbitrary point in stress space, a state of loading is specified by

$$\text{(loading)} \qquad f = \kappa, \qquad \dot{\kappa} \neq 0, \qquad \text{and} \qquad \frac{\partial f}{\partial \sigma_{ij}} \dot{\sigma}_{ij} > 0$$

and under this condition $\dot{\varepsilon}''_{ij}$ is given by (3.3).

During neutral loading

$$\text{(neutral loading)} \qquad f = \kappa, \qquad \dot{\kappa} = 0, \qquad \text{and} \qquad \frac{\partial f}{\partial \sigma_{ij}} \dot{\sigma}_{ij} = 0$$

and we have $\dot{\varepsilon}''_{ij} = 0$.

Finally, during unloading we have

$$\text{(unloading)} \qquad f = \kappa, \qquad \dot{\kappa} = 0, \qquad \text{and} \qquad \frac{\partial f}{\partial \sigma_{ij}} \dot{\sigma}_{ij} < 0$$

and again $\dot{\varepsilon}''_{ij} = 0$.

With explicit forms for $f(\ )$ and $F_{ij}(\ )$ the constitutive forms with rates must be integrated to obtain stress and deformation histories. The central problem then is the specification of the constitutive functions $f(\ )$ and $F_{ij}(\ )$, as well as κ.

Yield Functions

First we consider the initial yield function. The two most common yield functions are those of the Mises and Tresca forms. The Tresca yield function admits yielding when the maximum shear stress reaches a specified value, as

$$\sigma_1 - \sigma_3 = 2k \tag{3.4}$$

where k is the yield stress in simple shear. To avoid the use of labels, to specify the maximum shear stress the Tresca condition can be written in general form as

$$4J_2^3 - 27J_3^2 - 36k^2J_2^2 + 96k^4J_2 - 64k^6 = 0 \tag{3.5}$$

where J_2 and J_3 are the invariants of the deviatoric stress tensor, defined in Section 1.1. These invariants are given by

$$J_2 = \tfrac{1}{2}s_{ij}s_{ij}$$

$$J_3 = |s_{ij}|$$

The Mises criteria is more simple than that of Tresca, being stated simply as

$$J_2 = \tfrac{1}{2}s_{ij}s_{ij} = k^2 \tag{3.6}$$

where again k is the yield stress in simple shear. The Mises criteria is interpreted so that the strain energy associated with shear deformation has a limiting value.

For an elastic-perfectly plastic material, all that are needed are the yield functions, as previously given, and the entire theory of deformation can be formulated directly. The absence of work hardening makes this case particularly simple. The governing equations, known as the *Prandtl-Reuss relations*, are given many places, as for example, in [1.7].

Flow Rule

For a work hardening form of the stress-strain relations, equation (3.3) is taken in the particular form

$$\dot{\varepsilon}_{ij}'' = \Lambda \frac{\partial f}{\partial \sigma_{ij}} \tag{3.7}$$

where

$$\Lambda = \Lambda(\sigma_{kl}, \dot{\sigma}_{kl}, \varepsilon''_{kl})$$

and Λ must be homogeneous of degree one in $\dot{\sigma}_{kl}$. Relation (3.7) is known as the flow rule. It is an expression of the condition that the plastic strain rate vector be normal to the yield surface. This latter result follows from Drucker's postulate, which is a basic work inequality. It also follows that the loading surface must be convex. Before we consider the determination of the parameter Λ in (3.7), we discuss the two most common types of work hardening.

Work Hardening

After initial yielding, the subsequent loading surfaces are said to be harder because of the plastic deformation. The hardening rule specifies the manner in which the loading surface changes size and shape during deformation. The simplest type of hardening is that of so-called isotropic hardening. In this case the loading function (3.1) is taken in the special form

$$f(J_2, J_3) = \kappa \tag{3.8}$$

where the entire hardening effect occurs solely through the parameter κ. Under this condition there can be no anisotropy of effects during the plastic deformation. Isotropic hardening does not exhibit a Bauschinger effect, and it has not been found to provide a widely useful material model.

The other common type of work hardening is that of kinematic hardening. In the case of kinematic work hardening the loading function (3.1) is taken in the special form

$$f(\sigma_{ij} - \alpha_{ij}) = k^2 \tag{3.9}$$

where k is a constant and α_{ij} are the hardening parameters, which necessarily depend on the basic variables of the problem. It is seen that the form (3.9) represents a loading surface that translates in stress space, without change in shape. There are two common means of specifying the hardening parameters, α_{ij}, namely, Prager's rule, and Zeigler's rule. In the former, α_{ij} is specified through

$$\dot{\alpha}_{ij} = c\dot{\varepsilon}''_{ij} \tag{3.10}$$

where c is a scalor to be determined, whereas in Zeigler's rule

$$\dot{\alpha}_{ij}=\dot{\mu}(\sigma_{ij}-\alpha_{ij}) \tag{3.11}$$

The latter form has been shown to be invariant with respect to a reduction in the dimensions of the problem. That is, the form (3.11) applies in all stress subspaces, whereas (3.10) does not. We employ the form (3.11) in our later work. It can be shown that $\dot{\mu}$ in (3.11) can be evaluated as

$$\dot{\mu}=\frac{(\partial f/\partial\sigma_{ij})\dot{\sigma}_{ij}}{(\partial f/\partial\sigma_{k1})(\sigma_{k1}-\alpha_{k1})} \tag{3.12}$$

Returning to the flow rule for kinematic hardening, we write (3.9) as

$$f(\sigma_{ij},\varepsilon''_{ij})=k^2 \tag{3.13}$$

Using this form with the flow rule (3.7) we can evaluate Λ as

$$\Lambda=\frac{-(\partial f/\partial\sigma_{ij})\dot{\sigma}_{ij}}{(\partial f/\partial\sigma_{ij})\dfrac{\partial f}{\partial\varepsilon''_{ij}}} \tag{3.14}$$

The preceding forms give a complete specification of the stress-strain relations for the theory of inviscid plasticity. Of course, more general types of work hardening can be formulated. The remaining conditions to form the complete set of field equations—strain displacement relations, compatibility equations, and the equations of motion—are identical with those of linear elasticity. The stress-strain relations of plasticity theory are inherently nonlinear; thus the complete theory is nonlinear. Despite this complication, many important problems have been solved, and we employ the general theory to solve two important composite material plastic deformation problems in Chapter VIII.

1.4 ESHELBY'S FORMULA

There is a basic result in elasticity theory that is of great usefulness to us in heterogeneous material analysis. This result is that due to Eshelby [1.12] concerning the calculation of strain energy in systems containing inhomo-

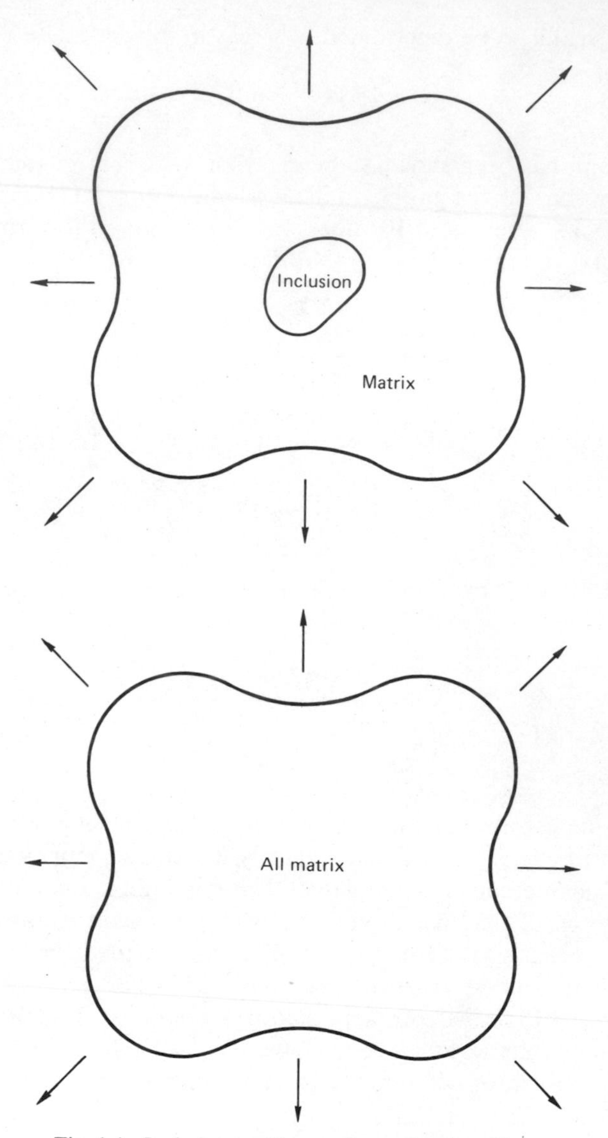

Fig. 1.4 Inclusion problem and associated problem.

geneities. Because this result is not available in textbooks on elasticity, it is derived here, in the form suitable for our purposes.

The formula derived by Eshelby reduces the usual volume integrations for calculating strain energy to a particular type of surface integration. This simplification is found to be of great advantage. Consider a homogeneous media, subjected to surface tractions, in which there exists a single inclusion of another material, as shown schematically in Fig. 1.4*a*. The elastic strain energy in the heterogeneous body is defined by

$$U=\frac{1}{2}\int_V \sigma_{ij}\varepsilon_{ij}\,dv \tag{4.1}$$

where V is the volume of the region. Next we define the strain energy in a body identical with that just specified, but with the inclusion replaced by homogeneous matrix material. Further, take this second problem to be subjected to the same set of surface tractions as in the preceding problem. The problem is as shown in Fig. 1.4*b*, and the strain energy is given by

$$U_0=\frac{1}{2}\int_V \sigma^0_{ij}\varepsilon^0_{ij}\,dv \tag{4.2}$$

where now the field variables must be distinguished as

$$\left.\begin{matrix}\sigma_{ij}\\ \varepsilon_{ij}\\ u_i\end{matrix}\right\}\qquad \text{Problem Fig. 1.4}a$$

$$\left.\begin{matrix}\sigma^0_{ij}\\ \varepsilon^0_{ij}\\ u^0_i\end{matrix}\right\}\qquad \text{Problem Fig. 1.4}b$$

With (4.2), (4.1) now can be written as

$$U=U_0+\frac{1}{2}\int_V\left(\sigma_{ij}\varepsilon_{ij}-\sigma^0_{ij}\varepsilon^0_{ij}\right)dv \tag{4.3}$$

Applying the divergence theorem, along with $\sigma_{ij,j}=0$ and $\sigma^0_{ij,j}=0$ gives (4.3) as

$$U=U_0+\frac{1}{2}\int_S\left(\sigma_i u_i-\sigma^0_i u^0_i\right)ds \tag{4.4}$$

where S is the surface of the body. On the surface

$$\sigma_i=\sigma^0_i \qquad \text{on } S \tag{4.5}$$

since both problems are subjected to the same surface tractions. Using (4.5) in (4.4) gives

$$U=U_0+\frac{1}{2}\int_S \sigma^0_i\left(u_i-u^0_i\right)ds \tag{4.6}$$

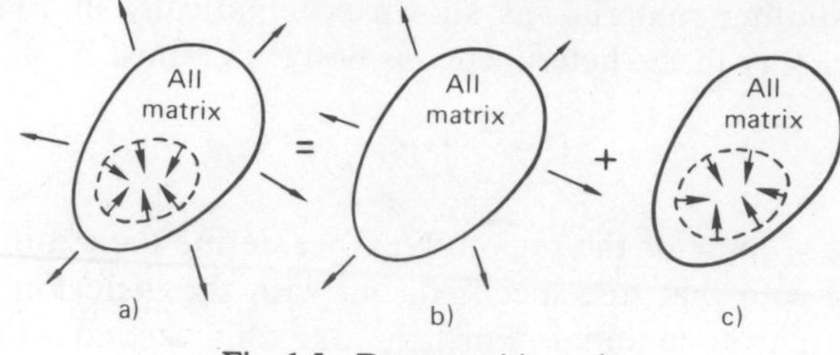

Fig. 1.5 Decomposition scheme.

Relation (4.6) is an intermediate form of the result we are seeking. Specifically, we wish to convert the integral term in (4.6) to another form. To this end we now consider an auxiliary problem.

Consider the problem of a body of the same shape as that in Fig. 1.4*a*, but composed entirely of matrix material, with the effect of the inclusion accounted for by a particular distribution of body forces. This problem is shown schematically in Fig. 1.5*a*. It is easy to show that the problem of Fig. 1.5*a*, under the action of a particular set of body forces taken over the surface of the inclusion of the problem in Fig. 1.4*a*, produces exactly the same state of field variables in the region outside that of the inclusion. Next the problem in Fig. 1.5*a* is decomposed into the two problems shown in Figs. 1.5*b* and *c*. Designate the field variables in the problems of Fig. 1.5 by

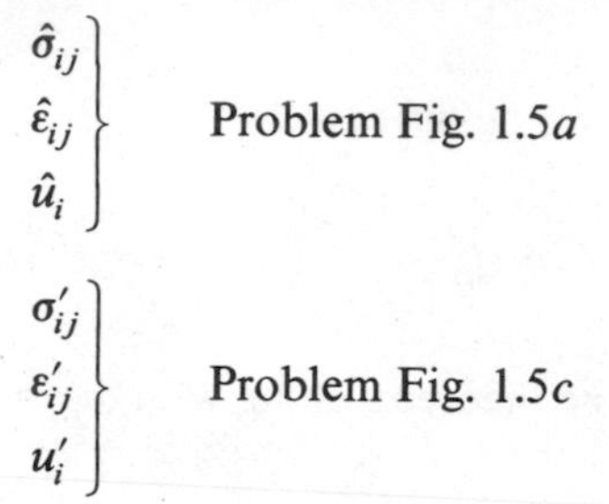

$$\left.\begin{matrix}\hat{\sigma}_{ij}\\ \hat{\varepsilon}_{ij}\\ \hat{u}_i\end{matrix}\right\} \quad \text{Problem Fig. 1.5}a$$

$$\left.\begin{matrix}\sigma'_{ij}\\ \varepsilon'_{ij}\\ u'_i\end{matrix}\right\} \quad \text{Problem Fig. 1.5}c$$

Of course, the problem in Fig. 1.5*b* is the same as that in Fig. 1.4*b*. Later use is made of the fact that $\hat{\sigma}_{ij}$, $\hat{\varepsilon}_{ij}$, $\hat{u}_i$ are identical with σ_{ij}, ε_{ij}, u_i, respectively, outside the region enclosed by the body forces, that is, outside the region of the inclusion.

With the decomposition shown in Fig. 1.5 we can now write

$$\hat{\sigma}_{ij} = \sigma^0_{ij} + \sigma'_{ij}$$

$$\hat{\varepsilon}_{ij} = \varepsilon^0_{ij} + \varepsilon'_{ij}$$

$$\hat{u}_i = u^0_i + u'_i \tag{4.7}$$

With (4.7), the elastic energy in the problem of Fig. 1.5*a* can be written as

$$\hat{U} = \frac{1}{2}\int_V \left(\sigma_{ij}^0 + \sigma_{ij}'\right)\left(\varepsilon_{ij}^0 + \varepsilon_{ij}'\right) dv \tag{4.8}$$

Expanding the terms in (4.8) gives

$$\hat{U} = U_0 + U' + U_{\text{INT}} \tag{4.9}$$

where U_0 is given by (4.2) and

$$U' = \frac{1}{2}\int_V \sigma_{ij}'\varepsilon_{ij}'\, dv \tag{4.10}$$

and

$$U_{\text{INT}} = \frac{1}{2}\int \left(\sigma_{ij}^0\varepsilon_{ij}' + \sigma_{ij}'\varepsilon_{ij}^0\right) dv \tag{4.11}$$

The term U_{INT} is just the interaction energy effect of the two stress states shown in Fig. 1.5*b* and *c*.

The interaction energy (4.11) can be put into a more convenient form. Consider the second term in the integrand of (4.11) and use the stress-strain relations to write it as

$$\sigma_{ij}'\varepsilon_{ij}^0 = C_{ijkl}\varepsilon_{kl}'\varepsilon_{ij}^0 \tag{4.12}$$

Using the symmetry, $C_{ijkl} = C_{klij}$, (4.12) becomes

$$\sigma_{ij}'\varepsilon_{ij}^0 = C_{ijkl}\varepsilon_{kl}^0\varepsilon_{ij}' \tag{4.13}$$

But, the right-hand side of (4.13) is just $\sigma_{ij}^0\varepsilon_{ij}'$; thus

$$\sigma_{ij}'\varepsilon_{ij}^0 = \sigma_{ij}^0\varepsilon_{ij}' \tag{4.14}$$

Using (4.14) in (4.11) gives

$$U_{\text{INT}} = \int_V \sigma_{ij}^0\varepsilon_{ij}'\, dv \tag{4.15}$$

Now use the divergence theorem and $\sigma_{ij,j}^0 = 0$ to put (4.15) into the form

$$U_{\text{INT}} = \int_S \sigma_i^0 u_i'\, ds \tag{4.16}$$

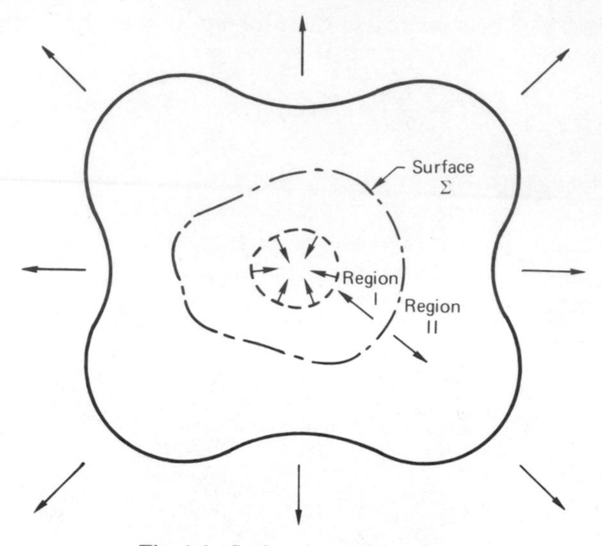

Fig. 1.6 Surface integral location.

This form of U_{INT} relates to the integral term in (4.6). To show this, use the last of (4.7) to write (4.6) as

$$U = U_0 + \frac{1}{2}\int_S \sigma_i^0 u_i' \, ds \tag{4.17}$$

where $u_i = \hat{u}_i$ on S has been used. Comparing the integral term in (4.17) with the form in (4.16) we see we can write

$$U = U_0 + \tfrac{1}{2} U_{\mathrm{INT}} \tag{4.18}$$

At this point we finally develop the desired form for U_{INT}, for use in (4.18). Returning to the form (4.15) for U_{INT}, decompose the region of volume integration into two parts as shown in Fig. 1.6, with the surface of separation Σ. This surface Σ is taken outside the region of the body forces. With this decomposition, (4.15) has the form

$$U_{\mathrm{INT}} = \int_{V_{\mathrm{I}}} \sigma_{ij}^0 \varepsilon_{ij}' \, dv + \int_{V_{\mathrm{II}}} \sigma_{ij}^0 \varepsilon_{ij}' \, dv \tag{4.19}$$

Using (4.14) in the last term of (4.19) then gives

$$U_{\mathrm{INT}} = \int_{V_{\mathrm{I}}} \sigma_{ij}^0 \varepsilon_{ij}' \, dv + \int_{V_{\mathrm{II}}} \sigma_{ij}' \varepsilon_{ij}^0 \, dv \tag{4.20}$$

Relation (4.20) can be written as

$$U_{\mathrm{INT}}=\int_{\Sigma}\sigma_i^0 u_i'\,ds-\int_{\Sigma}\sigma_i' u_i^0\,ds+\int_{S}\sigma_i' u_i^0\,ds \tag{4.21}$$

where the divergence theorem again has been used along with $\sigma_{ij,j}^0=0$ in region V_{I} and $\sigma_{ij,j}'=0$ in region V_{II}. The minus sign enters (4.21) since the positive direction of the unit vector normal to Σ is taken to be outward.

Recalling that $\sigma_i'=0$ on S, (4.21) reduces to

$$U_{\mathrm{INT}}=\int_{\Sigma}(\sigma_i^0 u_i'-\sigma_i' u_i^0)\,ds \tag{4.22}$$

Using relations (4.7), (4.22) takes the form

$$U_{\mathrm{INT}}=\int_{\Sigma}(\sigma_i^0 \hat{u}_i-\hat{\sigma}_i u_i^0)\,ds \tag{4.23}$$

Now recall that in the basic problem of Fig. 1.5a the variables $\hat{\sigma}_i$ and $\hat{u}_i$ are identical with the respective variables σ_i and u_i outside the region enclosed by the body forces; thus (4.23) becomes

$$U_{\mathrm{INT}}=\int_{\Sigma}(\sigma_i^0 u_i-\sigma_i u_i^0)\,ds \tag{4.24}$$

Finally, substituting (4.24) into (4.18) gives the final form

$$U=U_0+\frac{1}{2}\int_{S_i}(\sigma_i^0 u_i-\sigma_i u_i^0)\,ds \tag{4.25}$$

where Σ has been taken over the surface of the inclusion.

Formula (4.25) has been developed for the heterogeneous problem involving applied tractions. It can be shown that the corresponding result in the problem involving specified displacements on the outer boundary is

$$U=U_0+\frac{1}{2}\int_{S_i}(\sigma_i u_i^0-\sigma_i^0 u_i)\,ds \tag{4.26}$$

These results readily can be generalized to the case of multiple inclusions.

These formulas show a remarkable simplicity. The basic form for the total strain energy involves complicated quadratic form integrations over the volumetric region. However, these results of Eshelby show that the strain energy can be calculated by performing the simple surface integrations shown in (4.25) and (4.26). This result is of great utility to us in application to heterogeneous media behavior.

REFERENCES

1.1 I. S. Sokolnikoff, *Mathematical Theory of Elasticity*, McGraw-Hill, New York, 1956.

1.2 A. E. Green and W. Zerna, *Theoretical Elasticity*, 2nd ed., Oxford University Press, New York, 1968.

1.3 G. K. Batchelor, *An Introduction to Fluid Dynamics*, Cambridge University Press, New York, 1967.

1.4 B. Gross, *Mathematical Structure of the Theories of Viscoelasticity*, Hermann, Paris, 1953.

1.5 R. M. Christensen, *Theory of Viscoelasticity*, Academic, New York, 1971.

1.6 A. C. Pipkin, *Lectures on Viscoelasticity Theory*, Springer-Verlag, New York, 1972.

1.7 W. Prager and P. G. Hodge, Jr., *Theory of Perfectly Plastic Solids*, Wiley, New York, 1961.

1.8 R. Hill, *The Mathematical Theory of Plasticity*, Oxford University Press, New York, 1960.

1.9 L. M. Kachanov, *Foundations of the Theory of Plasticity*, North-Holland, Amsterdam, 1971.

1.10 A. Mendelson, *Plasticity: Theory and Application*, Macmillan, London, 1968.

1.11 P. M. Naghdi, "Stress-strain relations in plasticity and thermoplasticity," in *Symposium on Naval Structural Mechanics*, Pergamon, Elmsford, N. Y., 1960.

1.12 J. D. Eshelby, "The continuum theory of lattice defects," in *Progress in Solid State Physics*, Vol. 3, F. Seitz and D. Turnbull, Eds., Academic, New York, 1956, p. 79.

CHAPTER II
EFFECTIVE MODULI: SPHERICAL INCLUSIONS

Materials come in a variety of forms with a very wide range of properties. Rarely, however, does a single material suffice to provide the perfect balance of properties required for a particular application. As a practical expedient, it has been found, mixtures of materials often provide an advantageous blend of properties. It is perhaps inevitable that engineering science would be called on to explain accidental discoveries and empirical results. More importantly, engineering science is being utilized to gain a fundamental understanding of materials behavior in heterogeneous systems, so that further improvements can be achieved. Ultimately, of course, only a rigorous discipline of heterogeneous material behavior can provide the key to optimizing material utilization.

Two examples will help provide the motivation and thrust for the further developments. Many polymers in homogeneous form are glassy and brittle. It was found by trial and error, in a chemical laboratory, that a uniform dispersion of rubber spheres can greatly improve the impact sensitivity of the material. A significant proportion of the glassy polymers now manufactured contain the rubber "toughening" agent, as a means of upgrading the material behavior. The second example is that of fiber reinforced composite materials. Some fibers are composed of materials that are not even available in bulk form, or, if available, the fiber form has completely different and improved properties from those of the bulk form. In either case it has been found to be of great advantage to retain the material in fiber form, and bind the fibers together with a compliant matrix phase. In composite form the system still retains many of the properties associated with the unique fiber form of material. These examples are typical of the kinds of systems that may be considered in heterogeneous material combinations. It must be kept in mind that combinations of materials that enhance a particular property often involve the degradation of another property. Therefore, all relevant properties must be considered and trade-offs must often be made.

In this chapter we begin our detailed study of the behavior of heterogeneous material systems. As in any beginning it is necessary to focus on a particular aspect of the field. Accordingly, it is the stiffness, or effective modulus, aspect of the theory that is explored in this chapter and the following two chapters. Stiffness and strength are by far the most important of all the engineering properties. Indeed, the mathematical characterization of the stiffness property is the basis of the methods of analysis that ultimately lead to design using engineering materials. The developments in the present chapter assume conditions of linear elastic behavior, an assumption that provides the backbone of modern methods of engineering design.

A variety of different types of heterogeneous media are considered in the next three chapters. Primary emphasis is on the cases of spherical, cylindrical, and lamellar geometric forms of inclusions embedded in a continuous matrix phase. Very different types of reinforcement are found to result from these forms. The cylindrical and lamellar type problems are deferred until the next chapter; only spherical inclusions are treated in this chapter.

We focus here on determining the effective stiffness properties for these types of heterogeneous media. By effective stiffness properties, we mean an average measure of the stiffness of the material, taking into account the properties of all phases of the heterogeneous media and their interaction. The actual averaging process to be used to obtain these properties must be specified very carefully. This will be done in the next section. In some cases it is possible to obtain exact solutions for the effective moduli. When this is not possible, it is at least shown to be possible to obtain bounds on the effective moduli.

2.1 THE CONCEPT OF EQUIVALENT HOMOGENEITY

On some sufficiently small scale all materials are heterogeneous. One need only start at the scale of atoms and molecules to be convinced of the high degree of disorder and variability that can occur. If engineering materials were to be designed at this level of observation, the task would be insurmountable. To overcome this difficulty the continuum hypothesis is introduced. This hypothesis involves a statistical averaging process whereby the actual constitution and structure of the material is idealized as one in which the material is taken as a continuum. This continuum hypothesis involves the existence of certain measures associated with properties that govern the deformability of the media. These properties reflect averages of necessarily very complicated interactions on the molecular or atomic scale. Once the continuum model is taken to exist, the concept of homogeneity is of relevance. For a homogeneous medium the inherent properties that characterize it are taken to be the same at all points of the medium. Next we consider various types of inhomogeneous media.

Conditions of heterogeneity can occur either as an idealized continuous variation of properties with position, or as an abrupt change in properties across interfaces. In this work we are concerned exclusively with the latter case, involving the combination of various phases that remain distinct. The type of heterogeneity involving discrete phases is of by far more technological importance than the naturally occurring heterogeneity involving a continuous change in properties. Conditions of homogeneity and isotropy are assumed here to prevail within individual phases. To proceed further it is necessary to say something about the scale of the inhomogeneity.

We assume the existence of a characteristic dimension of the inhomogeneity. For example, in a fiber system this could be the mean distance between fibers. Of course, this characteristic dimension is a gross idealization of a necessarily statistical description of an actual heterogeneous system. Now, quite obviously there is a length scale over which the properties can be averaged in some meaningful way. The length scale of averaging, say δ, must be of a dimension much larger than that of the characteristic dimension of the inhomogeneity. A very advantageous circumstance occurs if there exists such a length scale of averaging, δ, that is still small compared with the characteristic dimension of the body. Under this condition the material can be idealized as being effectively homogeneous, and the problem of the load bearing body can be solved using the average properties associated with length scale δ.

In all further considerations we assume that the scale of properties averaging exists and is meaningful. That is, the scale of the inhomogeneity is assumed to be orders of magnitude smaller than the characteristic dimension of the problem of interest, such that there exists the intermediate dimension over which the properties averaging can be legitimately performed. The condition just described is said to be that of *effective* or *equivalent homogeneity*. Other terms in common usage with the same implication are those of *macroscopic homogeneity* and *statistical homogeneity*. The alternative to this proposition of equivalent homogeneity would be to account explicity for each and every separate region of homogeneity, with analyses providing continuity of stress and displacement across all interfaces. This would be an unthinkable task in typical composite systems with thousands or millions of separate particle or fiber regions.

With the admissibility of equivalent homogeneity, the fundamental problem in heterogeneous material behavior can now be posed. The basic problem is to utilize the averaging process to predict the effective properties of the idealized homogeneous medium in terms of the properties of the individual phases and some information on the interfacial geometry. The resulting effective properties are those that would be employed in the analysis of a load bearing body composed of the composite material. The relationship between the effective properties and the individual phase properties plus the analysis of the structural problem of interest then provides the means of optimizing structural performance by varying individual phase properties or characteristics.

Volumetric Averaging

The averaging process must be specified at this point. We introduce a volume element of the heterogeneous material having a characteristic dimension identical with that of the averaging dimension δ, defined earlier, and shown schematically in Fig. 2.1. Let this volume element be referred to as the representative volume element and let V designate its volume. Under conditions of an imposed macroscopically homogeneous stress or deformation field on the representative volume element, the average stress is defined by

$$\langle\sigma_{ij}\rangle = \int_V \sigma_{ij}(x_i)\,dv \tag{1.1}$$

and the average strain by

$$\langle\varepsilon_{ij}\rangle = \int_V \varepsilon_{ij}(x_i)\,dv \tag{1.2}$$

where ε_{ij} is the infinitesimal strain tensor. Relations (1.1) and (1.2) are very general in the sense that no restrictions whatsoever are placed on the interfacial geometry of the heterogeneous combination of phases. With nothing more than the preceding definitions, we can now define the effective properties of the heterogeneous media. The effective, linear stiffness, or modulus type properties, designated by the tensor C_{ijkl} are defined through their presence in the relation

$$\langle\sigma_{ij}\rangle = C_{ijkl}\langle\varepsilon_{kl}\rangle \tag{1.3}$$

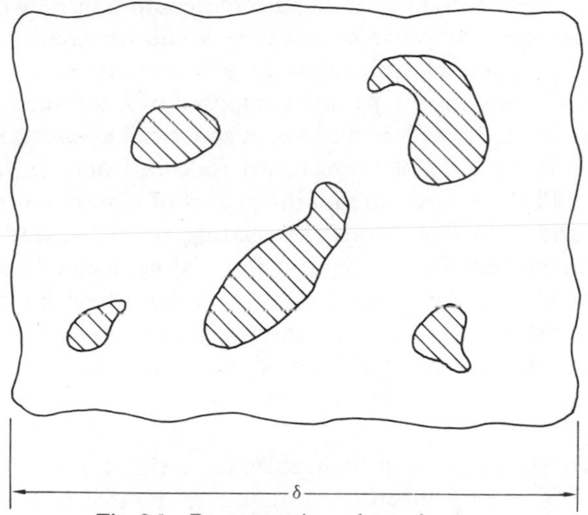

Fig. 2.1 Representative volume element.

Thus to solve for the effective properties of the heterogeneous media, we need to perform the averaging process specified by (1.1) and (1.2) and thence solve for C_{ijkl} from (1.3).* Although this process sounds simple in outline, it is complicated in detail, and great care must be exercised in performing the operation indicated. To perform this operation rigorously, we need exact solutions for the stress and strain fields, $\sigma_{ij}(x_i)$ and $\varepsilon_{ij}(x_i)$ in the heterogeneous media. Our intent is to perform this procedure with a minimum of approximation, and with no recourse to empirical procedures. As such, our results are limited to a few idealized geometric models of typical heterogeneous media combinations. Remarkably, it will be found that the few geometric models employed here cover a wide range of material types, and do so with a preservation of the essential aspects of the material types.

To be more specific about the possible geometric features, heterogeneous media can be classified into at least five broad geometric types. The first type is that typified by the crystal grain structure of common metals. Each grain is anisotropic and different grains have different orientations of their planes or axes of symmetry. This is a one phase system, and it is the only single phase system of the five types. In the second type, having two or more distinct phases, all phases are continuous and there is no distinguishing geometric characteristic of the interfaces that can be used to tell which material is on which side of the interface. This system is

*Similar volume averaging definitions apply in the nonlinear case; see Cowin [2.1] for a description.

sometimes said to be composed of an interpenetrating network. Rigorous criteria to distinguish a system of this type would necessarily be statistical.

The remaining three classifications of geometric types are variants of a common form. These three types are composed of a continuous phase, said to be the matrix phase, with inclusions of spherical, cylindrical, or lamellar shape. The commonality of these forms follows from the fact that in a general sense all three forms are limiting cases of ellipsoidal inclusions, the cylindrical and lamellar forms associating with prolate and oblate ellipsoids, respectively. Of course an ellipsoidal geometry offers an infinite gradation in between the limiting cases; however, these limiting cases are of prime importance. Indeed in some polymeric systems where kinetic driving forces are present and thermodynamical equilibrium is achieved, one of the three forms always results: spherical, cylindrical, or lamellar structure.

We shall consider all five of these basic geometric types of heterogeneity. More general types of geometry can sometimes be treated as combinations of two or more of these basic types. Otherwise such cases must be treated individually. At the most elementary level composites can be classified into one of two types: (1) those systems containing one continuous phase with discrete inclusions of one or more other materials, and (2) all other systems. At this simple level of classification, the first type includes the spherical, cylindrical, and lamellar inclusions of the above groupings. We will now derive an explicit formula that will be used to obtain the effective properties in the type of heterogeneous media involving discrete inclusions.

The problem of interest now is that of a two-material heterogeneous system, one material of which is continuous, the other of which is in the form of discrete inclusions. Both materials will be taken to be isotropic. The stress strain relations for the two materials are given by

$$\sigma_{ij}=\lambda_I\delta_{ij}\varepsilon_{kk}+2\mu_I\varepsilon_{ij} \tag{1.4}$$

for the inclusion phase and by

$$\sigma_{ij}=\lambda_m\delta_{ij}\varepsilon_{kk}+2\mu_m\varepsilon_{ij} \tag{1.5}$$

for the matrix phase, where λ and μ are the Lamé constants. The average stress formula (1.1) can be written as

$$\langle\sigma_{ij}\rangle=\frac{1}{V}\int_{V-\Sigma_{n=1}^{N}V_n}\sigma_{ij}\,dv+\frac{1}{V}\sum_{n=1}^{N}\int_{V_n}\sigma_{ij}\,dv \tag{1.6}$$

where there are taken to be N inclusions within the representative volume element, with volumes designated by V_n, and $V-\Sigma_{n=1}^{N} V_n$ designates the region excluding the inclusions. Using (1.5), (1.6) can be expressed as

$$\langle\sigma_{ij}\rangle = \frac{1}{V}\int_{V-\Sigma_{n=1}^{N} V_n} (\lambda_m \delta_{ij}\varepsilon_{kk} + 2\mu_m \varepsilon_{ij})\, dv + \frac{1}{V}\sum_{n=1}^{N}\int_{V_n} \sigma_{ij}\, dv \tag{1.7}$$

The first integral in (1.7) can be decomposed into the following two integrals:

$$\langle\sigma_{ij}\rangle = \frac{1}{V}\int_{V} (\lambda_m \delta_{ij}\varepsilon_{kk} + 2\mu_m \varepsilon_{ij})\, dv - \frac{1}{V}\sum_{n=1}^{N}\int_{V_n} (\lambda_m \delta_{ij}\varepsilon_{kk} + 2\mu_m \varepsilon_{ij})\, dv$$

$$+ \frac{1}{V}\sum_{n=1}^{N}\int_{V} \sigma_{ij}\, dv \tag{1.8}$$

The average stress term in (1.8) is rewritten with the use of (1.3) and the first integral in (1.8) can be explicitly written in terms of average strains to give

$$C_{ijkl}\langle\varepsilon_{kl}\rangle = \lambda_m \delta_{ij}\langle\varepsilon_{kk}\rangle + 2\mu_m \langle\varepsilon_{ij}\rangle$$

$$+ \frac{1}{V}\sum_{n=1}^{N}\int_{V_n} (\sigma_{ij} - \lambda_m \delta_{ij}\varepsilon_{kk} - 2\mu_m \varepsilon_{ij})\, dv \tag{1.9}$$

where the isotropic form of (1.3) has been used. This formula is of use to us in later derivations of effective properties. It is seen that only the conditions within the inclusions are needed for the evaluation of the effective properties tensor, C_{ijkl}. The derivation of (1.9) follows that of Russel and Acrivos [2.2].

Dilute Suspensions

Next (1.9) is specialized to the case of dilute suspension conditions, appropriate to the behavior of a single inclusion in an infinite matrix phase. Note that the dilute suspension problem as posed, involving a single particle in an infinite medium, violates the condition that the representative volume element be small compared with the scale of the problem of intended application. However, the physical meaning of this idealization is simply that the particles are so small and so far apart that all interaction between particles can be neglected, no matter what the size of the representative volume element may be.

Consider the dilute suspension, in a state of imposed simple shear deformation at large distances from the inclusion; then (1.9) takes the form

$$2\mu\langle e\rangle = 2\mu_m\langle e\rangle + \frac{1}{V}\int_{V_i}(\lambda_I\delta_{ij}\varepsilon_{kk} + 2\mu_I\varepsilon_{ij} - \lambda_m\delta_{ij}\varepsilon_{kk} - 2\mu_m\varepsilon_{ij})\,dv \tag{1.10}$$

where (1.4) has been used and $\langle e\rangle$ is the average shear strain. Now Eshelby [2.3] has shown that a single ellipsoidal inclusion embedded in an infinite medium undergoes a very special behavior. Namely, the inclusion is proved to be in a state of homogeneous deformation, corresponding to that imposed in the infinite media, at large distances from the inclusion. With this result (1.10) can be written as

$$\mu\langle e\rangle = \mu_m\langle e\rangle + \frac{V_i}{V}(\mu_I - \mu_m)\langle e_I\rangle \tag{1.11}$$

where $\langle e_I\rangle$ is the uniform shear strain in the inclusion. Finally, (1.11) can be written as

$$\frac{\mu - \mu_m}{\mu_I - \mu_m} = c\frac{\langle e_I\rangle}{\langle e\rangle} \tag{1.12}$$

where c is the volume fraction of the inclusion

$$c = \frac{V_i}{V} \tag{1.13}$$

In a similar manner the effective bulk modulus can be derived as

$$\frac{k - k_m}{k_I - k_m} = c\frac{\langle \varepsilon_I\rangle}{\langle \varepsilon\rangle} \tag{1.14}$$

where $\langle\varepsilon\rangle$ is the average dilatational strain and $\langle\varepsilon_I\rangle$ is the uniform dilatational strain in the inclusion. These remarkably simple formulas show that it is only necessary to know the state of uniform strain in the inclusion in order to derive the effective properties. Of course, these results (1.12) and (1.14) are valid only for dilute suspension conditions, whereas the formula (1.9) is valid for nondilute conditions.

Energy Methods

Up to this point the effective properties have been defined in terms of the explicit formula (1.3) involving average stress and average strain in the representative volume element. To conclude the developments in this section, an alternate means is derived for defining the effective properties through energy equivalence.

Contract expression (1.3) with the average strain tensor $\langle\varepsilon_{ij}\rangle$, to give

$$\langle\sigma_{ij}\rangle\langle\varepsilon_{ij}\rangle=C_{ijkl}\langle\varepsilon_{ij}\rangle\langle\varepsilon_{kl}\rangle \tag{1.15}$$

The average stress and strain terms in (1.15) are due to the imposed conditions on the boundary of the representative volume element. Since the states of stress and deformation are macroscopically homogeneous within the representative volume element, the averages can be obtained equivalently from boundary values rather than through volume integrations. Similarly then, the work term on the left-hand side of (1.15) is equivalent to the work integral calculated from surface stresses and displacements; thus (1.15) can be written as

$$\tfrac{1}{2}\int_S \sigma_i u_i\, ds=\tfrac{1}{2}C_{ijkl}\langle\varepsilon_{ij}\rangle\langle\varepsilon_{kl}\rangle \tag{1.16}$$

where u_i is the displacement vector and the $\frac{1}{2}$ factor has been added for convenience of later interpretation. Finally, using the divergence theorem and $\sigma_{ij,j}=0$, we can write (1.16) as

$$\tfrac{1}{2}\int_V \sigma_{ij}\varepsilon_{ij}\, dv=\tfrac{1}{2}C_{ijkl}\langle\varepsilon_{ij}\rangle\langle\varepsilon_{kl}\rangle \tag{1.17}$$

Relation (1.17) then defines the effective properties C_{ijkl} through the equality of the strain energy stored in the heterogeneous media to that stored in the equivalent homogeneous media.

Having derived the alternative forms (1.3) and (1.17) for the effective property tensor C_{ijkl}, one might expect a similar energy form to that of expression (1.9), applicable in the case of a two phase composite with discrete inclusions. To show this equivalence, contract (1.9) with $\langle\varepsilon_{ij}\rangle$, giving

$$\begin{aligned}\langle\sigma_{ij}\rangle\langle\varepsilon_{ij}\rangle&=(\lambda_m\delta_{ij}\langle\varepsilon_{kk}\rangle+2\mu_m\langle\varepsilon_{ij}\rangle)\langle\varepsilon_{ij}\rangle\\&\quad+\frac{1}{V}\sum_{n=1}^{N}\int_{V_n}\left[\sigma_{ij}\langle\varepsilon_{ij}\rangle-(\lambda_m\delta_{ij}\varepsilon_{kk}+2\mu_m\varepsilon_{ij})\langle\varepsilon_{ij}\rangle\right]dv\end{aligned} \tag{1.18}$$

Consider the last term in the integrand in (1.18), which written in general anisotropic form gives

$$C^m_{ijkl}\langle\varepsilon_{kl}\rangle\langle\varepsilon_{ij}\rangle$$

Now using the symmetry $C^m_{ijkl}=C^m_{klij}$ we can write

$$C^m_{ijkl}\langle\varepsilon_{kl}\rangle\langle\varepsilon_{ij}\rangle=\sigma^0_{ij}\langle\varepsilon_{ij}\rangle \tag{1.19}$$

where

$$\sigma^0_{ij}=C^m_{ijkl}\langle\varepsilon_{kl}\rangle \tag{1.20}$$

with stress σ^0_{ij} having the interpretation as the stress in the region of the inclusions, but under the condition that the inclusions have been replaced by matrix material. Using (1.19), (1.18) then becomes

$$\tfrac{1}{2}C_{ijkl}\langle\varepsilon_{ij}\rangle\langle\varepsilon_{kl}\rangle=\frac{U_0}{V}+\frac{1}{2V}\sum_{n=1}^{N}\int_{V_n}\left(\sigma_{ij}\varepsilon^0_{ij}-\sigma^0_{ij}-\sigma^0_{ij}\varepsilon_{ij}\right)dv \tag{1.21}$$

where $\langle\varepsilon_{ij}\rangle$ in (1.18) is written in alternate form as ε_{ij}, which is simply the strain in a uniform medium composed entirely of the matrix material, and U_0 is just the corresponding strain energy for the medium composed entirely of matrix material. Now, using (1.17) on the left-hand side of (1.21), and applying the divergence theorem to the integral term, it is shown that this expression is identical to Eshelby's formula (I-4.26) when the latter is generalized to the case of N inclusions. Thus (1.21) is the appropriate and operational energy form to be used to solve for the effective properties C_{ijkl}. Note that only the solution for the stress and strain fields within the inclusions are needed for the evaluation of (1.21). Finally, note that although the formula (1.21) was derived in this section using only the definition of the averaging process without recourse to Eshelby's formula, the latter is a more general result than the specific formula (1.21). That is, Eshelby's formula(s) (I-4.25, 4.26) have significance in elasticity theory beyond just the particular application involving averages in the heterogeneous media context.

2.2 DILUTE SUSPENSION, SPHERICAL INCLUSIONS

In this section we derive the solution for the effective shear modulus of a dilute suspension of elastic spherical particles in a continuous phase of another elastic material.

Shear Modulus

Consider first the case of a homogeneous medium in a state of simple shear deformation. With reference to an x, y, z rectangular Cartesian coordinate system, the displacement components are specified as

$$u_x = cx$$

$$u_y = -cy$$

$$u_z = 0 \tag{2.1}$$

where the maximum shear strain has the value c. We convert to a spherical polar coordinate system, r, θ, ϕ, and the displacement expressions corresponding to (2.1) have the forms

$$u_r = cr\sin^2\theta\cos 2\phi$$

$$u_\theta = cr\sin\theta\cos\theta\cos 2\phi$$

$$u_\phi = -cr\sin\theta\sin 2\phi \tag{2.2}$$

Guided by the preceding forms for homogeneous media deformation, a general solution for the heterogeneous problem is assumed here in the form

$$u_r = U_r(r)\sin^2\theta\cos 2\phi$$

$$u_\theta = U_\theta(r)\sin\theta\cos\theta\cos 2\phi$$

$$u_\phi = U_\phi(r)\sin\theta\sin 2\phi \tag{2.3}$$

where $U_r(r)$, $U_\theta(r)$, and $U_\phi(r)$ are unknown functions of r to be solved from the equilibrium equations. The equilibrium equations in spherical coordinates have the forms

$$\frac{\lambda+2\mu}{\mu}\left(\frac{1}{r}\frac{\partial^2 u_r}{\partial r\,\partial\theta}+\frac{2}{r^2}\frac{\partial u_r}{\partial\theta}+\frac{1}{r^2}\frac{\partial^2 u_\theta}{\partial\theta^2}-\frac{1}{r^2}u_\theta-\frac{\cot^2\theta}{r^2}u_\theta\right.$$

$$\left.\frac{\cot\theta}{r^2}\frac{\partial u_\theta}{\partial\theta}-\frac{\cot\theta}{r^2\sin\theta}\frac{\partial u_\phi}{\partial\phi}+\frac{1}{r^2\sin\theta}\frac{\partial^2 u_\phi}{\partial\theta\,\partial\phi}\right)-\frac{1}{r^2\sin\theta}\frac{\partial^2 u_\phi}{\partial\theta\,\partial\phi}$$

$$-\frac{\cot\theta}{r^2\sin\theta}\frac{\partial u_\phi}{\partial\phi}+\frac{1}{r^2\sin^2\theta}\frac{\partial^2 u_\theta}{\partial\phi^2}+\frac{1}{r}\frac{\partial u_\theta}{\partial r}+\frac{1}{r^2}u_\theta-\frac{1}{r^2}\frac{\partial u_r}{\partial\theta}+\frac{\partial^2 u_\theta}{\partial r^2}$$

$$-\frac{1}{r^2}u_\theta+\frac{1}{r}\frac{\partial u_\theta}{\partial r}+\frac{1}{r^2}\frac{\partial u_r}{\partial\theta}-\frac{1}{r}\frac{\partial^2 u_r}{\partial r\,\partial\theta}=0 \tag{2.4}$$

$$\frac{(\lambda+2\mu)}{\mu}\left(\frac{\partial^2 u_r}{\partial r^2}+\frac{2}{r}\frac{\partial u_r}{\partial r}-\frac{2}{r^2}u_r+\frac{1}{r}\frac{\partial^2 u_\theta}{\partial r\,\partial\theta}-\frac{1}{r^2}\frac{\partial u_\theta}{\partial\theta}+\frac{\cot\theta}{r}\frac{\partial u_\theta}{\partial r}\right.$$

$$\left.-\frac{\cot\theta}{r^2}u_\theta+\frac{1}{r\sin\theta}\frac{\partial^2 u_\phi}{\partial r\,\partial\phi}-\frac{1}{r^2\sin\theta}\frac{\partial u_\phi}{\partial\phi}\right)-\frac{1}{r}\frac{\partial^2 u_\theta}{\partial r\,\partial\theta}-\frac{\cot\theta}{r}\frac{\partial u_\theta}{\partial r}$$

$$-\frac{1}{r^2}\frac{\partial u_\theta}{\partial\theta}-\frac{\cot\theta}{r^2}u_\theta+\frac{1}{r^2}\frac{\partial^2 u_r}{\partial\theta^2}+\frac{\cot\theta}{r^2}\frac{\partial u_r}{\partial\theta}+\frac{1}{r^2\sin^2\theta}\frac{\partial^2 u_r}{\partial\phi^2}$$

$$-\frac{1}{r\sin\theta}\frac{\partial^2 u_\phi}{\partial r\,\partial\phi}-\frac{1}{r^2\sin\theta}\frac{\partial u_\phi}{\partial\phi}=0 \tag{2.5}$$

and

$$\frac{(\lambda+2\mu)}{\mu}\frac{1}{r\sin\theta}\left(\frac{\partial^2 u_r}{\partial r\,\partial\phi}+\frac{2}{r}\frac{\partial u_r}{\partial\phi}+\frac{1}{r}\frac{\partial^2 u_\theta}{\partial\theta\,\partial\phi}+\frac{\cot\theta}{r}\frac{\partial u_\theta}{\partial\phi}+\frac{1}{r\sin\theta}\frac{\partial^2 u_\phi}{\partial\phi^2}\right)$$

$$+\frac{1}{r}\frac{\partial u_\phi}{\partial r}-\frac{1}{r\sin\theta}\frac{\partial^2 u_r}{\partial r\,\partial\phi}+\frac{\partial^2 u_\phi}{\partial r^2}+\frac{1}{r}\frac{\partial u_\phi}{\partial r}+\frac{1}{r^2}\frac{\partial^2 u_\phi}{\partial\theta^2}-\frac{1}{r^2}u_\phi-\frac{\cot^2\theta}{r^2}u_\phi$$

$$+\frac{\cot\theta}{r^2}\frac{\partial u_\phi}{\partial\theta}+\frac{1}{r^2}\frac{\cot\theta}{\sin\theta}\frac{\partial u_\theta}{\partial\phi}-\frac{1}{r^2\sin\theta}\frac{\partial^2 u_\theta}{\partial\theta\,\partial\phi}=0 \tag{2.6}$$

Substituting (2.3) into (2.4)–(2.6) and equating to zero the coefficients of

$\sin^2\theta$ and the terms independent of θ gives the three governing equations

$$2(1-\nu)\left(U_r''+\frac{2}{r}U_r'-\frac{2}{r^2}U_r-\frac{3}{r}U_\theta'+\frac{3}{r^2}U_\theta\right)$$
$$+(1-2\nu)\left(-\frac{6}{r^2}U_r+\frac{3}{r}U_\theta'+\frac{3}{r^2}U_\theta\right)=0 \qquad (2.7)$$

$$2(1-\nu)\left(\frac{2}{r}U_r'+\frac{4}{r^2}U_r-\frac{6}{r^2}U_\theta\right)+(1-2\nu)\left(-\frac{2}{r}U_r'+U_\theta''+\frac{2}{r}U_\theta'\right)=0 \qquad (2.8)$$

$$U_\theta+U_\phi=0 \qquad (2.9)$$

where the prime designates derivatives with respect to r. The solution of (2.7)–(2.9) may be shown to be given by

$$U_r=A_1r-\frac{6\nu}{1-2\nu}A_2r^3+\frac{3A_3}{r^4}+\frac{(5-4\nu)}{(1-2\nu)}\frac{A_4}{r^2}$$

$$U_\theta=A_1r-\frac{(7-4\nu)}{(1-2\nu)}A_2r^3-\frac{2A_3}{r^4}+\frac{2A_4}{r^2}$$

$$U_\phi=-U_\theta \qquad (2.10)$$

At this point, solutions of the form of (2.10) are taken for the inclusion phase and the matrix phase; thus

$$U_{ri}=A_1r-\frac{6\nu_i}{1-2\nu_i}A_2r^3$$

$$U_{\theta i}=A_1r-\frac{(7-4\nu_i)}{1-2\nu_i}A_2r^3 \qquad (2.11)$$

and

$$U_{rm}=B_1r+\frac{3B_3}{r^4}-\frac{5-4\nu_m}{1-2\nu_m}\frac{B_4}{r^2}$$

$$U_{\theta m}=B_1r-\frac{2B_3}{r^4}+\frac{2B_4}{r^2} \qquad (2.12)$$

where the U_ϕ solution in each phase follows from (2.10). The particular terms in (2.10) that are missing in (2.11) and (2.12) are taken with vanishing coefficients to avoid unbounded or singular conditions.

Coefficient B_1 is considered as being a given quantity since it specifies the state of imposed simple shear deformation at $r \to \infty$. Thus within (2.11) and (2.12) there are four constants to be determined.

The interface conditions require the continuity of

$$\sigma_{rr} \quad \sigma_{r\theta} \quad \sigma_{r\phi}$$

$$u_r \quad u_\theta \quad u_\phi$$

at $r = a$, the radius of the spherical inclusion. However, only four of these conditions are independent. The resulting continuity conditions across the inclusion-matrix interface become

$$aA_1 - \frac{6\nu_i}{1-2\nu_i} a^3 A_2 = aB_1 - \frac{6\nu_m}{1-2\nu_m} a^3 B_2 + \frac{3}{a^4} B_3 + \frac{(5-4\nu_m)}{(1-2\nu_m)} \frac{B_4}{a^2} \tag{2.13}$$

$$aA_1 - \frac{7-4\nu_i}{1-2\nu_i} a^3 A_2 = aB_1 - \frac{7-4\nu_m}{1-2\nu_m} a^3 B_2 - \frac{2}{a^4} B_3 + \frac{2}{a^2} B_4 \tag{2.14}$$

$$21\lambda_i a^2 A_2 + 2\mu_i\left(A_1 - \frac{18\nu_i}{1-2\nu_i} a^2 A_2\right) = \lambda_m\left(21 a^2 B_2 - \frac{6}{a^3} B_4\right)$$
$$+ 2\mu_m\left(B_1 - \frac{18\nu_m}{1-2\nu_m} a^3 B_2 - \frac{12}{a^5} B_3 - 2\frac{(5-4\nu_m)}{(1-2\nu_m)} \frac{B_4}{a^3}\right) \tag{2.15}$$

$$\mu_i\left(A_1 - \frac{7+2\nu_i}{1-2\nu_i} a^2 A_2\right) = \mu_m\left(B_1 - \frac{7+2\nu_m}{1-2\nu_m} a^2 B_2 + \frac{8}{a^5} B_3 + 2\frac{(1+\nu_m)}{(1-2\nu_m)} \frac{B_4}{a^3}\right) \tag{2.16}$$

Use is now made of Eshelby's formula (I-4.26) and the criterion of the preceding section that the equivalent homogeneous medium store the same level of strain energy as the suspension. First, Eshelby's formula (I-4.26) takes the form

$$U = U_0 + \tfrac{1}{2}\int_S \left[-\sigma_{rr}^0 u_r - \sigma_{r\theta}^0 u_\theta - \sigma_{r\phi}^0 u_\phi \right.$$

$$\left. + \sigma_{rr} u_r^0 + \sigma_{r\theta} u_\theta^0 + \sigma_{r\phi} u_\phi^0 \right] ds \tag{2.17}$$

where it is recalled that the variables with the "0" designation refer to the case in which the entire medium is composed of the matrix phase. For this homogeneous problem involving just the matrix phase we have

$$u_r^0 = B_1 r \sin^2\theta \cos 2\theta$$

$$u_\theta^0 = B_1 r \sin\theta \cos\theta \cos 2\phi$$

$$u_\phi^0 = -B_1 r \sin\theta \sin 2\phi \tag{2.18}$$

and

$$\sigma_{rr}^0 = 2\mu_m B_1 \sin^2\theta \cos 2\phi$$

$$\sigma_{r\theta}^0 = 2\mu_m B_1 \sin\theta \cos\theta \cos 2\phi$$

$$\sigma_{r\phi}^0 = -2\mu_m B_1 \sin\theta \sin 2\phi \tag{2.19}$$

The strain energy stored in the homogeneous sphere of matrix material is given by

$$U_0 = \tfrac{4}{3}\pi b^3 \mu_m B_1^2 \tag{2.20}$$

where b is the radius of the sphere. The strain energy stored in the equivalent homogeneous medium is given by

$$U_{\text{COMP}} = \tfrac{4}{3}\pi b^3 \mu B_1^2 \tag{2.21}$$

where μ is the effective shear modulus. Finally, setting the strain energy in the suspension equal to that of the equivalent homogeneous medium, as $U = U_{\text{COMP}}$, and thence using (2.20) and (2.21) in (2.17) results in the form

$$\mu = \mu_m + \frac{3}{4\pi b^3 B_1^2} \int_0^{2\pi}\int_0^{\pi} \left[-\sigma_{rr}^0 u_r - \sigma_{r\theta}^0 u_\theta - \sigma_{r\phi}^0 u_\phi + \sigma_{rr} u_r^0 \right.$$

$$\left. + \sigma_{r\theta} u_\theta^0 + \sigma_{r\phi} u_\phi^0 \right] a^2 \sin\theta \, d\theta \, d\phi \tag{2.22}$$

In evaluating the integrals in (2.22) the stresses and displacements are those at the interface, $r=a$. The variables with the "0" superscript are from (2.18) and (2.19) whereas the other variables are those of the solution of the suspension problem. These displacement variables follow from (2.3), (2.10), and (2.13)–(2.16). The stresses follow directly from the linear elasticity stress-strain relations. The stress and displacements for use in (2.22) may be evaluated in either phase at the interface, $r=a$.

After performing the integrations in (2.22) the final evaluation is completely algebraic, although lengthy and tedious. Using the dilute suspension condition $(a/b)^3 \ll 1$ (2.22) finally becomes

$$\frac{\mu}{\mu_m} = 1 - \frac{15(1-\nu_m)\left[1-(\mu_i/\mu_m)\right]c}{7-5\nu_m+2(4-5\nu_m)\dfrac{\mu_i}{\mu_m}} \tag{2.23}$$

where $c=(a/b)^3$, the volume fraction of the spherical particles under dilute conditions. This result apparently was first derived by Dewey [2.4], based on an elasticity solution given by Goodier.

It is of interest to examine the solution (2.23) in the special case where the inclusion is perfectly rigid and the matrix material is incompressible, $\nu_m = \frac{1}{2}$. The result is

$$\frac{\mu}{\mu_m} = 1 + \tfrac{5}{2}c \tag{2.24}$$

This formula is of basic importance beyond that of an elastic suspension of rigid particles. As discussed in Section 1.1 there is a parallel structure to the mathematics governing linear elastic behavior and that of a Newtonian viscous fluid, with inertia term effects neglected in both cases. In fact, the effective properties characterization can be posed for a fluid suspension, as well as for a solid suspension. The criterion for the determination of the effective viscosity is that of equivalence between the rate of dissipation of energy in the fluid suspension and in the equivalent homogeneous medium. Thus in the fluid case the rate of dissipation of energy plays the role of the strain energy in elasticity. In both cases the mathematical structure of the theory and the procedure is identical whether the matrix material is fluid or solid. This equivalence is, however, restricted at this point to the present consideration of a dilute suspension of rigid particles. Accordingly, for the fluid suspension problem the counterpart of (2.24) is given by

$$\frac{\eta}{\eta_m} = 1 + \tfrac{5}{2}c \tag{2.25}$$

where η is the effective viscosity of the dilute suspension of spherical particles in a carrier fluid of viscosity η_m. Relation (2.25) is the celebrated Einstein formula, first devised by him in 1906 [2.5].

The analogy between viscous fluid behavior and elastic solid behavior has important limitations. Although it is completely correct and rigorous in this problem involving rigid particles, as we later see, a suspension of elastic particles in a fluid has no simple relationship to the behavior of the completely elastic suspension studied in this section.

Bulk Modulus

The procedure followed in this section to determine the effective shear modulus for a dilute elastic suspension can also be followed to determine the effective bulk modulus. The determination of the effective bulk modulus k is much simpler than that just outlined for μ, since the problem has spherical symmetry and thus is one-dimensional. It is left as an exercise to show that the effective bulk modulus is given by

$$k = k_m + \frac{(k_i - k_m)c}{1 + \left[(k_i - k_m)/\left(k_m + \frac{4}{3}\mu_m\right)\right]} \tag{2.26}$$

where dilute suspension conditions prevail.

2.3 COMPOSITE SPHERES MODEL

We have determined the effective shear and bulk moduli for a dilute elastic suspension of spherical particles. We now wish to proceed to the nondilute case. The logical way to make this step is to introduce a particular geometric model, and then to solve for its properties. Naturally, we wish the geometric model to be realistic and free of artificiality. On the other hand, we realize that no single geometric model will cover all cases. With this situation in mind we proceed now to the most common geometric model in use.

The composite spheres model was introduced by Hashin [2.6]. The model is composed of a gradation of sizes of spherical particles embedded in a continuous matrix phase. The size distribution, however, is not random, but rather has a very particular characteristic, as now described. The composite spheres model is shown in Fig. 2.2. The broken curves shown in Fig. 2.2 are taken to define a region of the matrix phase associated with each particular particle. The ratio of radii a/b is taken to be a constant for each composite sphere, independent of its absolute size.

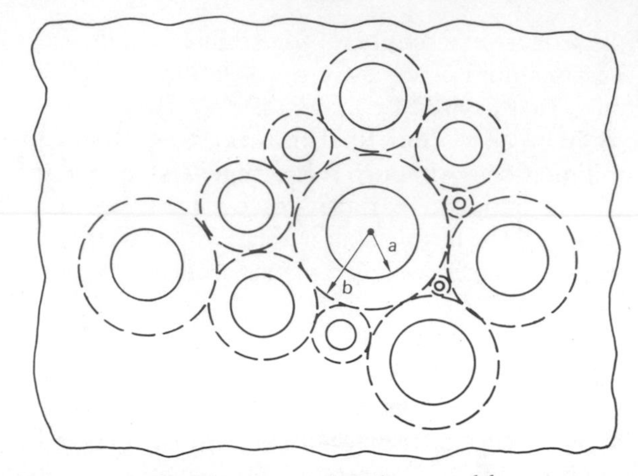

Fig. 2.2 Composite spheres model.

Thus there must be a specific gradation of sizes of particles such that each composite sphere has a/b=constant, while still having a volume filling configuration. Obviously, this distribution requires particle sizes down to infinitesimal. This model would be expected to provide reasonable results for actual systems that do have a rather fine gradation of sizes. Quite obviously this model would not be expected to provide reasonable results for systems containing single size particles, at high concentrations. Later we consider the general applicability of the model; first, however, we proceed to derive the approximate effective moduli.

Bulk Modulus

The analysis begins with the specific analysis of a single composite sphere. Then the means of generalizing the result to the representative volume element is considered. A single composite sphere is subjected to a hydrostatic stress p on its outer boundary; thus

$$\text{at} \quad r=b, \quad \sigma_{rr}=p \tag{3.1}$$

An equivalent homogeneous material sphere is subjected to the same stress.

The displacements at the outer boundaries of the composite sphere and the equivalent homogeneous sphere are equated to provide the same average state of dilatation within each. This procedure then results in the solution for the effective bulk modulus of the single composite sphere.

The single equilibrium equation to be satisfied is given by

$$\frac{\partial \sigma_{rr}}{\partial r} + \frac{2}{r}(\sigma_{rr} - \sigma_{\theta\theta}) = 0 \tag{3.2}$$

where $\sigma_{\phi\phi} = \sigma_{\theta\theta}$. In terms of displacements, (3.2) assumes the form

$$\frac{\partial^2 u_r}{\partial r} + \frac{2}{r}\frac{\partial u_r}{\partial r} - \frac{2}{r^2}u_r = 0 \tag{3.3}$$

The solution of (3.3) is given by

$$u_r = Ar + \frac{B}{r^2}$$

Adopting this solution to the separate inclusion and matrix phases gives

$$\begin{aligned} u_{ri} &= A_i r \\ u_{rm} &= A_m r + \frac{B_m}{r^2} \end{aligned} \tag{3.4}$$

where the coefficient B_i in the inclusion phase must be taken to vanish to avoid a singularity. The corresponding stresses are found to be

$$\begin{aligned} \sigma_{rri} &= (3\lambda_i + 2\mu_i)A_i \\ \sigma_{rrm} &= (3\lambda_m + 2\mu_m)A_m - 4\mu_m\frac{B_m}{r^3} \end{aligned} \tag{3.5}$$

The three constants of integration are evaluated from the continuity conditions

$$\text{At} \quad r = a, \quad \begin{aligned} u_{ri} &= u_{rm} \\ \sigma_{rri} &= \sigma_{rrm} \end{aligned} \tag{3.6}$$

along with the boundary condition (3.1). It is found that

$$A_m = L\frac{p}{4\mu_m} \tag{3.7}$$

$$\frac{B_m}{b^3} = -\frac{p}{4\mu_m} + \left(\frac{3k_m}{4\mu_m}\right)L\left(\frac{p}{4\mu_m}\right) \tag{3.8}$$

and

$$A_i = A_m + \frac{B_m}{a^3} \tag{3.9}$$

where

$$L = \frac{3k_i + 4\mu_m}{3(k_i - k_m)c + \frac{9}{4}(k_i k_m / \mu_m) + 3k_m} \tag{3.10}$$

The displacement at the outer boundary of the composite sphere follows as

$$u_{rm}|_{r=b} = A_m b + \frac{B_m}{b^2} \tag{3.11}$$

For the equivalent homogeneous sphere it is easily shown that

$$u_r|_{r=b} = \frac{pb}{3\hat{k}} \tag{3.12}$$

where $\hat{k}$ is the effective bulk modulus of the homogeneous sphere. Equating (3.11) to (3.12) gives the solution for $\hat{k}$ as

$$\hat{k} = k_m + \frac{c(k_i - k_m)}{1 + (1-c)\left[(k_i - k_m)/\left(k_m + \frac{4}{3}\mu_m\right)\right]} \tag{3.13}$$

Now it remains to prove that the solution (3.13) applies to the entire representative volume element, rather than just to the single composite sphere. First, note that the same pressure p could have been applied to all composite spheres in the representative volume element. Thus the stress state is continuous throughout the entire representative volume element and satisfies the equations of equilibrium. From the point of view of the theorem of minimum complementary energy, Section 1.1, the stress state just described is an admissible stress state. Accordingly, the energy calculated from this stress state is an upper bound to that of the actual stress state in the representative volume element. Now for an isotropic material in a state of dilatation, the local complementary energy is given by

$$U = \frac{p^2}{2k} \tag{3.14}$$

where p is the hydrostatic stress state. Since the complementary energy associated with the admissible stress solution (3.13) is an upper bound to the actual energy, it follows that

$$\frac{1}{k} \leqslant \frac{1}{\hat{k}} \tag{3.15}$$

where now k is the effective bulk modulus. Having found a lower bound to k, it is reasonable to expect to find an upper bound, and this can be done directly.

Reverse the procedure; rather than imposing the stress boundary condition on the single composite sphere, impose a displacement condition. Then solve for an effective bulk modulus $\tilde{k}$ of the composite sphere by requiring that the average strain state be identical in the composite sphere and in the equivalent homogeneous sphere. The details of this are left as an exercise. Now observe that with the addition of a rigid body component of displacement, the displacement field for the single composite sphere can be viewed as an admissible displacement field for the representative volume element; admissible in the sense of the theorem of minimum potential energy, see Section 1.1. It now follows from the theorem that the effective bulk modulus $\tilde{k}$ of the single composite sphere is an upper bound to that of the representative volume element. Thus

$$k \leqslant \tilde{k} \tag{3.16}$$

In carrying out the process just described, fortunately it is found that $\hat{k} = \tilde{k}$; thus the bounds coincide, giving the effective bulk modulus as

$$\frac{k - k_m}{k_i - k_m} = \frac{c}{1 + \left[(1-c)(k_i - k_m)/\left(k_m + \frac{4}{3}\mu_m\right)\right]} \tag{3.17}$$

where (3.13) has been put into an alternate form.

Shear Modulus

Having obtained the effective bulk modulus for the composite spheres model, we now turn to the problem of obtaining the effective shear modulus for the same model. This problem of determining μ is expected to be more complicated than that for determining k because the former is a problem of three-dimensional elasticity whereas that for k was just seen to be one-dimensional. Nevertheless, the procedure follows that just outlined, and, in fact, employs the elasticity solutions obtained in the preceding section. The complete procedure is given by Hashin [2.6]. As with k, the minimum theorems of elasticity are used to obtain upper and lower bounds on μ. In constrast to the situation with the bulk modulus, the bounds on the shear modulus do not coincide, except at very small and very large volume concentrations. The small volume concentration result is just the result for the dilute solution case. These details may be found in [2.6]. The large volume concentration result is found as

$$\frac{\mu}{\mu_i}=1-\frac{(1-\mu_m/\mu_i)[7-5\nu_m+2(4-5\nu_m)\mu_i/\mu_m]}{15(1-\nu_m)}c' \tag{3.18}$$

where

$$c'=1-c$$

and relation (3.18) is valid to the first order in c'.

The fact that the bounds do not coincide in the shear modulus problem is disappointing, but not surprising. In going through the details of the procedure it is found that in the case where simple shear type displacement components are prescribed on the surface of the composite sphere, the resulting boundary stresses are not those corresponding to a state of simple shear stress. Correspondingly, when the simple shear stresses are prescribed on the boundary, the resulting surface deformation state is not that of simple shear deformation. Accordingly, only bounds on μ result, and the greater the disparity between μ_m and μ_i, the greater is the gap between the bounds. Obviously, a different approach is needed to gain an exact solution or even an estimate for the effective shear modulus of the composite spheres model. At this time the exact solution is not known. A reasonable approach to this problem is given in the next section.

2.4 A THREE-PHASE MODEL

Not having succeeded, in the previous section, in finding the solution for the effective shear modulus of the composite spheres model, we now take a different approach.

Consider again the composite spheres model of Fig. 2.2. Now replace all except one of the individual composite spheres by equivalent homogeneous media, as shown in Fig. 2.3. Consider that the infinite region is subjected to homogeneous deformation conditions at large distances from the origin. The outer layer of material, being the equivalent homogeneous phase, has as its properties the unknown effective properties, μ and k. The composite configuration of Fig. 2.3 is rendered equivalent to a completely homogeneous material by requiring that they both store the same strain energy, under conditions of identical average strain. In contrast to the preceding

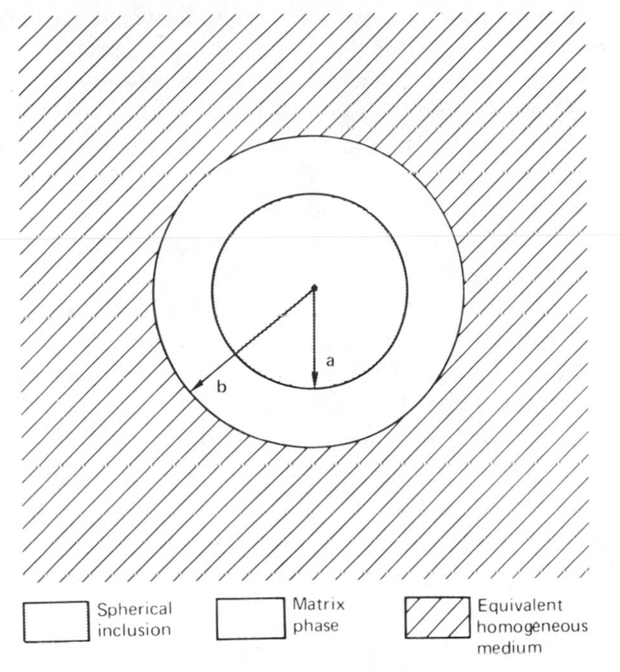

Fig. 2.3 Three phase model.

problems, the effective properties to be determined now enter both problems of the equivalence criteria, rather than just entering the equivalent homogeneous media problem. Nevertheless, the problem as stated is well posed, and we shall find the solution for it. Before doing this, however, we must consider the relationship of the problem just posed to that of the composite spheres model.

There is no assurance that the solutions for the composite media problem of Fig. 2.3 are the effective properties for the composite spheres model of Fig. 2.2. We can consider this to be a hypothesis, and test it. Specifically, in the case of the effective bulk modulus k, we know the solution for the composite spheres model. It is given by relation (3.17). The heterogeneous media problem of Fig. 2.3 can be solved for states of dilatation, and thereby solved for k in an equivalent homogeneous media. The details of this are left as an exercise. Reassuringly, it is found that the effective bulk modulus k for the model of Fig. 2.3 is identical to that of the composite spheres model. With this reinforcing result we therefore proceed to obtain the solution for the effective shear modulus of the model of Fig. 2.3. It must be emphasized, however, that it remains an open question whether or not the solution for μ for the model of Fig. 2.3 is actually the solution for the composite spheres model. The best that can be said is that the results we find are exact for the model of Fig. 2.3, which is a geometry of interest in its own right, whether or not it represents the composite spheres model in the case of shear. Finally, before proceeding to the analysis, it should be mentioned that the model of Fig. 2.3 was first introduced by Kerner [2.7] and van der Pol [2.8]. However, the results found here are different from those of these two references, and, as is discussed later, there are some errors or unjustified assumptions in these latter two approaches.

The method of solution to the problem of Fig. 2.3 closely follows that developed in the dilute suspension case treated in Section 2.2. Specifically, solutions in the form of (2.3) are assumed for the condition of imposed simple shear deformation at large distances from the origin. The functions $U_r(r)$, $U_\theta(r)$, and $U_\phi(r)$ have the form (2.10), as found by solving the equilibrium equations. Separate solutions must be taken for all three regions shown in Fig. 2.3. Specifically, take

$$
\begin{aligned}
U_{ri} &= A_1 r - \frac{6\nu_i}{1-2\nu_i} A_2 r^3 \\
U_{\theta i} &= A_1 r - \frac{7-4\nu_i}{1-2\nu_i} A_2 r^3
\end{aligned} \tag{4.1}
$$

$$
\begin{aligned}
U_{rm} &= B_1 r - \frac{6\nu_m}{1-2\nu_m} B_2 r^3 + \frac{3B_3}{r^4} + \frac{5-4\nu_m}{1-2\nu_m}\frac{B_4}{r^2} \\
U_{\theta m} &= B_1 r - \frac{7-4\nu_m}{1-2\nu_m} B_2 r^3 - \frac{2B_3}{r^4} + \frac{2}{r^2} B_4
\end{aligned} \tag{4.2}
$$

and finally in the third phase composed of the equivalent homogeneous media

$$
\begin{aligned}
U_{re} &= D_1 r + \frac{3D_3}{r^4} + \frac{5-4\nu}{1-2\nu}\frac{D_4}{r^2} \\
U_{\theta e} &= D_1 r - \frac{2D_3}{r^4} + \frac{2D_4}{r^2}
\end{aligned} \tag{4.3}
$$

where it is recalled from Section 2.2 that $U_\phi = -U_\theta$. The Poisson's ratio property ν entering (4.3) is just the effective property of that type for the equivalent homogeneous media. Constant D_1 in (4.3) is considered as given, by the imposed state of simple shear deformation at large distances from the origin. There remain eight constants to be determined. The continuity conditions on stress and displacement across the two interfaces provide the requisite eight conditions. The four conditions that result from the continuity conditions at $r=a$ are already stated as equations (2.13)–(2.16). The remaining four conditions result from continuity conditions at $r=b$, the interface between the matrix material and the effective homogeneous phase. These equations are given by

$$bB_1-\frac{6\nu_m}{1-2\nu_m}b^3B_2+\frac{3}{b^4}B_3+\frac{5-4\nu_m}{1-2\nu_m}\frac{B_4}{b^2}=bD_1+\frac{3}{b^4}D_3+\frac{5-4\nu}{1-2\nu}\frac{D_4}{b^2} \tag{4.4}$$

$$bB_1-\frac{7-4\nu_m}{1-2\nu_m}b^3B_2-\frac{2}{b^4}B_3+\frac{2}{b^2}B_4=dD_1-\frac{2}{b^4}D_3+\frac{2}{b^2}D_4 \tag{4.5}$$

$$\lambda_m\left(21b^2B_2-\frac{6}{b^3}B_4\right)+2\mu_m\left(B_1-\frac{18\nu_m}{1-2\nu_m}b^2B_2+\frac{12}{b^5}B_3-2\frac{(5-4\nu_m)}{1-2\nu_m}\frac{B_4}{b^3}\right)$$
$$=-6\lambda\frac{D_4}{b^3}+2\mu\left(D_1-\frac{12}{b^5}D_3-\frac{2(5-4\nu)}{1-2\nu}\frac{D_4}{b^3}\right) \tag{4.6}$$

and

$$\mu_m\left(B_1-\frac{7+2\nu_m}{1-2\nu_m}b^2B_2+\frac{8}{b^5}B_3+\frac{2(1+\nu_m)}{1-2\nu_m}\frac{B_4}{b^3}\right)=$$
$$\mu\left(D_1+\frac{8D_3}{b^5}+\frac{2(1+\nu)}{1-2\nu}\frac{D_4}{b^3}\right) \tag{4.7}$$

where λ, μ, and ν are the effective properties, only two of which are independent. The solution of these equations is considered shortly.

Next, Eshelby's formula (I-4.26), in the form for imposed displacements is used to evaluate the strain energy stored in the configuration of Fig. 2.3. Rewriting formula (I-4.26) here for convenience gives

$$U_{\text{COMP}}=U_0+\tfrac{1}{2}\int_S\left(\sigma_i u_i^0-\sigma_i^0 u_i\right)ds \tag{4.8}$$

where surface S is the interface $r=b$ and U_0 is the strain energy stored in the homogeneous material problem composed of the material outside of the inclusion. In this problem U_0 is thus the strain energy stored in the equivalent homogeneous media. Our criterion for determining the effective properties is to set

$$U_{\text{COMP}} = U_{\text{EQUIV HOMOG MED}} \tag{4.9}$$

but as discussed earlier the right-hand side of (4.9) is just U_0; using (4.9) in this form reduces (4.8) to

$$\int_S (\sigma_i u_i^0 - \sigma_i^0 u_i)\, ds = 0 \tag{4.10}$$

Following exactly the same procedure as in Section 2.2, using the proper stress and displacement expressions in (4.10) reduces it to the very simple result

$$D_4 = 0 \tag{4.11}$$

The task at hand then is to solve the eight simultaneous equations (2.13)–(2.16) and (4.4)–(4.7) for D_4 and thence set the result equal to zero. This then gives the criterion for determining the effective shear modulus μ. It is found that this process eliminates the appearance of all other effective properties, such as λ and ν, leaving only μ involved. The final solution for μ is given by the solution of the quadratic equation

$$A\left(\frac{\mu}{\mu_m}\right)^2 + 2B\left(\frac{\mu}{\mu_m}\right) + C = 0 \tag{4.12}$$

where

$$A = 8\left(\frac{\mu_i}{\mu_m} - 1\right)(4 - 5\nu_m)\eta_1 c^{10/3} - 2\left[63\left(\frac{\mu_i}{\mu_m} - 1\right)\eta_2 + 2\eta_1\eta_3\right]c^{7/3}$$
$$+ 252\left(\frac{\mu_i}{\mu_m} - 1\right)\eta_2 c^{5/3} - 25\left(\frac{\mu_i}{\mu_m} - 1\right)(7 - 12\nu_m + 8\nu_m^2)\eta_2 c + 4(7 - 10\nu_m)\eta_2\eta_3 \tag{4.13}$$

$$B = -2\left(\frac{\mu_i}{\mu_m} - 1\right)(1 - 5\nu_m)\eta_1 c^{10/3} + 2\left[63\left(\frac{\mu_i}{\mu_m} - 1\right)\eta_2 + 2\eta_1\eta_3\right]c^{7/3}$$
$$- 252\left(\frac{\mu_i}{\mu_m} - 1\right)\eta_2 c^{5/3} + 75\left(\frac{\mu_i}{\mu_m} - 1\right)(3 - \nu_m)\eta_2\nu_m c + \tfrac{3}{2}(15\nu_m - 7)\eta_2\eta_3 \tag{4.14}$$

$$C=4\left(\frac{\mu_i}{\mu_m}-1\right)(5\nu_m-7)\eta_1 c^{10/3}-2\left[63\left(\frac{\mu_i}{\mu_m}-1\right)\eta_2+2\eta_1\eta_3\right]c^{7/3}$$

$$-252\left(\frac{\mu_i}{\mu_m}-1\right)\eta_2 c^{5/3}+25\left(\frac{\mu_i}{\mu_m}-1\right)(\nu_m^2-7)\eta_2 c-4(7+5\nu_m)\eta_2\eta_3 \tag{4.15}$$

with

$$\eta_1=(49-50\nu_i\nu_m)\left(\frac{\mu_i}{\mu_m}-1\right)+35\frac{\mu_i}{\mu_m}(\nu_i-2\nu_m)+35(2\nu_i-\nu_m)$$

$$\eta_2=5\nu_i\left(\frac{\mu_i}{\mu_m}-4\right)+7\left(\frac{\mu_i}{\mu_m}+4\right)$$

$$\eta_3=\frac{\mu_i}{\mu_m}(8-10\nu_m)+(7-5\nu_m) \tag{4.16}$$

and where as always $c=(a/b)^3$, the volume fraction of inclusions.

The final result, (4.12), may be shown to reduce to the previously derived result (2.23) under dilute suspension conditions. This derivation shows the great utility of the Eshelby formulas (I-4.25, 4.26) for calculating strain energy from just a surface integration. Note that in this problem it would not have been possible to use the simple formula (1.12) to calculate the effective shear modulus since the inclusion in this problem is composed of two materials and it does not experience a uniform strain state. Finally, it is important to observe the internal consistency in the present derivation. Recall that in determining the effective bulk modulus k for the composite spheres model, the solution for k was found in a form uncoupled from the second property, say μ. A completely parallel situation was found in the present derivation; μ was determined in uncoupled form from k. This provides indirect support for the view that the present result may be the exact solution for the effective shear modulus of the composite spheres model. As it stands, however, all that can be said with certainty is that we have found the solution for the model of Fig. 2.3.

Experimental data for a suspension of spherical particles has been given by Richard [2.9]. The measurements are of the effective uniaxial modulus E. For purposes of comparing the data with theoretical predictions, the effective shear modulus from (4.12) and the effective bulk modulus from (3.17) have been combined to give

$$E=\frac{9k\mu}{3k+\mu}$$

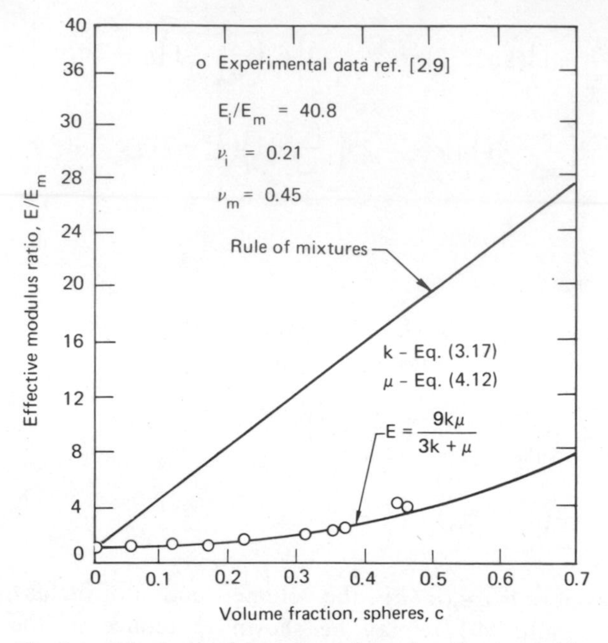

Fig. 2.4 Effective modulus, glass spheres in polyester matrix.

The results are shown in Fig. 2.4. The composite material contains a suspension of glass microspheres embedded in a polyester matrix material. The size range of the glass spheres is 210–297 μm in diameter. Also shown in Fig. 2.4 is the rule of mixtures prediction, which is seen to be meaningless.

The work just described is from that of Christensen and Lo [2.10]. A very similar approach to [2.10] was given by Smith [2.11], [2.12]. In these works, however, an unjustified assumption was introduced that was not made in the present approach; specifically, in [2.11] and [2.12] it was assumed that coefficient D_4 in equations (4.3) vanishes. The present approach makes no such assumption; rather that result is proven. Also, as mentioned earlier, this same problem has been studied by Kerner [2.7] and by van der Pol [2.8]. Both of these results are different from that obtained here. A specific error is present in the manipulations in the latter reference. In the work of Kerner no error can be pinpointed because of the brevity of the derivation. However, the final form of Kerner's solution suggests that an assumption was made that the inclusion is in a homogeneous state of simple shear deformation, an assumption the present work disproves.

2.5 THE SELF-CONSISTENT SCHEME

The two most commonly used models for the behavior of macroscopically isotropic composite media are those of the composite spheres model and the "self-consistent scheme" model. The latter model is now described and evaluated.

The method of the self-consistent scheme was first devised by Hershey [2.13] and Kröner [2.14] as a means to model the behavior of polycrystalline materials. Such materials are just one phase media, but because of the random or partially random orientation of the crystals, discontinuities in properties exist across crystal interfaces. Thus the properties vary with position, and this is certainly a particular type of heterogeneous media.

In the application of the method to polycrystalline aggregates, a single anisotropic crystal is viewed as a spherical or ellipsoidal inclusion embedded in an infinite medium of the unknown isotropic properties of the aggregate. Then the system is subjected to uniform stress or strain conditions at large distances from the inclusion. Next the orientation average of the stress or strain in the inclusion is set equal to the corresponding applied value of stress or strain. This procedure is where the name originates. There results a condition from which the isotropic effective properties can be solved. The method actually reduces to coupled equations to be solved for μ and k. The entire procedure requires numerical evaluation at some stage, and has primarily been done for the spherical boundary case.

The extension of the self-consistent scheme to multiphase media was given by Hill [2.15] and Budiansky [2.16]. As discussed by Budiansky [2.16] the method has a very simple geometric interpretation. Specifically, each phase of the composite is alternatively viewed as being lumped as a single ellipsoidal inclusion in an infinite matrix of the unknown effective properties of the problem. Applying uniform stress or strain conditions at infinity then allows the determination of the average conditions in the inclusion. After this operation is performed for all phases, the average conditions are known in all phases, in terms of the individual phase properties and the effective properties. Thence, average conditions in the entire composite are known and the effective moduli can be calculated from the averages. As with the polycrystalline case, coupled equations are obtained, to be solved for μ and k in terms of the properties of the individual phases and their volume fractions. Although the method for multiphase media seems straightforward, there are some problems with it.

A symptom of the difficulty can be observed in the cases of rigid inclusions and cavities. As noted in Budiansky [2.16] in the case of cavities, the predicted effective shear modulus is given by

$$\mu = \frac{3(1-2c)}{1-c}\mu_m \tag{5.1}$$

where c is the volume fraction of the voids, whereas for rigid inclusions in an incompressible matrix phase

$$\begin{aligned} \mu &= \frac{\mu_m}{1-\frac{5}{2}c} \qquad 0 \leqslant c \leqslant \tfrac{2}{5} \\ &= \infty \qquad \tfrac{2}{5} \leqslant c \leqslant 1 \end{aligned} \tag{5.2}$$

where now c is the volume fraction of the rigid inclusions. Thus μ becomes zero at a volume fraction of $c = \frac{1}{2}$ in the case of voids and μ becomes infinite at a volume fraction of $c = \frac{2}{5}$ in the case of rigid inclusions.

Clearly, there is a very strange behavior associated with the application of the self-consistent scheme to multiphase media. This result cannot be too surprising. The method takes enormous liberties with the geometry of the material combination. To calculate the averages needed for each phase, the geometry is successively rearranged to view the phase under consideration as an inclusion, even if in reality the phase is completely continuous. Of course, the scheme is not a geometric model at all; it is merely a convenient and appealing operational scheme, which, unfortunately, gives erroneous results.* Despite these inconsistencies, the self-consistent scheme rather frequently has been applied to model multiphase media, with the only apparent justification being its simplicity. However, the point of view adopted here is that the self-consistent scheme does not provide a realistic model of multiphase composite media behavior, and no further reference to it is made in the context of multiphase behavior.

The situation with respect to single phase media is quite different from that described for multiphase media. For single phase polycrystalline media, the self-consistent method does not take great liberties with the geometry, and at that level it is, in fact, a geometric model. Certainly in application to polycrystalline materials it appears to be a very reasonable approach. In fact, the three phase geometric model of the preceding section is the counterpart for multiphase media of the self-consistent scheme model for single phase media. The three phase model also is

*A different scheme, which gives better results, has been discussed by McLaughlin [2.17], based on earlier work.

sometimes referred to as the self-consistent scheme, to complicate matters further. We reserve the term *self consistent scheme* for the procedure described in this section.

2.6 A CONCENTRATED SUSPENSION MODEL

Next a model for a concentrated suspension is developed. In particular, we consider a suspension of single size spheres in a continuous matrix phase, at a volume concentration near the maximum value. This is a particularly difficult problem, and we cannot develop an exact solution as we have done with the previous models. In fact, we have to make use of asymptotic methods and use several approximations. Nevertheless, the method is completely rational, rather than empirical, and we find valuable results.

Our restrictions begin with an idealization. We consider only the case of perfectly rigid spherical particles. The problem is that of a suspension of single size, perfectly rigid spherical particles in an incompressible continuous matrix phase, and we wish to determine the effective shear type property. This problem is equally well posed for either an elastic matrix material or a viscous fluid. We develop the results here in the formalization of the elastic problem. The derivation follows that given by Frankel and Acrivos [2.18], in which the fluid type problem was treated. These alternate derivations illustrate the duality of the two approaches. Some necessary restrictions must be appended to the interpretation of results in the case of the fluid. This situation is fully discussed in the next section.

In considering a concentrated suspension, it is necessary to specify the packing arrangement, to pose the problem in a mathematical context. We consider a cubical packing of the single size spherical particles. The reason for picking the cubical packing geometry is that its maximum volume fraction of $c_{\text{MAX}} = \pi/6$ corresponds closely to the observation that in the important, practical case of loosely packed systems, there results a particle volume fraction of slightly more than one-half; see, for example, Happel and Brenner [2.19].

Consider the arrangement shown in Fig. 2.5. At volume fractions near the maximum value, the spherical particles are nearly touching. In this condition the effective region over which high states of deformation occur are just those regions where the spheres are nearly in contact. First, the basic geometry of the model must be established. Referring to Fig. 2.5, it is seen that

$$\frac{\Delta}{a} = \frac{l}{a} - 2 \tag{6.1}$$

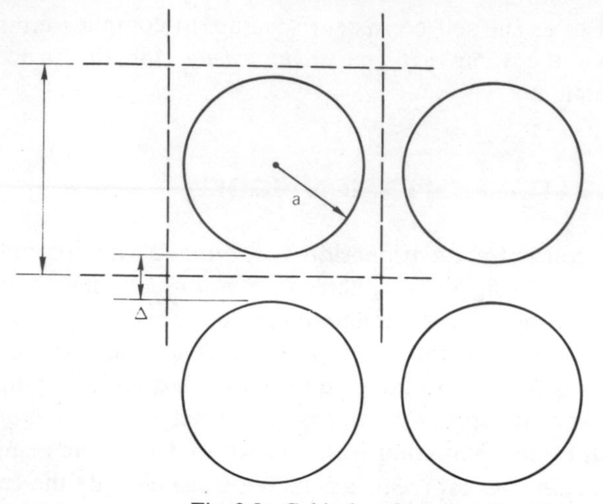

Fig. 2.5 Cubical packing arrangement.

where l is the dimension of the cube or cell used to model the material. The parameter ratio (l/a) can be expressed in terms of the volume fraction of the particles c through

$$\left(\frac{a}{l}\right)=\frac{1}{2}\left(\frac{c}{c_{\text{MAX}}}\right)^{1/3} \tag{6.2}$$

where the maximum volume fraction for cubical packing is

$$c_{\text{MAX}}\big|_{\text{CUBICAL}}=\frac{\pi}{6}$$

The state of shear strain is specified by

$$\begin{aligned} \varepsilon_{xx}&=\varepsilon \\ \varepsilon_{yy}&=0 \\ \varepsilon_{zz}&=-\varepsilon \end{aligned} \tag{6.3}$$

In the context of the suspension this state of average deformation is imposed by requiring that the spheres undergo a relative displacement of value u in one direction, while they undergo a relative displacement of $-u$ in the orthogonal direction. The average strain then is just $\varepsilon=u/l$. Furthermore, the maximum shear strain also is given by $e=u/l$.

As before, our approach in determining the effective shear modulus is to equate the strain energy stored in the suspension and in the equivalent homogeneous medium. Relative to the cubical unit in Fig. 2.5 the strain energy is given by

$$U = \mu l^3 \left(\frac{u}{l} \right)^2 \tag{6.4}$$

where μ is the effective shear modulus. Next we must obtain the strain energy in the suspension.

It would be very difficult to obtain an exact analytical solution to the problem pictured in Fig. 2.5. We use the lubrication approximation effectively to replace the problem of Fig. 2.5 involving curved boundaries with a much more simple problem. Speaking in terms of fluid behavior for a moment, the shearing flow of a fluid between nonparallel boundaries with relative translation is idealized to be that of simple shearing flow locally, as determined by the distance between surfaces at each location. The nature of the approximation involves the violation of one of the two balance of momentum equations in a two-dimensional problem. So long as the variation in the gap thickness is changing over large distances compared with the gap thickness, the approximation is found to be satisfactory. See

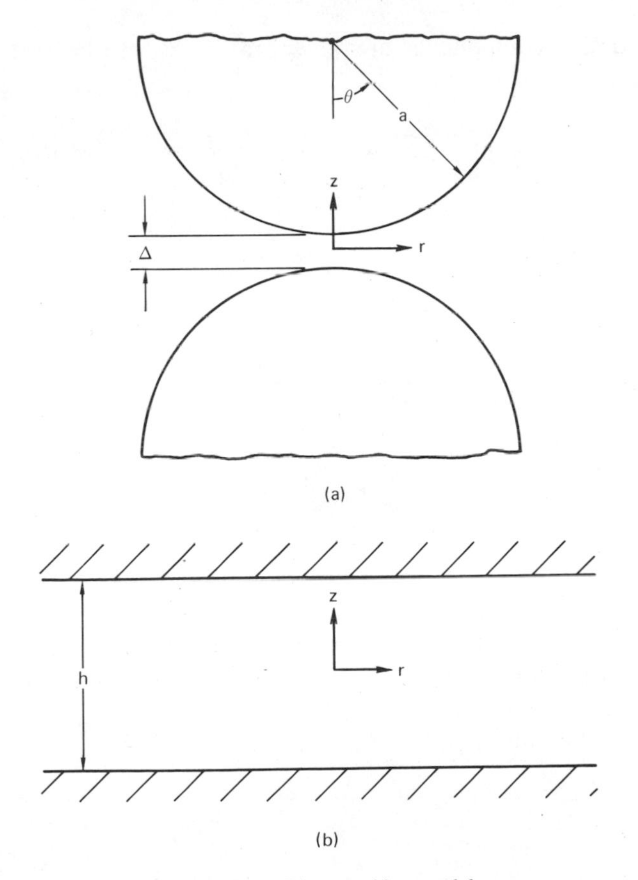

Fig. 2.6 Close packing model.

Batchelor [2.20] for a more detailed description. The general method of the lubrication approximation can be profitably employed in the present problem, involving the relative coaxial displacement of spheres constrained by an elastic medium.

The problem to be solved is that shown in Fig. 2.6*a*, but in the spirit of the lubrication approximation we begin by solving the parallel boundary problem of Fig. 2.6*b*. Employing cylindrical coordinates, the problem is specified by

$$\text{at} \quad z = \pm\frac{h}{2} \quad u_z = \pm\frac{u}{2}$$

$$u_r = 0 \tag{6.5}$$

where u is the relative displacement of the boundaries. The material is incompressible and therefore must satisfy the condition

$$\varepsilon_{rr} + \varepsilon_{\theta\theta} + \varepsilon_{zz} = 0 \tag{6.6}$$

where the strain components are given in terms of the displacement components by

$$\varepsilon_{rr} = \frac{\partial u_r}{\partial r}$$

$$\varepsilon_{\theta\theta} = \frac{u_r}{r}$$

$$\varepsilon_{zz} = \frac{\partial u_z}{\partial z}$$

$$\varepsilon_{rz} = \frac{1}{2}\left(\frac{\partial u_z}{\partial r} + \frac{\partial u_r}{\partial z}\right) \tag{6.7}$$

The equilibrium equations are given by

$$-\frac{\partial p}{\partial r} + \mu_m \nabla^2 u_r = 0 \tag{6.8}$$

$$-\frac{\partial p}{\partial z} + \mu_m \nabla^2 u_z = 0 \tag{6.9}$$

where p is the hydrostatic pressure supported by the incompressible material and μ_m is the shear modulus of the material. Following Landau and Lifshitz [2.21] in this type of problem $u_z \ll u_r$ and $\partial u_r/\partial r \ll \partial u_r/\partial z$; thus the equations of equilibrium become

$$\mu_m \frac{\partial^2 u_r}{\partial z^2} = \frac{\partial p}{\partial r} \tag{6.10}$$

and

$$\frac{\partial p}{\partial z}=0 \tag{6.11}$$

The solution of the governing equations (6.6), (6.10), and (6.11) are given by

$$u_r=\frac{3}{4}\left(\frac{u}{h}\right)r\left(\frac{4z^2}{h^2}-1\right)$$

$$u_z=-\frac{3}{2}\left(\frac{u}{h}\right)\left(\frac{4z^3}{3h^2}-z\right) \tag{6.12}$$

where the boundary conditions (6.5) have been satisfied.

Designate the local strain energy by w, which in general form is given by

$$w=\mu_m\varepsilon_{ij}\varepsilon_{ij} \tag{6.13}$$

It will be convenient for later purposes to form the integral of w over the thickness. Using the strain relations (6.7) and the solution (6.12) it can be shown that

$$\int_0^{h/2} w\,dz=\frac{9\mu_m}{10}\left(\frac{u}{h}\right)^2 h+\frac{3\mu_m}{8}\left(\frac{u}{h}\right)^2\frac{r^2}{h} \tag{6.14}$$

where the last term follows from the shear strain term.

The elastic counterpart of the lubrication approximation assumes that the solution (6.14) represents the corresponding solution of the problem in Fig. 2.6a, so long as thickness h is taken to be the total gap dimension in Fig. 2.6a; thus $h=h(r)$. The total energy in four of the half gaps (four half gaps per cube) is then given by

$$U=8\pi\mu_m\int_0^a\left[\frac{9}{10}\left(\frac{u}{h}\right)^2 h+\frac{3}{8}\left(\frac{u}{h}\right)^2\frac{r^2}{h}\right]r\,dr \tag{6.15}$$

where

$$\frac{h}{2}=\left(\frac{\Delta}{2}+a\right)-a\cos\theta \tag{6.16}$$

and θ is the angle designated in Fig. 2.6a. Coordinate r also has the relation

$$r = a\sin\theta \tag{6.17}$$

Our objective is to determine the character of the energy term U in the limit of vanishingly small gap Δ between spheres. It can be seen from (6.15) that the first term in the integrand of (6.15) is of higher order than the second term, as $\Delta \to 0$ for $\theta = 0$; thus the first term will be neglected. Combining the modified relation (6.15) with (6.16) and (6.17) then gives

$$U = \frac{3\pi}{8}\mu_m u^2 a \int_0^{\pi/2} \frac{\sin^2\theta\cos\theta\, d\theta}{(\Delta/2a + 1 - \cos\theta)^3} \tag{6.18}$$

Now the integration in (6.18) can be performed, and retaining only the terms giving a singular type behavior as $\Delta \to 0$ gives

$$U = \frac{3\pi}{4}\mu_m u^2 a\left(\frac{a}{\Delta}\right) \tag{6.19}$$

In the final operation we equate the energy in the suspension (6.19) to the energy in the equivalent homogeneous material (6.4) to obtain

$$\frac{\mu}{\mu_m} = \frac{3\pi}{4}\left(\frac{a}{l}\right)\left(\frac{a}{\Delta}\right) \tag{6.20}$$

The terms a/Δ and a/l are given by (6.1) and (6.2), which when combined with (6.20) form

$$\frac{\mu}{\mu_m} = \frac{3\pi}{16\left[1 - \left(\frac{c}{c_{\text{MAX}}}\right)^{1/3}\right]} \tag{6.21}$$

where (6.21) only has meaning as $c \to c_{\text{MAX}}$ in view of the approximations introduced relating to the behavior as $\Delta \to 0$.

We thus have found the nature of the effective shear modulus for the suspension of rigid spherical particles, as $c \to c_{\text{MAX}}$. Under this type of idealization we, of course, knew that $\mu \to \infty$ as $c \to c_{\text{MAX}}$, but the relation (6.21) gives us the order of the singularity. In a strict sense the solution (6.21) only applies to the specified deformation state, since the cubical packing arrangement is not isotropic. This distinction, however, is probably unimportant, in a practical assessment of (6.21), in view of the other approximations that have been introduced.

The corresponding problem for the fluid suspension is only trivially different. The displacements are to be reinterpreted as velocities, and the shear modulus corresponds to viscosity. Whether for a solid or a fluid, the result (6.21) applies only to the cubical packing geometry, with $c_{MAX}=\pi/6$. Frankel and Acrivos [2.18] considered other cell geometries.

The practical assessment of the result (6.21) is given in the next section.

2.7 SOME GENERAL OBSERVATIONS

The previous sections provide the models developed here for use with spherical particle composite systems. As has been seen, the composite spheres model and the related model of Section 2.4 are the only ones for which we have a reasonably complete and exact analytical characterization. These models cover the entire volume fraction range in the sense that the inclusions can exist at any level $0 \leqslant c \leqslant 1$. The type of composite material, involving a combination of different size spherical filler particles, is well represented by such models. What about other cases, such as single size spherical particles; can these be represented by the composite spheres model? The answer must be qualified. It is yes, provided the volume fraction of particles is not close to the maximum volume fraction of the particles in aggregate form. As shown in Fig. 2.4, the composite spheres model and the associated model of Section 2.4 give a very close prediction of the experimentally measured effective uniaxial modulus up to a volume fraction between $c=0.45$ and 0.50. Considering that this is only slightly below the volume fraction for the loose (cubical) packing of single size spheres, the observed range of validity of the model(s) is impressive. However, no precise limits can be stated for the range of validity of such models to all composite systems. Any relative difference depends not only on the volume fraction, but also on the properties of each phase. Nevertheless, the composite spheres model does represent the behavior of many systems of practical interest.

It should also be pointed out that the composite spheres model is representative of the behavior of a wide variety of composite systems, not just those with exactly spherical particles. So long as the particles are not greatly different in shape from the spherical configuration, the model would be expected to give a reasonable prediction. This flexibility of the model is due to the fact that the effective properties relate to global averages of stress and strain, which themselves are more dependent on the volume fractions of various phases, rather than on the fine details of the local geometry of phases. Of course, this flexibility of application has limits, and there is no substitute for common sense, in the applications.

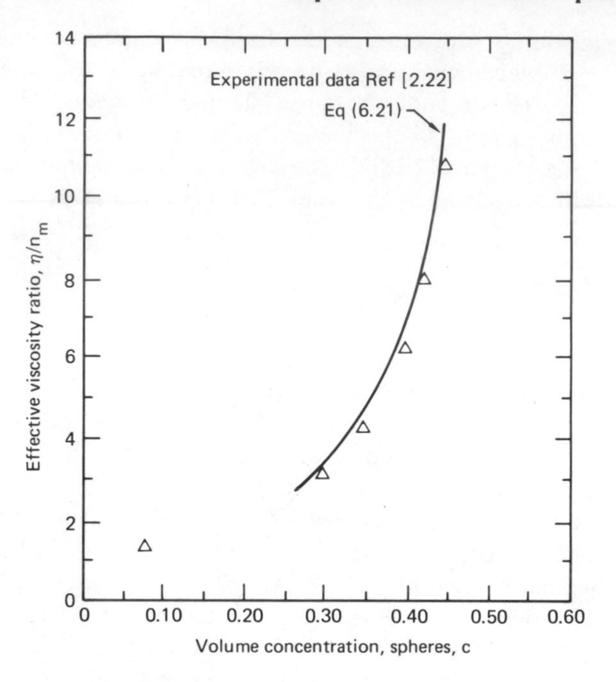

Fig. 2.7 Effective viscosity single size spherical particles.

It is unlikely that there will ever be a closed form, theoretically derived result for a model involving a suspension of single size spheres that is applicable over the full volume fraction range. In the preceding section we obtained the behavior of the effective shear modulus for a suspension near the condition of maximum volume fraction of the spherical particles. It is of interest, however, to see how well the resulting expression (6.21) actually models data. As mentioned before, the result (6.21) applies in both the elastic solid case and the viscous fluid case. With the idealization of perfectly rigid inclusions, probably the fluid case provides the more stringent test of the prediction of the effective property in shear.

Carefully screened data on the effective viscosity have been given by Krieger [2.22]. His effective viscosity data are shown in Fig. 2.7. The prediction of the result (6.21) for cubical packing with $c_{MAX}=\pi/6$ are also given. The comparison is remarkably good. The formula was derived by approximate means that have significance as a rigorous result only in the limit $c \rightarrow c_{MAX}$. For this reason the result (6.21) is only meaningful in the region of volume fractions near to that for maximum packing. Therefore, the close agreement between the data and the prediction (6.21) at volume fractions much lower than c_{MAX} is merely fortuitous.

There remains the problem of finding an expression to fit data such as those in Fig. 2.7 over the entire range. Necessarily, an expression of this type must be empirical. Consider, for example, the empirical expression

$$\frac{\mu}{\mu_m}=\frac{2+c}{2(1-2c)} \tag{7.1}$$

This expression becomes unbounded at $c=\frac{1}{2}$, a value near that for cubic packing, which seems to be realistic. Furthermore, expression (7.1) properly reduces to the Einstein result under dilute suspension conditions. The comparison between expression (7.1) and experimental data of Fig. 2.7 can be made with very favorable results. This, or any other empirical expression, can be employed for convenience, but caution must be exercised, for such expressions have no fundamental meaning and may be very misleading in particular applications.

Finally, we must mention some of the realities in dealing with fluid suspensions. We have made use of the duality of the suspension problem for elastic solids and for viscous fluids. Recall that it is fundamental to the linear elastic analysis that only infinitesimal deformations occur. A similar restriction must be placed on the fluid suspension problem for the results of the two like analyses to be interchangeable. Otherwise, in the fluid suspension problem there is a fundamental indeterminacy concerning the relative positions of the particles. The problem of the continuously flowing suspension has been studied by Batchelor and Green [2.23] at the level involving the interaction of two spherical particles, being transported by the Newtonian viscous fluid. They find the fundamental indeterminacy mentioned earlier, with consequent results that the suspension is not isotropic and further that non-Newtonian effects are possible. These non-Newtonian effects imply that the effective viscosity is shear rate dependent, and also imply the existence of normal stress effects in simple shearing flow. Despite these complications, Batchelor and Green estimate the zero shear rate effective viscosity to order c^2 and obtain

$$\frac{\mu}{\mu_m}=1+\frac{5}{2}c+(5.2\pm0.3)c^2+O(c^3) \tag{7.2}$$

In an extensive study Chen and Acrivos [2.24, 2.25] have refined the result shown in (7.2) to give the c^2 coefficient as 5.01. It is interesting to note that the empirical expression (7.1) has the expansion

$$\frac{\mu}{\mu_m}=1+\frac{5}{2}c+5c^2+O(c^3) \tag{7.3}$$

This supports the view that expression (7.1) may at least have utilitarian value.

The non-Newtonian effects mentioned earlier play a large role in suspension behavior. The data of Krieger shown in Fig. 2.7 are effectively zero shear rate results, but as shear rate increases, the effective viscosity is found to undergo shear thinning and takes on reduced values. So far, we have only discussed suspensions of spherical particles, and we find great complications. In suspensions involving particles of shape other than spherical, non-Newtonian effects appear to predominate the behavior. Of course, non-Newtonian effects are not basic in themselves, but rather are symptoms of basic effects. Particle agglomeration and flow alignment of particles with concurrent anisotropy are also symptoms of other basic effects. These basic effects may be strictly hydrodynamic or of other origin. For example, Brownian motion and electrical forces comprise underlying effects that typically become more important as the size of the particles decreases. On the colloidal scale, effects such as these are of premier importance. However, our interests here are in systems of a length scale larger than colloidal and we are not explicitly concerned with such effects.

PROBLEMS

1. Derive Eshelby's formula (I-4.26) for the case of imposed displacement boundary conditions.
2. Prove that in the derivation of Eshelby's formulas (I-4.25, 4.26), the body force in the problem with energy designated by U_o is only distributed over the surface of the inclusion region, and is not interior to the region.
3. Solve the imposed displacement problem to determine the effective bulk modulus for the composite spheres model.
4. Solve the three phase model of Section 2.4 for the effective bulk modulus. Verify that the result is identical to that of the composite spheres model.
5. Can the composite spheres model be extended to account for N different material phases? Consider both the exact solutions for the effective properties and bounds on the effective properties.
6. Use the composite spheres model to predict the effective bulk modulus for a void containing foam type material. What model would you use to obtain the effective shear modulus for a foam? Do these models account for bending of the cell walls? If not, how would you account for such effects?

7. Derive the formula for the volume fraction of single size spheres in hexagonal packing. Compare the result with that for cubic packing. How would one determine the size distribution of spheres required for the composite spheres model?
8. Adapt the analysis of Section 2.6 to obtain the effective shear modulus for the hexagonal packing of single size rigid spheres in an incompressible matrix phase. Compare the result with the experimental data of [2.22].

REFERENCES

2.1 S. C. Cowin, "Effective stress strain relations for finitely deformed inhomogeneous bodies," *Mech. Res. Commun.*, vol. 4, 163 (1977).

2.2 W. B. Russel and A. Acrivos, "On the effective moduli of composite materials: slender rigid inclusions at dilute concentrations," *Z. Angw. Math. Phys.*, vol. 23, 434 (1972).

2.3 J. D. Eshelby, "The determination of the elastic field of an ellipsoidal inclusion, and related problems," Pro. R. Soc. Lond., vol. A 241, 376 (1957).

2.4 J. M. Dewey, "The elastic constants of materials loaded with non-rigid fillers," *J. Appl. Phys.*, vol. 18, 578 (1947).

2.5 A. Einstein, *Investigations of the Theory of Brownian Movement*, Dover, New York, 1956.

2.6 Z. Hashin, "The elastic moduli of heterogeneous materials," *J. Appl. Mech.*, vol. 29, 143 (1962).

2.7 E. H. Kerner, "The elastic and thermoelastic properties of composite media," *Proc. Phys. Soc.*, vol. 69, 808 (1956).

2.8 C. van der Pol, "On the rheology of concentrated dispersions," *Rheol. Acta*, vol. 1, 198 (1958).

2.9 T. G. Richard, "The mechanical behavior of a solid microsphere filled composite," *J. Comp. Mater.*, vol. 9, 108 (1975).

2.10 R. M. Christensen and K. H. Lo, "Solutions for effective shear properties in three phase sphere and cylinder models," *J. Mech. Phys. Solids*, vol. 27, No. 4 (1979).

2.11 J. C. Smith, "Correction and extension of van der Pol's method for calculating the shear modulus of a particulate composite," *J. Res. Natl. Bur. Stand.*, vol. 78A, 355 (1974).

2.12 J. C. Smith, "Simplification of van der Pol's formula for the shear modulus of a particulate composite," *J. Res. Natl. Bur. Stand.*, vol. 79A, 419 (1975).

2.13 A. V. Hershey, "The elasticity of an isotropic aggregate of anisotropic cubic crystals," *J. Appl. Mech.*, vol. 21, 236 (1954).

2.14 E. Kröner, "Berechnung der elastischen konstanten des vielkristalls aus den konstanten des einkristalls," *Z. Phys.*, vol. 151, 504 (1958).

2.15 R. Hill, "A self-consistent mechanics of composite materials," *J. Mech. Phys. Solids*, vol. 13, 213 (1965).

2.16 B. Budiansky, "On the elastic moduli of some heterogeneous materials," *J. Mech. Phys. Solids*, vol. 13, 223 (1965).

2.17 R. McLaughlin, "A study of the differential scheme for composite materials," *Int. J. Eng. Sci.*, vol. 15, 237 (1977).

2.18 N. A. Frankel and A. Acrivos, "On the viscosity of a concentrated suspension of solid spheres," *Chem. Eng. Sci.*, vol. 22, 847 (1967).

2.19 J. Happel and H. Brenner, *Low Reynolds Number Hydrodynamics*, Prentice-Hall, Englewood Cliffs, N.J., 1965.

2.20 G. K. Batchelor, *An Introduction to Fluid Dynamics*, Cambridge University Press, New York, 1967.

2.21 L. D. Landau and E. M. Lifshitz, *Fluid Mechanics*, Pergamon, London, 1959.

2.22 I. M. Krieger, "Rheology of monodisperse lattices," *Adv. Colloid Interface Sci.*, vol. 3, 111 (1972).

2.23 G. K. Batchelor and J. T. Green, "The hydrodynamic interaction of two small freely-moving spheres in a linear flow field," *J. Fluid Mech.*, vol. 56, 375 (1972).

2.24 H.-S. Chen and A. Acrivos, "The solution of the equations of linear elasticity for an infinite region containing two spherical inclusions," *Int. J. Solids Structures*, vol. 14, 331 (1978).

2.25 H.-S. Chen and A. Acrivos, "The effective elastic moduli of composite materials containing spherical inclusions at non-dilute concentrations," *Int. J. Solids Structures*, vol. 14, 349 (1978).

CHAPTER III

EFFECTIVE MODULI: CYLINDRICAL AND LAMELLAR SYSTEMS

We introduced the concept of effective homogeneity in the preceding chapter and studied suspensions of spherical particles. In this chapter we extend the geometric forms of material combination by studying cylindrical and lamellar systems. The previous method for determining effective moduli is employed extensively in this chapter, and recourse must be made to Section 2.1 as the starting point for the present work.

We are still only at a preliminary stage in laying out the complete theory of behavior for heterogeneous media. Nevertheless, it is well to keep in mind our objectives in considering different systems. For example, rigid spherical inclusions are normally considered to degrade the strength of the matrix material into which they are inserted. Although such inclusions have a stiffening effect, it may not be a large effect, and such inclusions are often employed for other reasons, such as to reduce the cost of the material or to improve its dynamic response characteristics. In contrast, the use of fiber reinforced materials has a very direct rationale to improve both the stiffness and strength properties. Just how much stiffening effect can be achieved is seen in the present chapter. Initial consideration is also given in this chapter to materials containing platelet or lamellar inclusions.

In the next chapter platelet systems are shown to have a very different behavior from that of fiber systems.

It is useful to consider two separate geometric means of specifying cylindrical and lamellar inclusions. In one approach these types of inclusions can be obtained as limiting cases of prolate and oblate ellipsoids, or even as cases in which the ratios of the maximum to the minimum dimension of the ellipsoids are very large. In the other approach the cylindrical or lamellar inclusions are taken to be of infinite extent. For the fiber case this simply means that the fibers are idealized as uniform circular cylinders of infinite length.

An assessment of the relative efficiency of reinforcement for different types of inclusions is, for the most part, deferred until the next chapter. In Chapter IV minimum theorems are employed to deduce the potential limits of reinforcement efficiency. The present results are indispensable in assessing reinforcement efficiency.

3.1 TRANSVERSELY ISOTROPIC MEDIA

In both the fiber and the platelet case our objective is to characterize the effective stiffness properties of the heterogeneous media. Consider for example a system of aligned fibers. So long as the fibers are randomly

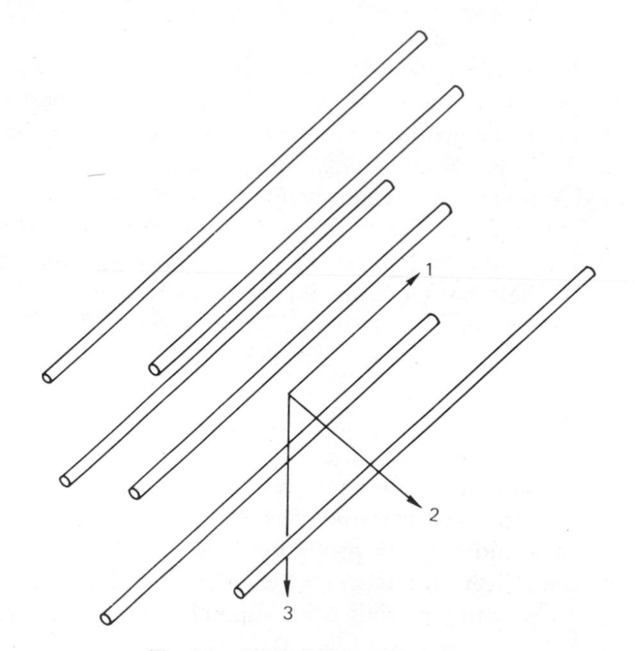

Fig. 3.1 Aligned fiber system.

positioned in space, the media can be idealized as being effectively homogeneous. Further, media of this type, as shown in Fig. 3.1, have symmetry properties in the plane normal to that of the fiber direction. Such media are characterized as being transversely isotropic. It will be useful in our later work, therefore, to have a complete characterization for transversely isotropic media. The moduli type properties with which we are dealing are those of the equivalent homogeneous characterization of the heterogeneous media. In the next section we begin determining these effective properties in terms of the properties of the individual phases.

Modulus Properties

Let axis number 1 designate the axis of symmetry. The stress-strain relations can be written in the form

$$\begin{aligned}
\sigma_{11} &= C_{11}\varepsilon_{11} + C_{12}\varepsilon_{22} + C_{12}\varepsilon_{33} \\
\sigma_{22} &= C_{12}\varepsilon_{11} + C_{22}\varepsilon_{22} + C_{23}\varepsilon_{33} \\
\sigma_{33} &= C_{12}\varepsilon_{11} + C_{23}\varepsilon_{22} + C_{22}\varepsilon_{33} \\
\sigma_{12} &= 2C_{66}\varepsilon_{12} \\
\sigma_{23} &= (C_{22} - C_{23})\varepsilon_{23} \\
\sigma_{31} &= 2C_{66}\varepsilon_{31}
\end{aligned} \tag{1.1}$$

The five constants, C_{11}, C_{12}, C_{22}, C_{23}, and C_{66}, designate the five independent effective properties of the media. Although these properties give a complete description of the stiffness of the media, they are not necessarily the most convenient form for the properties characterization. One sees the difficulty, when considering how to determine the properties by direct experimental means. For this reason the so-called engineering properties are introduced.

Consider a state of uniaxial stress specified by

$$\sigma_{11} \neq 0$$

$$\sigma_{22} = \sigma_{33} = \sigma_{12} = \sigma_{23} = \sigma_{31} = 0$$

Using these conditions in (1.1), it is readily found that

$$\sigma_{11} = E_{11}\varepsilon_{11}$$

where

$$E_{11}=C_{11}-\frac{2C_{12}^2}{C_{22}+C_{23}} \tag{1.2}$$

Modulus E_{11} is called the *uniaxial modulus*, and it is directly measurable. The lateral contraction that accompanies the state of uniaxial stress is used to define Poisson's ratios through

$$\nu_{12}=\frac{-\varepsilon_{22}}{\varepsilon_{11}}$$

and

$$\nu_{13}=\frac{-\varepsilon_{33}}{\varepsilon_{11}}$$

where in the convention ν_{ij}, the first index i refers to the coordinate of imposed stress or strain and the second index j refers to the response direction. It is found from (1.1) for the state of uniaxial stress in the 1 direction that

$$\nu_{12}=\nu_{13}=\frac{C_{12}}{C_{22}+C_{23}} \tag{1.3}$$

Next consider the state specified by

$$\varepsilon_{11}=0, \qquad \varepsilon_{22}=\varepsilon_{33}=\varepsilon$$

Let

$$\sigma_{22}=\sigma_{33}=\sigma$$

From (1.1) it is found that

$$\sigma=2K\varepsilon$$

where

$$K=\tfrac{1}{2}(C_{22}+C_{23}) \tag{1.4}$$

Modulus, K is designated as the *plane strain bulk modulus*.

Finally, the directly measurable shear moduli are defined as

$$\mu_{12}=\mu_{31}=C_{66} \tag{1.5}$$

and

$$\mu_{23} = \tfrac{1}{2}(C_{22} - C_{23}) \tag{1.6}$$

Relations (1.2)–(1.6) can be inverted to give

$$\begin{aligned} C_{11} &= E_{11} + 4\nu_{12}^2 K_{23} \\ C_{12} &= 2K_{23}\nu_{12} \\ C_{22} &= \mu_{23} + K_{23} \\ C_{23} &= -\mu_{23} + K_{23} \\ C_{66} &= \mu_{12} \end{aligned} \tag{1.7}$$

By no means are the five properties defined by (1.2)–(1.6) the only possible, measurable properties. For example, one can imagine that a state of uniaxial tension normal to the axis of symmetry can easily be imposed. To accomplish this, specify

$$\sigma_{22} \neq 0$$

$$\sigma_{11} = \sigma_{33} = \sigma_{12} = \sigma_{23} = \sigma_{31} = 0$$

From (1.1) it is found that

$$\sigma_{22} = E_{22}\varepsilon_{22}$$

where

$$E_{22} = C_{22} + \frac{C_{12}^2(-C_{22} + C_{23}) + C_{23}(-C_{11}C_{23} + C_{12}^2)}{C_{11}C_{22} - C_{12}^2} \tag{1.8}$$

Defining the Poisson's ratios, ν_{21} and ν_{23} by

$$\nu_{21} = \frac{-\varepsilon_{11}}{\varepsilon_{22}}$$

$$\nu_{23} = \frac{-\varepsilon_{33}}{\varepsilon_{22}}$$

it is found that

$$\nu_{21} = \frac{C_{12}(C_{22} - C_{23})}{C_{11}C_{22} - C_{12}^2}$$

$$\nu_{23} = \frac{C_{11}C_{23} - C_{12}^2}{C_{11}C_{22} - C_{12}^2} \tag{1.9}$$

From the symmetry properties it also follows that

$$\nu_{31} = \nu_{21}$$

$$\nu_{32} = \nu_{23} \tag{1.10}$$

It must be noted that $\nu_{12} \neq \nu_{21}$, but the following relation between the two properties can easily be verified, using the previously defined forms

$$\frac{\nu_{12}}{E_{11}} = \frac{\nu_{21}}{E_{22}} \tag{1.11}$$

Some further property interrelations that are useful are as follows:

$$E_{22} = \frac{4\mu_{23}K_{23}}{K_{23} + \mu_{23} + 4\nu_{12}^2 \mu_{23}K_{23}/E_{11}} \tag{1.12}$$

$$\nu_{23} = \frac{K_{23} - \mu_{23} - 4\nu_{12}^2 \mu_{23}K_{23}/E_{11}}{K_{23} + \mu_{23} + 4\nu_{12}^2 \mu_{23}K_{23}/E_{11}} \tag{1.13}$$

$$\nu_{21} = \frac{4\nu_{12}^2 \mu_{23}K_{23}}{E_{11}(K_{23} + \mu_{23}) + 4\nu_{12}^2 \mu_{23}K_{23}} \tag{1.14}$$

$$\nu_{12}^2 = \left(-\nu_{23} - \frac{1}{4}\frac{E_{22}}{K_{23}} + \frac{1}{4}\frac{E_{22}}{\mu_{23}} \right) \frac{E_{11}}{E_{22}} \tag{1.15}$$

Compliance Properties

It is useful to express the inverted form of the stress-strain relations (1.1). Writing (1.1) in matrix notation

$$\{\sigma_i\} = [C_{ij}]\{\varepsilon_j\} \tag{1.16}$$

where $\{\sigma_i\}$ and $\{\varepsilon_i\}$ are six element vectors and $[C_{ij}]$ is a 6×6 matrix. Relations (1.16) may be inverted to give

$$\{\varepsilon_i\} = [S_{ij}]\{\sigma_j\} \tag{1.17}$$

where $[S_{ij}]$ is defined as

$$[S_{ij}] = \frac{\begin{bmatrix} \text{cofactor matrix} \\ \text{of } C_{ij} \end{bmatrix}^{\mathrm{T}}}{|C_{ij}|}$$

with the transpose operation being designated for the term in the numerator. From the symmetry $C_{ij}=C_{ji}$ it follows that $S_{ij}=S_{ji}$.

We now wish to express (1.17) in explicit form. To this end note that if σ_{11} were the only existing stress, then ε_{11} would be given by σ_{11}/E_{11}. Alternatively, if only σ_{22} were nonzero, then $\varepsilon_{22}=\sigma_{22}/E_{22}$; but in this case $\varepsilon_{11}=-\nu_{21}\varepsilon_{22}$; thus $\varepsilon_{11}=-\nu_{21}\sigma_{22}/E_{22}$. Superimposing effects of these types we get

$$\varepsilon_{11}=\frac{1}{E_{11}}\sigma_{11}-\frac{\nu_{21}}{E_{22}}\sigma_{22}-\frac{\nu_{21}}{E_{22}}\sigma_{33}$$

Two more relations of this type allow $[S_{ij}]$ to be written as

$$[S_{ij}]=\begin{bmatrix} \frac{1}{E_{11}} & -\frac{\nu_{21}}{E_{22}} & -\frac{\nu_{21}}{E_{22}} & 0 & 0 & 0 \\ -\frac{\nu_{12}}{E_{11}} & \frac{1}{E_{22}} & -\frac{\nu_{32}}{E_{22}} & 0 & 0 & 0 \\ -\frac{\nu_{12}}{E_{11}} & -\frac{\nu_{23}}{E_{22}} & \frac{1}{E_{22}} & 0 & 0 & 0 \\ 0 & 0 & 0 & \frac{1}{2\mu_{12}} & 0 & 0 \\ 0 & 0 & 0 & 0 & \frac{1}{2\mu_{23}} & 0 \\ 0 & 0 & 0 & 0 & 0 & \frac{1}{2\mu_{12}} \end{bmatrix} \tag{1.18}$$

Note that the symmetry property $S_{ij}=S_{ji}$ combined with the explicit terms in (1.18) reveals the symmetry relation (1.11).

Bounds on Poisson's Ratios

Finally, we wish to derive bounds on the Poisson's ratios ν_{ij} that will aid in their interpretation. First, with reference to relation (1.13) note that

$$\text{if } \mu_{23}\to\infty \qquad \text{then } \nu_{23}\to -1$$

also, from (1.13),

$$\text{if } K_{23}\to\infty, \qquad E_{11}\to\infty, \qquad \text{then } \nu_{23}\to 1$$

It can be shown from these results that

$$-1\leqslant\nu_{23}\leqslant 1 \tag{1.19}$$

otherwise some of the moduli measures would be negative. For an isotropic material, the upper bound upon ν is $\frac{1}{2}$; however, in the case of a transversely isotropic material, the upper bound of 1 can easily be rationalized. Consider an incompressible material that is infinitely rigid in the 1 direction. When a material of this type is extended in the 2 direction, it must contract by a like amount into the 3 direction to preserve incompressibility; thus in this case $\nu_{23}=1$.

Next consider bounds on ν_{12} and ν_{21}. From relation (1.12) solve for ν_{12}^2 to get

$$\nu_{12}^2=\frac{E_{11}}{E_{22}}-\frac{E_{11}}{4}\left(\frac{1}{K_{23}}+\frac{1}{\mu_{23}}\right) \tag{1.20}$$

All the moduli on the right-hand side of (1.20) are taken to be nonnegative. Clearly, the largest possible value for ν_{12}^2 is obtained when $K_{23}\to\infty$ and $\mu_{23}\to\infty$. This gives

$$|\nu_{12}|\leqslant\left(\frac{E_{11}}{E_{22}}\right)^{1/2} \tag{1.21}$$

Using (1.11) it also follows that

$$|\nu_{21}|\leqslant\left(\frac{E_{22}}{E_{11}}\right)^{1/2} \tag{1.22}$$

Consider for a moment the range of properties suggested by fiber reinforced materials. In such situations $E_{11}\gg E_{22}$ and from (1.21) and (1.22) it follows that ν_{12} can be larger than 1, as has been observed, whereas ν_{21} is typically very small compared with 1.

3.2 COMPOSITE CYLINDERS MODEL

We are now ready to seek solutions for the effective properties of fiber reinforced materials. As seen in the preceding section there are five independent properties to be determined. Our objective is to obtain an analytical representation for the five properties in terms of the fiber phase properties, the matrix phase properties, and the volume concentration of each phase.

A geometric model of the composite material must be introduced to accomplish this task. The most commonly used model is that of the

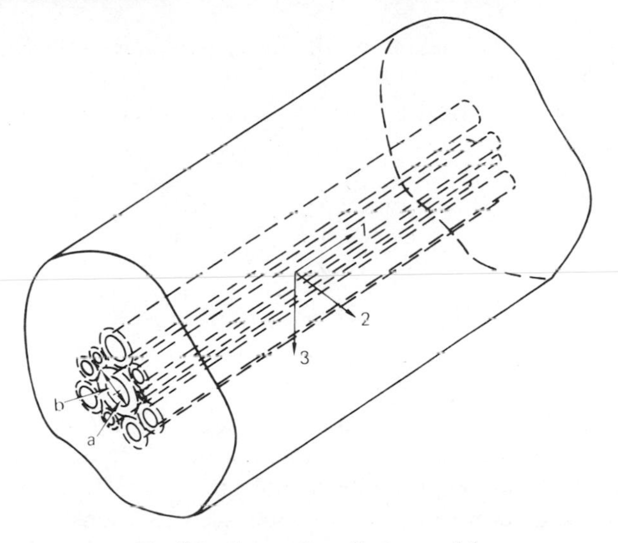

Fig. 3.2 Composite cylinders model.

composite cylinders model, introduced by Hashin and Rosen [3.1]. This model is the two-dimensional counterpart of the three-dimensional composite spheres model introduced in Section 2.3. The fiber phase is taken to be composed of infinitely long circular cylinders embedded in a continuous matrix phase. The model is shown in Fig. 3.2. With each individual fiber of radius a, there is associated an annulus of matrix material of radius b. Each individual cylinder combination of this type is referred to as a composite cylinder, and the absolute values of radii a and b vary with each composite cylinder such that a volume filling configuration is obtained. The ratio of radii a/b are, however, required to be constant for all composite cylinders. Naturally, the absolute size of the individual cylinders must vary, down to infinitesimal. The utility of this model results from the fact that the analysis of an individual composite cylinder will suffice to determine four of the five effective moduli for the representative volume element. At this point, we review the definition of effective properties in Section 2.1. The definitions and procedures delineated in that section are basic to the development of this chapter.

Uniaxial Modulus

The initial objective is to determine the effective uniaxial modulus E_{11}. To this end, take

$$\varepsilon_{11} = \varepsilon$$

$$\sigma_{22} = \sigma_{33} = \sigma_{12} = \sigma_{23} = \sigma_{31} = 0$$

where a rectangular Cartesian coordinate system is employed with coordinate 1 in the direction of the axis of the single composite cylinder. Converting to a cylindrical coordinate system, assume the following displacement field:

$$u_{rf}=A_f r$$

$$u_{rm}=A_m r+\frac{B_m}{r}$$

$$u_z=\varepsilon z \tag{2.1}$$

It is easy to show that these solution forms satisfy the equations of equilibrium, where A_f is a constant appropriate to the fiber phase and A_m and B_m are constants appropriate to the matrix phase. These three constants are to be determined from the interface conditions

$$\text{at} \quad r=a \qquad u_{rf}=u_{rm}$$

$$\sigma_{rf}=\sigma_{rm} \tag{2.2}$$

and the boundary condition

$$\text{at} \quad r=b \qquad \sigma_{rm}=0 \tag{2.3}$$

In the context of the present simple problem, the effective modulus E_{11} is defined by

$$E_{11}=\frac{\langle\sigma_{zz}\rangle}{\varepsilon}$$

which can be written as

$$E_{11}=\frac{1}{\pi b^2\varepsilon}\int\int_A \sigma_z(r)\,\mathrm{da} \tag{2.4}$$

where A is the cross-section area. Carrying out the operation in (2.4) gives

$$E_{11}=cE_f+(1-c)E_m+\frac{4c(1-c)(\nu_f-\nu_m)^2\mu_m}{\left[(1-c)\mu_m/(k_f+\mu_f/3)\right]+\left[c\mu_m/(k_m+\mu_m/3)\right]+1} \tag{2.5}$$

The result (2.5) is the effective uniaxial modulus for a single composite cylinder. It remains to show that it is, in fact, the proper result for the representative volume element. This problem is exactly the same one we faced in analyzing the composite spheres model, Section 2.3. We need not dwell on the procedure here; it is exactly the same as in Section 2.3. Namely, the theorem of minimum potential energy can be used to show that the result (2.5) is a lower bound for the corresponding result of the representative volume element. Alternatively, rather than imposing displacement boundary conditions, stresses can be imposed, and an upper bound for E_{11} can be found through the use of the theorem of minimum complementary energy. In so doing, it is found that the upper and lower bounds coincide, thus providing the exact solution, which then is given by (2.5).

Other Properties

The problem just outlined also provides a solution for the effective Poisson's ratio ν_{12}, where axis 1 is in the fiber direction and $\nu_{12} \neq \nu_{21}$. In the context of the present problem ν_{12} is defined by

$$\nu_{12} = -\frac{u_r|_{r=b}}{\varepsilon b} \tag{2.6}$$

which is minus the ratio of the lateral strain response to the imposed axial strain. It is found that

$$\nu_{12} = (1-c)\nu_m + c\nu_f + \frac{c(1-c)(\nu_f - \nu_m)[\mu_m/(k_m + \mu_m/3) - \mu_m/(k_f + \mu_f/3)]}{[(1-c)\mu_m/(k_f + \mu_f/3)] + [c\mu_m/(k_m + \mu_m/3)] + 1} \tag{2.7}$$

It is interesting to note that the formula (2.5) for E_{11} is very closely approximated by the rule of mixtures, simply by neglecting the last term in (2.5). To a lesser extent the formula (2.7) is also approximated by the rule of mixtures. It will be seen that the remaining three effective properties have values not well approximated by the rule of mixtures.

In an entirely similar manner to that just considered, two more problems can be posed to give exact solutions for the plane strain bulk modulus K_{23} and the shear modulus in the fiber direction μ_{12}. It is found that

$$K_{23} = k_m + \frac{\mu_m}{3} + \frac{c}{1/[k_f - k_m + \frac{1}{3}(\mu_f - \mu_m)] + (1-c)/(k_m + \frac{4}{3}\mu_m)} \tag{2.8}$$

and

$$\frac{\mu_{12}}{\mu_m} = \frac{\mu_f(1+c)+\mu_m(1-c)}{\mu_f(1-c)+\mu_m(1+c)} \tag{2.9}$$

These problems are equally simple to that outlined in the derivation of E_{11} and ν_{12}, in the sense that only one displacement component is involved in the field variable solutions. The explicit formulas (2.5), and (2.7)–(2.9) were derived by Hill [3.2] and Hashin [3.3].

Having found four of the five properties for the transversely isotropic media, our job would seem almost done. This unfortunately is not the case. The remaining property to be determined is μ_{23}, the transverse shear modulus, and this turns out to be a difficult matter for the composite cylinders model. Again the problem is completely parallel to the situation faced in Section 2.3 in seeking the effective shear modulus for the composite spheres model. Proceeding as described in Section 2.3, a displacement type problem can be posed to determine the upper bound for μ_{23}, and a stress type problem to determine the lower bound. As shown by Hashin and Rosen [3.1], the bounds do not coincide, except at very low and very high volume fractions. At this time no exact solution for μ_{23} has been presented. In the next section we present a solution for a different model that may also be the solution for μ_{23} of the composite cylinders model. The discussion of the range of validity of the composite cylinders model is deferred until Sections 4.3 and 4.6.

3.3 A MODEL FOR THE TRANSVERSE SHEAR OF A FIBER SYSTEM

The dilemma just faced in the preceding section inspires the use of yet another model. Specifically, it has proved to be very difficult to determine the transverse shear modulus for the composite cylinders model. Accordingly, we introduce a model that is intimately related to the composite cylinders model, but one for which we can determine the transverse shear modulus.

Consider again the composite cylinders model of Fig. 3.2. Replace all but a single composite cylinder by equivalent homogeneous media, as shown in Fig. 3.3. The determination of the effective transverse shear modulus for the model of Fig. 3.3 is now undertaken. The procedure is similar to that followed in Section 2.4, where a three phase spherical model was analyzed. The result to be found here may be the proper result for the

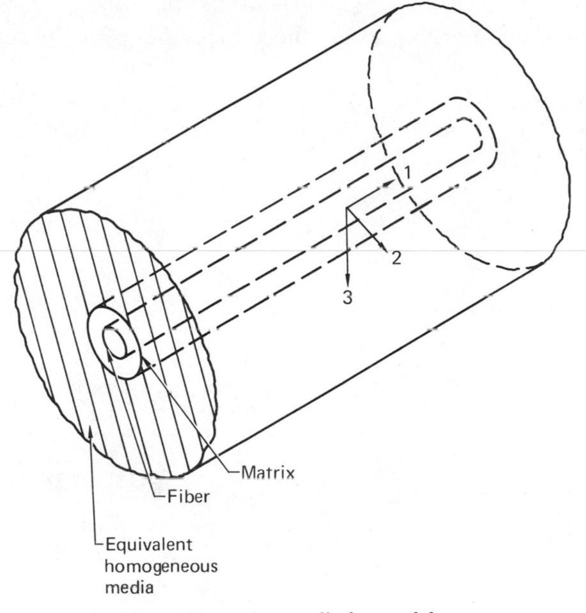

Fig. 3.3 Three phase cylinder model.

property μ_{23} of the composite cylinders model, but all we can say with certainty is that it is the exact result for the model of Fig. 3.3.

A state of shear deformation is imposed on the representative volume element. Specifically, the state of deformation in the equivalent homogeneous media phase is taken as

$$u_{re} = \frac{b}{4\mu_{23}}\left[\frac{2r}{b} + (\eta+1)\frac{b}{r}a_3 + \frac{b^3}{r^3}c_3\right]\cos 2\theta \tag{3.1}$$

$$u_{\theta e} = \frac{b}{4\mu_{23}}\left[-\frac{2r}{b} - (\eta-1)\frac{b}{r}a_3 + \frac{b^3}{r^3}c_3\right]\sin 2\theta \tag{3.2}$$

where

$$\eta = 3 - 4\nu_{23} \tag{3.3}$$

and where polar coordinates are employed, and as $r \to \infty$ relations (3.1, 3.2) represent an imposed state of simple shear deformation. Modulus μ_{23} and Poisson's ratio ν_{23} are the unknown effective properties of the equivalent homogeneous media. It is easy to show that relations (3.1, 3.2) satisfy the equations of equilibrium, as do those given here for the matrix phase

$$u_{rm}=\frac{b}{4\mu_m}\left[(\eta_m-3)\frac{r^3}{b^3}a_2+\frac{r}{b}d_2+(\eta_m+1)\frac{b}{r}c_3+\frac{b^3}{r^3}b_2\right]\cos 2\theta \quad (3.4)$$

$$u_{\theta m}=\frac{b}{4\mu_m}\left[(\eta_m+3)\frac{r^3}{b^3}a_2-\frac{r}{b}d_2-(\eta_m-1)\frac{b}{r}c_2+\frac{b^3}{r^3}b_2\right]\sin 2\theta \quad (3.5)$$

and for the fiber phase

$$u_{rf}=\frac{b}{4\mu_f}\left[(\eta_f-3)\frac{r^3}{b^3}a_1+\frac{r}{b}d_1\right]\cos 2\theta$$

$$u_{\theta f}=\frac{b}{4\mu_f}\left[(\eta_f+3)\frac{r^3}{b^3}a_1-\frac{r}{b}d_1\right]\sin 2\theta \quad (3.6)$$

where

$$\eta_m=3-4\nu_m$$

$$\eta_f=3-4\nu_f \quad (3.7)$$

There are eight unknown constants, a_1, a_2, a_3, b_2, c_2, c_3, d_1, d_2, to be determined.

The continuity conditions to be satisfied are given by

$$\sigma_r,\quad \sigma_{r\theta},\quad u_r,\quad u_\theta \qquad \text{continuous across} \quad r=\mathrm{a},b$$

These continuity conditions lead to the following eight independent equations

$$d_1-d_2+4\left(\frac{b}{a}\right)^2 c_2+3\left(\frac{b}{a}\right)^4 b_2=0$$

$$a_1-a_2+\left(\frac{b}{a}\right)c_2+\left(\frac{b}{a}\right)^6 b_2=0$$

$$[(\eta_m-3)\mu_f-(\eta_f-3)\mu_m]a_2+\left(\frac{b}{a}\right)^4[(\eta_m+1)\mu_f+(\eta_f+1)\mu_m]c_2$$

$$+(\mu_f-\mu_m)\left(\frac{b}{a}\right)^2 d_2+(\mu_f+\eta_f\mu_m)\left(\frac{b}{a}\right)^6 b_2=0$$

$$[(\eta_m+3)\mu_f-(\eta_f+3)\mu_m]a_2+\left(\frac{b}{a}\right)^4[(\eta_m-1)\mu_f-(\eta_f-1)\mu_m]c_2$$

$$-(\mu_f-\mu_m)\left(\frac{b}{a}\right)^2 d_2+(\mu_f+\eta_f\mu_m)\left(\frac{b}{a}\right)^6 b_2=0$$

$$2-4a_3-3c_3-d_2+4c_2+3b_2=0$$

$$-2-2a_3-3c_3-6a_2+d_2+2c_2+3b_2=0$$

$$\mu_m[2+(\eta+1)a_3+c_3]-\mu_{23}[(\eta_m-3)a_2+d_2+(\eta_m+1)c_2+b_2]=0$$

and

$$\mu_m[-2-(\eta-1)a_3+c_3]-\mu_{23}[(\eta_m+3)a_2-d_2-(\eta_m-1)c_2+b_2]=0 \qquad (3.8)$$

Our criterion for determining the effective properties requires the equality of strain energy in the heterogeneous media and in the equivalent homogeneous media. We write this condition as

$$U=U_{\text{EQUIV HOMOG}} \qquad (3.9)$$

We now employ Eshelby's formula (I-4.26) to obtain the energy stored in the composite material by integrating over the surface of the two layer inclusion of Fig. 3.3. But in formula (I-4.26) the term U_0 designates the strain energy in the media with the inclusion replaced by the material outside of it, which in this case is the material of equivalent homogeneous properties. Thus

$$U_0=U_{\text{EQUIV HOMOG}} \qquad (3.10)$$

Combining relations (3.9) and (3.10) with the energy formula (I-4.26) gives the appropriate form as

$$\int_0^{2\pi}\left(\sigma_{rre}u_r^0+\sigma_{r\theta e}u_\theta^0-\sigma_{rr}^0u_{re}-\sigma_{r\theta}^0u_{\theta e}\right)_{r=b}d\theta=0 \qquad (3.11)$$

where polar coordinates have been employed. The various terms in (3.11) are given by

$$\sigma_{rr}^{0}=\cos 2\theta$$

$$\sigma_{r\theta}^{0}=-\sin 2\theta$$

$$u_{r}^{0}=\frac{r}{2\mu_{23}}\cos 2\theta$$

$$u_{\theta}^{0}=\frac{r}{2\mu_{23}}\sin 2\theta$$

$$\sigma_{rre}\big|_{r=b}=\left(1-2a_3-\tfrac{3}{2}c_3\right)\cos 2\theta$$

$$\sigma_{r\theta e}\big|_{r=b}=-\left(1+a_3+\tfrac{3}{2}c_3\right)\sin 2\theta \tag{3.12}$$

and $u_{re}|_{r=b}$ and $u_{\theta e}|_{r=b}$ follow from (3.1,3.2). Using (3.1,3.2) and (3.12) in (3.11), and thence carrying out the integration gives the simple result

$$a_3=0 \tag{3.13}$$

Solving for a_3 from relations (3.8) and setting it equal to zero in accordance with (3.13) gives the determining form

$$A\left(\frac{\mu_{23}}{\mu_m}\right)^2+2B\left(\frac{\mu_{23}}{\mu_m}\right)+C=0 \tag{3.14}$$

where

$$A=3c(1-c)^2\left(\frac{\mu_f}{\mu_m}-1\right)\left(\frac{\mu_f}{\mu_m}+\eta_f\right)$$

$$+\left[\frac{\mu_f}{\mu_m}\eta_m+\eta_f\eta_m-\left(\frac{\mu_f}{\mu_m}\eta_m-\eta_f\right)c^3\right]\left[c\eta_n\left(\frac{\mu_f}{\mu_m}-1\right)-\left(\frac{\mu_f}{\mu_m}\eta_n+1\right)\right]$$

$$B=-3c(1-c)^2\left(\frac{\mu_f}{\mu_m}-1\right)\left(\frac{\mu_f}{\mu_m}+\eta_f\right)$$

$$+\frac{1}{2}\left[\eta_m\frac{\mu_f}{\mu_m}+\left(\frac{\mu_f}{\mu_m}-1\right)c+1\right]\left[(\eta_m-1)\left(\frac{\mu_f}{\mu_m}+\eta_f\right)\right.$$

$$\left.-2\left(\frac{\mu_f}{\mu_m}\eta_m-\eta_f\right)c^3\right]+\frac{c}{2}(\eta_m+1)\left(\frac{\mu_f}{\mu_m}-1\right)\left[\frac{\mu_f}{\mu_m}+\eta_f+\left(\frac{\mu_f}{\mu_m}\eta_m-\eta_f\right)c^3\right]$$

$$C=3c(1-c)^2\left(\frac{\mu_f}{\mu_m}-1\right)\left(\frac{\mu_f}{\mu_m}+\eta_f\right)$$

$$+\left[\eta_m\frac{\mu_f}{\mu_m}+\left(\frac{\mu_f}{\mu_m}-1\right)c+1\right]\left[\frac{\mu_f}{\mu_m}+\eta_f+\left(\frac{\mu_f}{\mu_m}\eta_m-\eta_f\right)c^3\right]$$

where the volume fraction of fiber phase c is given by

$$c=\left(\frac{a}{b}\right)^2$$

Under dilute conditions it may be shown that (3.14) reduces to the result

$$\frac{\mu_{23}}{\mu_m}=1+\frac{c}{\left[\mu_m/(\mu_f-\mu_m)\right]+\left(k_m+\frac{7}{3}\mu_m\right)/\left(2k_m+\frac{8}{3}\mu_m\right)} \tag{3.15}$$

Relation (3.14) is then the solution we seek. It yields the solution for the transverse shear modulus μ_{23} for the model of Fig. 3.3, in terms of the properties and volume fractions of each phase. It is reassuring and useful that the solution for μ_{23} (3.14) is independent of any of the other effective properties, even though ν_{23} was involved in the basic equations entering the problem.

Again it is emphasized that we do not know whether or not the result (3.14) is the solution for μ_{23} of the composite cylinders model. Nevertheless, for practical purposes we may proceed by using the four properties found in the preceding section, (2.5) and (2.7)–(2.9), along with (3.14) as the five independent properties of the transversely isotropic, fiber reinforced media. There is another purported solution, given by Hermans [3.4], of the same model, Fig. 3.3. However, there is an error in that work, which is discussed by Christensen and Lo [3.5]. The present work is taken from the latter reference.

The practical assessment of the fiber results obtained so far is deferred until later results are obtained for the behavior of systems involving a suspension of platelets. At this point it should be noted that the composite cylinders model and the model of this section allow the full range for the fiber phase, $0\leqslant c\leqslant 1$. The question of how well a model of this type represents the case of single size distribution cylinders is considered in Section 4.3.

3.4 FINITE LENGTH FIBER EFFECTS

The effective properties obtained in the preceding sections all assumed a condition of infinitely long cylinders. However, fiber composites are increasingly being used in forms involving so-called chopped fibers. Under this condition the fibers have a finite aspect ratio, and we would expect that under this condition the reinforcing properties of the fibers would be degraded. In this section we obtain a quantification of the effect. However, we restrict attention to a dilute condition, in which there is no interaction between fibers. The effective properties derived in the preceding sections had no restriction to dilute conditions, and in accordance with the geometry of the composite cylinders model the full volume fraction range for fiber phase $0 \leqslant c \leqslant 1$ was possible.

The derivation to be given here follows that of Russel [3.6], which in turn is based on the results obtained by Eshelby [3.7].

The problem to be solved is that of an ellipsoidal inclusion embedded in an infinite matrix phase, which is subjected to conditions of uniform deformation at large distances from the inclusion. From this solution we can determine the effective properties of the heterogeneous medium. Since our interest is in finite length fiber systems, we let the ellipsoid take the form of a slender prolate ellipsoid. The effective properties are determined as a function of the aspect ratio of the ellipsoid, as well as the properties of the two phases.

General Method

Unlikely as it may seem, we profitably begin by considering the problem of an infinite medium containing an ellipsoidal hole, rather than an inclusion of another material. An ellipsoidal plug of the same material, but of different size is to be fit into the hole; the situation is as shown in Fig. 3.4. Since the hole and the plug are of different size, a distribution of tractions on the surface of the plug is required to bring it to the same size

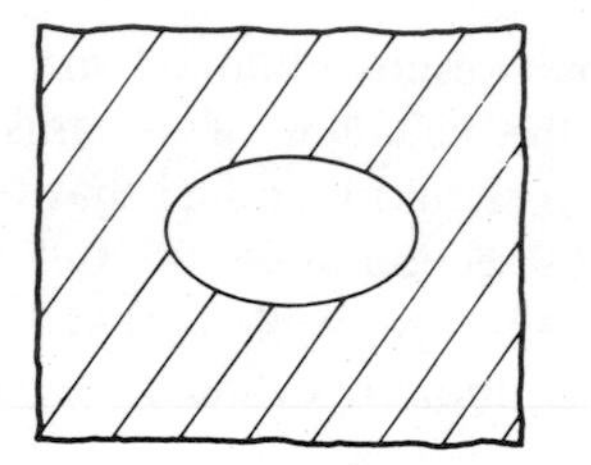

+

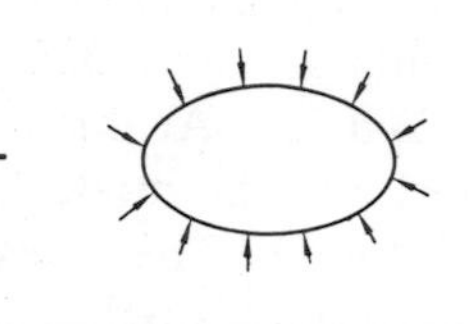

=

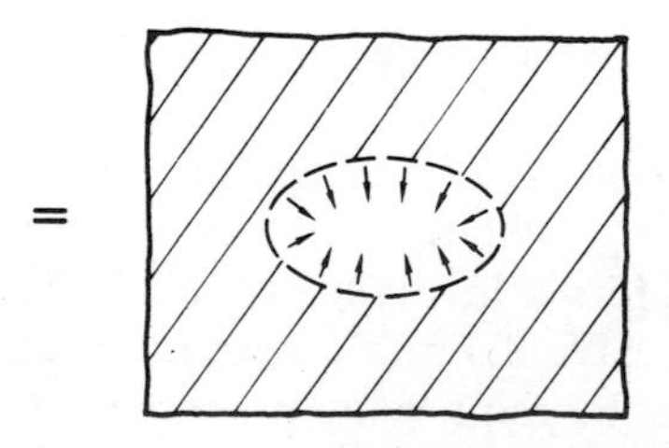

Fig. 3.4 Matching ellipsoidal plug to hole.

as the ellipsoidal hole. Then when combined, a continuous medium results, but with a distribution of body forces as shown in Fig. 3.4. In the final configuration of Fig. 3.4 there are no stresses or strains in the region outside that of the original hole.

Next an equal but opposite distribution of body forces is superimposed on the final configuration of Fig. 3.4, to remove the body forces. This alters the state of stress and strain everywhere in the medium. However, Eshelby [3.7] has shown that the state of stress and strain remains uniform inside the ellipsoidal region of the original hole. This result is crucial to further developments.

At this point we superimpose a uniform stress and strain over the entire region. In the region of the original hole, the the final state of stress has contributions from three sources: (1) the original body force application, (2) the superimposed opposite body forces on the continuous medium, and (3) the finally applied state of superimposed uniform stress. Relative to the original state of the unloaded medium with the ellipsoidal hole, the final state of strain in the region of the original hole has contributions from two sources,

$$\varepsilon_{ij}=\varepsilon_{ij}^{0}+\varepsilon_{ij}^{c} \tag{4.1}$$

where ε_{ij}^{0} is the applied state of uniform strain and ε_{ij}^{c} is the strain change resulting from the superimposed body forces, which negate the body forces needed to bring the plug to the size of the hole.

Now let ε_{ij}^{T} designate the original strain to bring the ellipsoidal plug to the size of the hole. The three sources of stress can be written as

$$\sigma_{ij}=\lambda_m\left(\varepsilon_{kk}^{0}+\varepsilon_{kk}^{c}-\varepsilon_{kk}^{T}\right)\delta_{ij}+2\mu_m\left(\varepsilon_{ij}^{0}+\varepsilon_{ij}^{c}-\varepsilon_{ij}^{T}\right) \tag{4.2}$$

where λ_m and μ_m are the properties of the isotropic matrix phase. The relationship between ε_{ij}^{c} and ε_{ij}^{T} can be written as a linear transformation

$$\varepsilon_{ij}^{c}=S_{ijkl}\varepsilon_{kl}^{T} \tag{4.3}$$

where the tensor S_{ijkl} can be found using Green's functions for homogeneous media. Eshelby [3.7] gives the solution for S_{ijkl}, and the results are stated later.

We are now finally in a position to consider the primary problem of interest, that involving an ellipsoidal inclusion of a material different from the matrix material. The stress-strain relation for the inclusion material is given by

$$\sigma_{ij}=\lambda_I\varepsilon_{kk}\delta_{ij}+2\mu_I\varepsilon_{ij} \tag{4.4}$$

Now we substitute the strain decomposition (4.1) into (4.4) and then substitute the resulting expression into (4.2). The result is

$$\lambda_I(\varepsilon_{kk}^0+\varepsilon_{kk}^c)\delta_{ij}+2\mu_I(\varepsilon_{ij}^0+\varepsilon_{ij}^c)=\lambda_m(\varepsilon_{kk}^0+\varepsilon_{kk}^c-\varepsilon_{kk}^T)\delta_{ij}$$
$$+2\mu_m(\varepsilon_{ij}^0+\varepsilon_{ij}^c-\varepsilon_{ij}^T) \tag{4.5}$$

Now substitute ε_{ij}^c from (4.3) into (4.5) and solve the resulting equation for ε_{ij}^T. We symbolically write this solution in the form

$$\varepsilon_{ij}^T=A_{ijkl}\varepsilon_{kl}^0 \tag{4.6}$$

Clearly, A_{ijkl} depends on the properties of both phases. With the solution for ε_{ij}^T from (4.6) then ε_{ij}^c is explicitly known from (4.3) and finally the total state of strain in the inclusion is known from (4.1). As we recall, knowledge of the state of strain in the inclusion leads us directly to the solution for the effective properties. We have used the solution for the homogeneous material problem to obtain the solution for the heterogeneous inclusion problem.

Now we explicitly obtain the expression for the effective properties in the present problem. In Section 2.1 we developed a formula for the effective properties, Eq. (II-1.9), which we repeat here:

$$C_{ijkl}\langle\varepsilon_{kl}\rangle=\lambda_m\delta_{ij}\langle\varepsilon_{kk}\rangle+2\mu_m\langle\varepsilon_{ij}\rangle$$
$$+\frac{1}{V}\sum_{n=1}^{N}\int_{V_N}(\sigma_{ij}-\lambda_m\varepsilon_{kk}\delta_{ij}-2\mu_m\varepsilon_{ij})\,dv \tag{4.7}$$

Substitute from (4.1) and (4.2) into (4.7) to obtain

$$C_{ijkl}\varepsilon_{kl}^0=\lambda_m\delta_{ij}\varepsilon_{kk}^0+2\mu_m\varepsilon_{ij}^0-\frac{1}{V}\sum_{n=1}^{N}\int_{V_n}(\lambda_m\varepsilon_{kk}^T\delta_{ij}+2\mu_m\varepsilon_{ij}^T)\,dv \tag{4.8}$$

where $\langle\varepsilon_{ij}\rangle=\varepsilon_{ij}^0$ has been used. At dilute concentration, and using the fact that ε_{ij}^T is uniform, (4.8) becomes

$$C_{ijkl}\varepsilon_{kl}^0=\lambda_m\delta_{ij}\varepsilon_{kl}^0+2\mu_m\varepsilon_{ij}^0-c(\lambda_m\varepsilon_{kk}^T\delta_{ij}+2\mu_m\varepsilon_{ij}^T) \tag{4.9}$$

where c is the volume concentration and ε_{ij}^T follows from the solution of (4.5) in the form of (4.6). The effective properties C_{ijkl} then are obtained from (4.9). The entire process follows once the particular state of strain ε_{ij}^0 is specified, and this leads to the corresponding properties.

Shear Properties

We begin the effective properties determination procedure by restricting attention to the transverse shear property μ_{23}, where axis 1 is in the fiber direction. Accordingly, let

$$\varepsilon_{23}^0 \neq 0 \qquad \text{all other} \quad \varepsilon_{ij}^0 = 0$$

From Eq. (4.5) and ε_{ij}^c from (4.3) we find that

$$\varepsilon_{23}^T = \frac{(\mu_m - \mu_I)\varepsilon_{23}^0}{\mu_m + 2(\mu_I - \mu_m)S_{2323}} \tag{4.10}$$

This result is then substituted into Eq. (4.9) to obtain

$$\frac{\mu_{23}}{\mu_m} = 1 + \frac{c[(\mu_I/\mu_m) - 1]}{1 + 2[(\mu_I/\mu_m) - 1]S_{2323}} \tag{4.11}$$

At this point we need the explicit formula for S_{2323} for use in (4.11). Later we need to use other forms of S_{ijkl}, so the appropriate forms from Eshelby [3.7] are now recorded. For purposes of writing S_{ijkl}, the summation convention is temporarily suspended, giving

$$S_{ijij} = \tfrac{1}{2}Q(r_i^2 + r_j^2)I_{ij} + \tfrac{1}{2}R(I_i + I_j)$$

$$S_{iijj} = Qr_j^2 I_{ij} - RI_i \qquad (i \neq j) \tag{4.12}$$

where

$$Q = \frac{3}{8\pi(1 - \nu_m)}$$

$$R = \frac{1 - 2\nu_m}{8\pi(1 - \nu_m)}$$

$$I_i = 2\pi a b_2 b_3 \int_0^\infty \frac{du}{(r_i^2 + u)\Delta}$$

$$I_{ij} = \tfrac{2}{3}\pi a b_2 b_3 \int_0^\infty \frac{du}{(r_i^2 + u)(r_j^2 + u)\Delta} \qquad (i \neq j)$$

$$I_{ii} = 2\pi a b_2 b_3 \int_0^\infty \frac{du}{(r_i^2 + u)^2\Delta}$$

$$\Delta = (a^2 + u)^{1/2}(b_2^2 + u)^{1/2}(b_3^2 + u)^{1/2} \tag{4.13}$$

with

$$r_1 = a \qquad r_2 = b_2 \qquad r_3 = b_3 \tag{4.14}$$

being the dimensions of the ellipsoid along the major and minor axes.

Interest here centers on the case of the prolate ellipsoid with $a > b_2 = b_3 = b$. Eshelby [3.7] shows that in this case

$$\begin{aligned}
I_1 &= 4\pi - I_2 - I_3 \\
I_2 = I_3 &= \frac{2\pi a b^2}{(a^2 - b^2)^{3/2}} \left[\frac{a}{b} \left(\frac{a^2}{b^2} - 1 \right)^{1/2} - \cosh^{-1} a/b \right] \\
I_{13} &= I_{12} \\
I_{22} = I_{33} &= 3 I_{23} \\
I_{11} &= \frac{4\pi}{3a^2} - 2 I_{12} \\
a^2 I_{11} &= I_1 - 2b^2 I_{12} \\
I_{12} + I_{22} + I_{23} &= \frac{4\pi}{3b^2}
\end{aligned} \tag{4.15}$$

The problem involved here is concerned with using the ellipsoid to model a finite length fiber; thus we make a slenderness approximation in accordance with

$$a \gg b$$

to obtain

$$\begin{aligned}
I_1 &\simeq 4\pi\kappa^2 \left(\ln \frac{2}{\kappa} - 1 \right) \\
I_2 = I_3 &\simeq 2\pi \left[1 - \kappa^2 \left(\ln \frac{2}{\kappa} - 1 \right) \right] \\
a^2 I_{11} &\simeq 4\pi\kappa^2 \left(\ln \frac{2}{\kappa} - \frac{4}{3} \right) \\
a^2 I_{12} = a^2 I_{13} &\simeq \frac{2\pi}{3} \left[1 - 3\kappa^2 \left(\ln \frac{2}{\kappa} - \frac{4}{3} \right) \right] \\
b^2 I_{22} = b^2 I_{33} &\simeq \pi \left(1 - \frac{\kappa^2}{2} \right) \\
b^2 I_{23} &\simeq \frac{\pi}{3} \left(1 - \frac{\kappa^2}{2} \right)
\end{aligned} \tag{4.16}$$

where

$$\kappa = \frac{b}{a} \ll 1 \tag{4.17}$$

Relations (4.16) are obtained using the expansion for $\cosh^{-1}(1/\kappa)$ in (4.15), namely,

$$\cosh^{-1}\left(\frac{1}{\kappa}\right) = \ln\frac{2}{\kappa} - \frac{\kappa^2}{4} - \frac{3}{32}\kappa^4 \ldots$$

With the use of (4.16) the S_{ijkl} forms in (4.12) can be written for slender prolate ellipsoids as

$$S_{1111} \simeq \frac{2-\nu_m}{1-\nu_m}\kappa^2\left[\ln\frac{2}{\kappa} - \frac{5-2\nu_m}{2(2-\nu_m)}\right]$$

$$S_{1122} \simeq -\frac{1-2\nu_m}{2(1-\nu_m)}\kappa^2\left[\ln\frac{2}{\kappa} - \frac{3-4\nu_m}{2(1-2\nu_m)}\right]$$

$$S_{2222} \simeq \frac{5-4\nu_m}{8(1-\nu_m)} - \frac{1-2\nu_m}{4(1-\nu_m)}\kappa^2\left[\ln\frac{2}{\kappa} - \frac{1-8\nu_m}{4(1-2\nu_m)}\right]$$

$$S_{2211} \simeq \frac{\nu_m}{2(1-\nu_m)} - \frac{1+\nu_m}{2(1-\nu_m)}\kappa^2\left[\ln\frac{2}{\kappa} - \frac{3+2\nu_m}{2(1+\nu_m)}\right]$$

$$S_{2233} \simeq -\frac{1-4\nu_m}{8(1-\nu_m)} + \frac{1-2\nu_m}{4(1-\nu_m)}\kappa^2\left[\ln\frac{2}{\kappa} - \frac{5-8\nu_m}{4(1-2\nu_m)}\right]$$

$$S_{1212} \sim \tfrac{1}{4}$$

$$S_{2323} \simeq \frac{3-4\nu_m}{8(1-\nu_m)} \tag{4.18}$$

Now, returning to the problem at hand, S_{2323} follows from (4.18) and when substituted into (4.11) the latter becomes

$$\frac{\mu_{23}}{\mu_m} = 1 + \frac{c(\mu_I/\mu_m - 1)}{1 + 2[(\mu_I/\mu_m) - 1]\{(3-4\nu_m)/[8(1-\nu_m)]\}} \tag{4.19}$$

Note that this result is identical with the prediction of the model of Section 3.3, Eq. (3.15). Thus we see that under the slenderness condition $b/a \ll 1$, the transverse shear modulus is, to that approximation, the same as the solution in the case of infinitely long cylinders. It can be shown that the slender prolate ellipsoid solution for μ_{12} also corresponds exactly to the infinitely long cylinder solution (2.9), under dilute conditions.

Uniaxial Modulus and Other Properties

The three remaining properties to be considered are E_{11}, ν_{12}, and K_{23}. With regard to the determination of these properties, we begin by specifying

$$\varepsilon_{11}^0 \neq \varepsilon$$

$$\varepsilon_{22}^0 = \varepsilon_{33}^0 = \varepsilon_{rr}^0 \neq 0$$

The latter strains can be evaluated to satisfy the condition of stress-free lateral surfaces in the determination of E_{11} and ν_{12}. For K_{23} we have $\varepsilon_{11}^0 = 0$. Equation (4.5) then gives

$$a_{11}\varepsilon_{11}^T + a_{12}\varepsilon_{rr}^T = b_{11}\varepsilon_{11}^A + b_{12}\varepsilon_{rr}^A$$

$$a_{21}\varepsilon_{11}^T + a_{22}\varepsilon_{rr}^T = b_{21}\varepsilon_{11}^A + b_{22}\varepsilon_{rr}^A \tag{4.20}$$

where

$$a_{11} = \Delta\lambda S_{ii11} + 2\Delta\mu S_{1111} + \lambda_m + 2\mu_m$$

$$a_{12} = 2\Delta\lambda S_{ii22} + 4\Delta\mu S_{1122} + 2\lambda_m$$

$$a_{21} = \Delta\lambda S_{ii11} + 2\Delta\mu S_{2211} + \lambda_m$$

$$a_{22} = 2\Delta\lambda S_{ii22} + 2\Delta\mu(S_{2222} + S_{2233}) + 2\lambda_m + 2\mu_m$$

$$b_{11} = -\Delta\lambda - 2\Delta\mu$$

$$b_{12} = -2\Delta\lambda$$

$$b_{21} = -\Delta\lambda$$

$$b_{22} = -2\Delta\lambda - 2\Delta\mu \tag{4.21}$$

with

$$\Delta\lambda = \lambda_I - \lambda_m$$

$$\Delta\mu = \mu_I - \mu_m \tag{4.22}$$

Solving (4.20) for ε_{ij}^T in the form of (4.6) gives the only nonzero elements of A_{ijkl} as

$$A_{1111}=\frac{1}{b}(b_{11}a_{22}-b_{21}a_{12})$$

$$A_{1122}=A_{1133}=\frac{1}{2D}(b_{12}a_{22}-a_{12}b_{22})$$

$$A_{2211}=A_{3311}=\frac{1}{D}(a_{11}b_{21}-b_{11}a_{21})$$

$$A_{2222}+A_{3333}=A_{3322}+A_{3333}=\frac{1}{D}(a_{11}b_{22}-a_{21}b_{12})$$

$$\text{other} \quad A_{ijkl}=0 \tag{4.23}$$

where

$$D=a_{11}a_{22}-a_{12}a_{12}$$

Now, employing relation (4.9) under the condition of uniaxial stress gives

$$\frac{E_{11}}{E_m}=1-c\left[A_{1111}-\nu_m(A_{1122}+A_{1133})\right] \tag{4.24}$$

where (4.6) and (4.23) have been used. Similarly, it is found that

$$\begin{aligned}\nu_{12}=\nu_m-c\{&A_{2211}+A_{3311}-\nu_m(A_{2222}+A_{2233})\\&+\nu_m[A_{1111}-\nu_m(A_{1122}+A_{1133})]\}\end{aligned} \tag{4.25}$$

$$K_{23}=K_m\{1-c[A_{2222}+A_{2233}+\nu_m(A_{1122}+A_{1133})]\} \tag{4.26}$$

The final solutions for E_{11}, ν_{12}, and K_{23} in the case of slender prolate ellipsoids are given by substituting (4.23) in (4.24)–(4.26), using (4.21) and (4.18).

We now write out the explicit final solution in the case of E_{11}; the expressions for ν_{12} and K_{23} are given in [3.6]. E_{11} is given by

$$\frac{E_{11}}{E_m}=1+c$$

$$\times\frac{\dfrac{\Delta\mu}{2(1-\nu_m)}(3\Delta\lambda+2\Delta\mu)+\dfrac{E_m(1-2\nu_m)}{2(1+\nu_m)}\Delta\lambda+\dfrac{E_m(1+2\nu_m^2)}{(1+\nu_m)(1-2\nu_m)}\Delta\mu}{\Delta\mu(3\Delta\lambda+2\Delta\mu)\left(\dfrac{1+\nu_m}{1-\nu_m}\right)\kappa^2\left[\ln\dfrac{2}{\kappa}-\dfrac{5-4\nu_m}{4(1-\nu_m)}\right]+\dfrac{E_m}{2(1-\nu_m)}(\Delta\lambda+\Delta\mu)+\mu_m(3\lambda_m+2\mu_m)} \tag{4.27}$$

It is helpful to examine this result in two special cases. To this end we classify the order of the terms in (4.27) according to their relationship to powers of the fiber modulus E_I. Writing (4.27) in symbolic form, the various terms in (4.27) have the following order:

$$\frac{E_{11}}{E_m} = 1 + c\left[\frac{O(E_I^2) + O(E_I) + O(E_I^0)}{O(E_I^2) + O(E_I) + O(E_I^0)}\right]$$

Consider first the special case where the fibers are nearly rigid, such that

$$\frac{E_I}{E_m}\kappa^2 \ln\frac{2}{\kappa} \gg 1$$

Then only the $O(E_I^2)$ terms in (4.27) need be retained, giving

$$\frac{E_{11}}{E_m} = 1 + c\left\{\frac{1/[2(1+\nu_m)]}{\kappa^2\{\ln(2/\kappa) - (5-4\nu_m)/[4(1-\nu_m)]\}}\right\} \tag{4.28}$$

Alternatively, take the opposite situation to that considered earlier, namely,

$$\frac{E_I}{E_m}\kappa^2 \ln\frac{2}{\kappa} \ll 1$$

In this case we retain only the lowest order terms in E_I in the numerator and denominator of (4.27) to obtain

$$\frac{E_{11}}{E_m} = 1 + c\left[\frac{E_I}{E_m} - 1 + \frac{2(\nu_I - \nu_m)^2}{(1+\nu_m) + (E_m/E_I)(1+\nu_I)(1-2\nu_I)}\right] \tag{4.29}$$

This form, (4.29), is identical with the dilute solution form of the infinitely long cylinder solution (II-2.5).

The general solution (4.27) and even the further approximation (4.28) show the strong dependence the effective property E_{11} has on the aspect ratio of the inclusion $\kappa = b/a$. Writing (4.27) in the form

$$\frac{E_{11}}{E_m} = 1 + cA \tag{4.30}$$

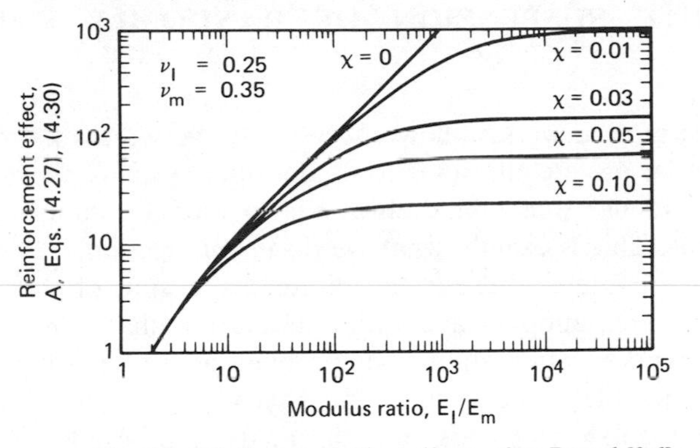

Fig. 3.5 Finite length fiber effects, after Russel [3.6].

the form of A is shown in Fig. 3.5 as a function of κ and E_I/E_m. The common curve in Fig. 3.5 at low values of E_I/E_m corresponds to (4.29), whereas the various asymptotes at high values of E_I/E_m correspond to (4.28). The particularly interesting range of $10 \leqslant E_I/E_m \leqslant 10^{+3}$ covers many fiber composites of practical interest. In this range there is a strong dependence of E_{11} on the aspect ratio κ. For example, for $E_I/E_m = 200$, with $\kappa = 0.01$ $(l/d = 100)$ the factor A in (4.30) has a value of about 82% of the infinitely long case, whereas at $\kappa = 0.1$ $(l/d = 10)$, A retains only about 11% of its value at the infinitely long asymptote. These results, of course, apply only to the dilute suspension case. In the nondilute case the degrading effect of a finite l/d value would probably be less severe, but still qualitatively the same as that shown here. Clearly, the fibers must be very long in comparison to the diameter, or significant stiffness effect is sacrificed.

Problems of these types have been approached using a method known as shear lag analysis. However, an examination of the method shows that it involves rather drastic assumptions, which are not in accordance with elasticity theory. More refined analyses utilizing a three-dimensional elasticity formulation have been given by Muki and Sternberg [3.8] and Sternberg and Muki [3.9]. An extension of these works has been given by Carne [3.10] in which the load distribution is obtained in two semi-infinite, interacting fibers, embedded in a three-dimensional elastic continuum.

Finally, mention should be made of the work of Levin [3.11]. In work similar to that just given, he utilized Eshelby's solution for an oblate ellipsoidal inclusion to obtain solutions for the effective properties of a material containing an array of nonintersecting cracks.

3.5 A DILUTE SUSPENSION OF RANDOMLY ORIENTED PLATELETS

In the previous section we saw how successfully we were able to use the analysis of slender prolate ellipsoids to model fiber systems. Obviously, we should be able to use thin oblate ellipsoids to model composite platelet lamellar systems. This is exactly what we do in this section.

The problem of interest here is the dilute suspension of thin platelets having random orientation in a continuous matrix phase. We defer the corresponding random orientation fiber problem as well as the consideration of aligned platelet systems until the next chapter. The platelets are modeled as thin oblate ellipsoids, and the analysis proceeds along much the same lines as in the preceding section concerned with prolate ellipsoids. The general outline here follows the work of Boucher [3.12].

A suspension of randomly oriented inclusions produces a composite material with isotropic effective properties. The proper form for the solution of μ and k is given by relations (II-2.14), rewritten here as

$$\frac{\mu-\mu_m}{\mu_I-\mu_m}=c\frac{\langle e_I\rangle}{e^0}$$

and

$$\frac{k-k_m}{k_I-k_m}=c\frac{\langle \varepsilon_I\rangle}{\varepsilon^0} \tag{5.1}$$

where the notation of the previous section is adopted with ε_{ij}^0 being the imposed uniform state of strain and $\langle e_I\rangle$ and $\langle \varepsilon_I\rangle$ being the volume averages over the inclusion of the shear strain and the dilatational strain, respectively. Thus we must determine the states of strain within the dilute suspension inclusion.

The inclusion strain can be written as

$$(\varepsilon_{ij})_{\text{INCL}}=\varepsilon_{ij}^0+\varepsilon_{ij}^c \tag{5.2}$$

where ε_{ij}^c is defined in the preceding section as

$$\varepsilon_{ij}^c=S_{ijkl}\varepsilon_{kl}^T \tag{5.3}$$

with ε_{ij}^T being the solution of Eq. (4.5) and S_{ijkl} are defined in the preceding section. The solution of Eq. (4.5) is written in symbolic form as

$$\varepsilon_{ij}^T=A_{ijkl}\varepsilon_{kl}^0 \tag{5.4}$$

Combining (5.3) and (5.4) gives

$$\varepsilon_{ij}^{c} = S_{ijkl} A_{klmn} \varepsilon_{mn}^{0} \tag{5.5}$$

Substitution of (5.5) into (5.2) allows the latter to be written as

$$(\varepsilon_{ij})_{\text{INCL}} = T_{ijkl} \varepsilon_{kl}^{0} \tag{5.6}$$

where

$$T_{ijkl} = \delta_{ik}\delta_{jl} + S_{ijmn} A_{mnkl} \tag{5.7}$$

The form (5.6) shows the direct transformation of the imposed uniform strain field to the strain conditions within the inclusion. It must be remembered that these strain conditions in the ellipsoidal inclusion are uniform.

Consider next an imposed state of dilatation, specified by

$$\varepsilon_{ij}^{0} = \frac{\varepsilon}{3}\delta_{ij} \tag{5.8}$$

Using (5.8) in (5.6) gives

$$(\varepsilon_{jj})_{\text{INCL}} = \frac{T_{jjkk}}{3}\varepsilon \tag{5.9}$$

Now, T_{jjkk} is an invariant of the tensor T_{ijkl}; thus relation (5.9) shows that the volume change in the ellipsoidal inclusion does not depend on its orientation, which, of course, is necessary in the present problem. Using (5.9) directly with (5.1) then gives

$$\frac{k - k_m}{k_I - k_m} = \frac{c}{3} T_{jjkk} \tag{5.10}$$

The problem thus reduces to finding T_{jjkk}, defined by (5.7). The tensor S_{ijkl} appears explicitly in (5.7) and also implicitly in A_{mnkl}. The next step is to deduce the appropriate form for S_{ijkl} in the case of thin oblate ellipsoids.

The tensor S_{ijkl}, in the special form needed, is defined by Eq. (4.12). It is first necessary to calculate I_i and thence I_{ij} for use in (4.12). From Eshelby [3.7], the appropriate form for I_2 and I_3 in the present problem is

$$I_2 = I_3 = \frac{2\pi a b^2}{(b^2 - a^2)^{3/2}}\left[\cos^{-1}\frac{a}{b} - \frac{a}{b}\left(1 - \frac{a^2}{b^2}\right)^{1/2}\right] \tag{5.11}$$

where the oblate ellipsoid has semi-axes a, $b=c$ with $a<b$, and coordinate direction 1 corresponds to the axis of symmetry of the ellipsoid. Interest here is in the case of very thin ellipsoids. In fact, we only explore the limiting case behavior as $a/b\to 0$. In this case, from (5.11)

$$\lim_{a/b\to 0} I_2=0$$

and from Eshelby [3.7] it follows that

$$\begin{aligned} I_1&=4\pi \\ I_2=I_3&=0 \end{aligned} \tag{5.12}$$

At this point the result (5.12) should be compared with the corresponding forms for slender prolate ellipsoids (4.16). We see that our objectives are more modest here: We are only examining the limiting case behavior as $a/b\to 0$, whereas in the work with the prolate ellipsoids we retained the first two terms in the relevant expansions. Thus we do not obtain results for platelets of finite aspect ratio, but just obtain the limiting case results for platelets with an infinite aspect ratio.

Having the result (5.12) the values of I_{ij} follow from the formulas in Eshelby [3.7], S_{jjkk} follows from (4.12), and finally T_{jjkk} follows from (5.7). The algebraic details of performing this operation are lengthy, but straightforward, and similar to those of the preceding section.
The final result is

$$\frac{k-k_m}{k_I-k_m}=\frac{c}{1+(k_I-k_m)/\left(k_m+\frac{4}{3}\mu_I\right)} \tag{5.13}$$

Now we must solve the corresponding problem for the effective shear modulus μ. To this end, take a state of shear deformation specified by

$$\varepsilon_{ij}^0=e_{ij}^0, \qquad \varepsilon_{kk}^0=0 \tag{5.14}$$

This form, when substituted into (5.6), gives the shear strain in the inclusion as

$$(e_{ij})_{\text{INCL}}=T_{ijkl}e_{kl}^0-\frac{T_{mmkl}}{3}e_{kl}^0\delta_{ij} \tag{5.15}$$

We need the volumetric average of $(e_{ij})_{\text{INCL}}$ over all equally possible orientations of the ellipsoidal inclusion. This procedure would require a rather involved integration, which we consider later in Section 4.5. For present purposes the integration can be avoided by the following simple procedure.

The average of $(e_{ij})_{\mathrm{INCL}}$ over all possible orientations of the inclusion is designated by $(e_{ij})_{\mathrm{INCL}}^{\mathrm{AVG}}$. This latter quantity must be a linear function of the invariants of first degree of T_{ijkl}, namely, T_{ijij} and T_{jjkk}. To find the linear function of these invariants we begin by writing specific forms for the equation

$$\frac{(e_{ij})_{\mathrm{INCL}}^{\mathrm{AVG}}}{e_{ij}^{0}} = C \tag{5.16}$$

where the fact that C is a constant follows from the original form of the effective property equation, (5.1). First, for a shear deformation in the 1,2 plane, (5.6) and (5.16) give

$$T_{1212}^{\mathrm{AVG}} = \frac{C}{2}$$

where T_{ijkl}^{AVG} is the integration of T_{ijkl} over all possible orientations of the ellipsoid. Two more equations of this type follow from similar shear deformations in the other two planes. Next a shear deformation of the type $\varepsilon_{22}^{A} = -\varepsilon_{11}^{A}$ is specified, and Boucher [3.12] shows that in this case (5.6) and (5.16) give

$$T_{1111}^{\mathrm{AVG}} + T_{2222}^{\mathrm{AVG}} - T_{1122}^{\mathrm{AVG}} - T_{2211}^{\mathrm{AVG}} = 2C$$

and two similar equations follow immediately. Boucher [3.12] shows that these six equations can be combined in linear combinations to give

$$3T_{ijij}^{\mathrm{AVG}} - T_{jjkk}^{\mathrm{AVG}} = 15C$$

This is the desired result since $T_{ijij}^{\mathrm{AVG}} = T_{ijij}$ and $T_{jjkk}^{\mathrm{AVG}} = T_{jjkk}$ because of the invariant properties of T_{ijkl}.

We can thus write

$$\frac{(e_{ij})_{\mathrm{INCL}}^{\mathrm{AVG}}}{e_{ij}^{0}} = \tfrac{1}{15}(3T_{ijij} - T_{jjkk}) \tag{5.17}$$

Proceeding as with the evaluation of k, $T_{ijij}T_{jjkk}$ are found for the limiting case of the oblate ellipsoid, and when combined with (5.1) and (5.17) there results

$$\frac{\mu - \mu_m}{\mu_I - \mu_m} = \frac{c}{1 + (\mu_I - \mu_m)/(\mu_m + \mu_d)} \tag{5.18}$$

where

$$\mu_d = \frac{\mu_I(9k_I + 8\mu_I)}{6(k_I + 2\mu_I)}$$

The final results (5.13) and (5.18) are basic results giving us the reinforcing effects of platelets under dilute conditions. These results are of great value to us beyond just directly giving reinforcing effect. In the next chapter we find that the results (5.13) and (5.18) have a special interpretation as providing the maximum possible reinforcing effect, but that conclusion must await developments in Chapter IV.

Finally, we note in passing that the development performed here could equally well be performed in the case of a random orientation model for fibers, using the prolate ellipsoid results of the preceding section. We do not pursue that line; an alternate means of studying the random orientation fiber problem is given in Section 4.4. In fact, in the next chapter we also reanalyze the randomly oriented platelet problem, by a method that allows us to remove the present restriction to dilute conditions.

PROBLEMS

1. For transversely isotropic materials prove that

$$\frac{\nu_{12}}{E_{11}} = \frac{\nu_{21}}{E_{22}}$$

2. Verify relations (1.12)–(1.15), which interrelate the properties of transversely isotropic materials.
3. Derive E_{11} and ν_{12} for the composite cylinders model.
4. Derive μ_{12} for the composite cylinders model.
5. The derivation of Section 3.4 gave effective properties for finite length fibers under dilute conditions. Under nondilute conditions would one expect the effect of fiber aspect ratio to be more severe or less severe than in the dilute case? Explain your reasoning.
6. At a given aspect ratio, would one expect the edge effects in finite platelets to be more significant or less significant than the end effect in fibers?
7. In the case of perfectly rigid fibers in an incompressible matrix phase, use the method of Section 2.6 (lubrication theory) to derive the transverse shear modulus under high volume fraction conditions.

8. Explain why the bounds on ν_{ij} for transversely isotropic media do not coincide with the bounds on ν for isotropic media.
9. Discuss the extension of the results given in Sections 3.2 and 3.3 to account for the effect of transversely isotropic fibers (see Hashin [3.13]).

REFERENCES

3.1 Z. Hashin and B. W. Rosen, "The elastic moduli of fiber-reinforced materials," *J. Appl. Mech.*, vol. 31, 223 (1964).

3.2 R. Hill, "Theory of mechanical properties of fibre-strengthened materials: I. Elastic behavior," *J. Mech. Phys. Solids*, vol. 12, 199 (1964).

3.3 Z. Hashin, "Viscoelastic fiber reinforced materials," *AIAA J.*, vol. 4, 1411 (1966).

3.4 J. J. Hermans, "The elastic properties of fiber reinforced materials when the fibers are aligned," *Proc. K. Ned. Akad. Wet.*, vol. B70, 1 (1967).

3.5 R. M. Christensen and K. H. Lo, "Solutions for effective shear properties in three phase sphere and cylinder models,"*J. Mech. Phys. Solids*, vol 27, no. 4 (1979).

3.6 W. B. Russel, "On the effective moduli of composite materials: effect of fiber length and geometry at dilute concentrations," *Z. Angew. Math. Phys.*, vol. 24, 581 (1973).

3.7 J. D. Eshelby, "The determination of the elastic field of an ellipsoidal inclusion, and related problems," *Pro. R. Soc. Lond.*, vol. A241, 376 (1957).

3.8 R. Muki and E. Sternberg, "On the diffusion of an axial load from an infinite cylindrical bar embedded in an elastic medium," *Int. J. Solids Structures*, vol. 5, 587 (1969).

3.9 E. Sternberg and R. Muki, "Load-absorption by a filament in a fiber-reinforced material," *Z. Angew. Math. Phys.*, vol. 21, 552 (1970).

3.10 T. G. Carne, "Load absorption and interaction of two adjacent filaments in a fiber-reinforced material," *J. Elasticity*, vol. 6, 1 (1976).

3.11 V. M. Levin, "Determination of effective elastic moduli of composite materials," *Sov. Phys. Dokl.*, vol. 20, 147 (1975).

3.12 S. Boucher, "On the effective moduli of isotropic two-phase elastic composites," *J. Comp. Mater.*, vol. 8, 82 (1974).

3.13 Z. Hashin, "Analysis of properties of fiber composites with anisotropic constituents," *J. Appl. Mech*, vol. 46, (1979).

CHAPTER IV BOUNDS ON EFFECTIVE MODULI AND FURTHER RESULTS

We have developed several important and practical results concerning the combination of materials having different geometric forms. Specifically, we have obtained some results on the effective stiffness properties for spherical, cylindrical, and lamellar type inclusions. The latter case, however, has been considered only at dilute concentrations. It would be interesting and revealing to compare the efficiency of reinforcement of these different forms. We assimilate enough results in this chapter, when combined with the previous results, to make a critical assessment of reinforcing efficiency for various types of interfacial geometry. We also seek to obtain information upon upper and lower bounds for effective properties, without specification of the interface geometry.

At first thought it might seem that information on upper and lower bounds for effective properties would be of only secondary importance. After all, a bound is no substitute for a carefully derived, direct prediction of the property of interest. Even when there is no acceptable derivation of a certain property, the bounds would seem to have limited usefulness. When a prediction is needed, should one use the upper bound, or the lower bound, or half the distance between them? These questions have no answer, and in this sense bounds are of little value. However, we find a much stronger use for the bounds. We succeed in identifying the upper and lower bounds with particular geometric forms for two phase material combinations. Thus from the upper bound behavior we gain a fundamental understanding of optimal geometric forms for reinforcement using a stiff inclusion phase in a compliant matrix phase. Alternatively, from the bounds characteristics we learn what geometric form of voids provides the least degrading effect on the effective stiffness properties. Thus, contrary to our first guess, we find that the bounds information is of great utility to us in interpreting the mechanical behavior of heterogeneous media.

Finally, we add just a few words at this point concerning the specific items to be covered in this chapter, other than the bounds work. In the preceding chapter we obtained the effective stiffness properties for a dilute suspension of randomly oriented platelets. In this chapter we provide the complement of that result by removing the restriction to dilute conditions. Also, we obtain the corresponding results for a randomly oriented distribution of fibers. The reason for placement of these results in separate chapters is as follows. The dilute suspension case for randomly oriented platelets was logically placed in the preceding chapter, in proximity to the finite length fiber analysis, since both studies were based on Eshelby's basic analysis of ellipsoidal inclusions. However, the analyses to be performed in the later sections here on randomly oriented platelets and fibers proceed by a totally different method from that involving prolate and oblate ellipsoids. As is shown, the analyses here have a logical proximity to the bounds studies.

4.1 BOUNDS ON ISOTROPIC EFFECTIVE MODULI

It is quite obvious that the minimum theorems of Section 1.1 can be used to obtain bounds on the effective properties. Proceed by using the simplest possible admissible fields involving uniform stress, or uniform strain, across all phases of the heterogeneous solid. To be more specific, we take the case of a macroscopically isotropic composite material containing N different material phases. Assuming a uniform state of strain across all phases imposed by a linearly varying displacement field, it immediately follows from the theorem of minimum potential energy and the definition of effective properties through energy equivalence that the effective moduli are bounded by

$$
\begin{aligned}
k &\leqslant \sum_{i=1}^{N} c_i k_i \\
\mu &\leqslant \sum_{i=1}^{N} c_i \mu_i
\end{aligned}
\tag{1.1}
$$

where c_i are the volume fractions of the respective phases, and in obtaining (1.1) separate admissible states of dilatation and simple shear must be considered. Alternatively, a uniform state of stress throughout the entire material, combined with the theorem of minimum complementary energy, gives

$$\frac{1}{k} \leqslant \sum_{i=1}^{N} \frac{c_i}{k_i}$$

$$\frac{1}{\mu} \leqslant \sum_{i=1}^{N} \frac{c_i}{\mu_i} \tag{1.2}$$

Combining relations (1.1) and (1.2) gives

$$\frac{1}{\sum_{i=1}^{N} c_i/k_i} \leqslant k \leqslant \sum_{i=1}^{N} c_i k_i$$

$$\frac{1}{\sum_{i=1}^{N} c_i/\mu_i} \leqslant \mu \leqslant \sum_{i=1}^{N} c_i \mu_i \tag{1.3}$$

These bounding relations were first noticed by Paul [4.1]. The lower bounds in (1.3) are referred to as the Voigt bounds, and the upper bounds are designated as the Reuss bounds. Both Reuss and Voigt noted relations of these types as being representative of models composed of series elements and parallel elements.

Unfortunately, the bounds predicted by (1.3) are so far apart for typical composite materials that they are of no practical value. Of course, we know how to obtain more explicit information on effective properties by specifying the geometry of the combination of materials. Our objective here, however, is to obtain the most explicit possible information on bounds without specifying phase geometry. The question is, can we obtain tighter bounds than those involved in (1.3) without specifying phase geometry? A first intuitive answer to this question would probably be no. How would we find the appropriate admissible field for use [other than the uniform fields already used in (1.3)] without specifying phase geometry? Well, the answer to the original question is yes. We shall indeed find tighter and more meaningful bounds than those in (1.3). That this can be done is one of the remarkable achievements in the field. The result was derived by Hashin and Shtrikman [4.2]. We here follow the derivation of Walpole [4.3], who provided a generalization of the original results. The latter derivation utilized much of the work of Hill, as noted therein.

Upper Bound Development

Consider a heterogeneous solid composed of N phases of individual volumes V_r and volume fractions c_r, subjected to displacement boundary conditions on its surface. We employ direct rather than tensor notation, as discussed in Section 1.1. The elastic moduli of phase r is designated by $\mathbf{C}_r$. We introduce a homogeneous comparison material that will be of value to us in the analysis. The comparison material is subjected to the same displacement boundary conditions as in the problem of interest. The situation is entirely similar to the one we encountered in Section 1.4 in deriving Eshelby's formula. Designate the properties of the comparison material by $\mathbf{C}_0$, to be specified later.

Let $\boldsymbol{\varepsilon}$ be any strain field in the comparison material that is derived from continuous displacements that satisfy the surface conditions. Let

$$\boldsymbol{\sigma}^* = \mathbf{C}_0\boldsymbol{\varepsilon} + \boldsymbol{\tau} \tag{1.4}$$

where we later insure that $\boldsymbol{\sigma}^*$ is self-equilibrated (satisfies equilibrium equations) and $\boldsymbol{\tau}$ is referred to as the *polarization stress*, which we specify presently. For

$$\boldsymbol{\tau} = (\mathbf{C}_r - \mathbf{C}_0)\boldsymbol{\varepsilon}$$

we would have the exact solution of the heterogeneous media problem of interest; otherwise $\boldsymbol{\sigma}^*$ and $\boldsymbol{\varepsilon}$ do not correspond to the solution.

Take a piecewise uniform polarization stress

$$\boldsymbol{\tau} = (\mathbf{C}_r - \mathbf{C}_0)\bar{\boldsymbol{\varepsilon}}_r \tag{1.5}$$

where $\bar{\boldsymbol{\varepsilon}}_r$ is the volume average over V_r of $\boldsymbol{\varepsilon}$. Thus

$$\boldsymbol{\sigma}^* = \mathbf{C}_r\bar{\boldsymbol{\varepsilon}}_r + \mathbf{C}_0\boldsymbol{\varepsilon}'_r \tag{1.6}$$

where

$$\boldsymbol{\varepsilon}'_r = \boldsymbol{\varepsilon} - \bar{\boldsymbol{\varepsilon}}_r \tag{1.7}$$

Variable $\boldsymbol{\varepsilon}'_r$ represents the deviation within V_r of $\boldsymbol{\varepsilon}$ from its average value.

The theorem of minimum potential energy, Section 1.1, is stated by

$$U \leqslant \frac{1}{2}\sum_{r=1}^{N}\int_{V_r}\boldsymbol{\varepsilon}\mathbf{C}_r\boldsymbol{\varepsilon}\,dv \tag{1.8}$$

where U is the strain energy associated with the exact solution and $\boldsymbol{\varepsilon}$ is any admissible strain field, already introduced in (1.4). We wish to add another term to (1.8), without changing its value; thus we write

$$U \leqslant \frac{1}{2} \sum_{r=1}^{N} \int_{V_r} \boldsymbol{\varepsilon} \mathbf{C}_r \boldsymbol{\varepsilon} \, dv + \frac{1}{2} \int_V \boldsymbol{\sigma}^* (\bar{\boldsymbol{\varepsilon}} - \boldsymbol{\varepsilon}) \, dv \tag{1.9}$$

where $\bar{\boldsymbol{\varepsilon}}$ is the volume average of $\boldsymbol{\varepsilon}$ and where we must now prove that the last term in (1.9) vanishes. Actually, the integrand in the last term in (1.9) vanishes

$$\boldsymbol{\sigma}^* (\bar{\boldsymbol{\varepsilon}} - \boldsymbol{\varepsilon}) = 0$$

This is true because $\boldsymbol{\sigma}^*$ is self-equilibrated, and therefore this expression is just a statement of the principle of virtual work.

From (1.6) we substitute for $\boldsymbol{\sigma}^*$ into (1.9) to get

$$U \leqslant \frac{1}{2} \sum_{r=1}^{N} \int_{V_r} \boldsymbol{\varepsilon} \mathbf{C}_r \boldsymbol{\varepsilon} \, dv + \frac{1}{2} \sum_{r=1}^{N} \int_{V_r} (\bar{\boldsymbol{\varepsilon}} - \boldsymbol{\varepsilon}) \left[\mathbf{C}_r (\bar{\boldsymbol{\varepsilon}} - \boldsymbol{\varepsilon}'_r) + \mathbf{C}_0 \boldsymbol{\varepsilon}'_r \right] dv \tag{1.10}$$

where (1.7) has been used.

Expression (1.10) can be put into the form

$$U \leqslant \frac{1}{2} \sum_{r=1}^{N} \int_{V_r} \bar{\boldsymbol{\varepsilon}} \mathbf{C}_r \bar{\boldsymbol{\varepsilon}}_r \, dv + \frac{1}{2} \sum_{r=1}^{N} \int_{V_r} \bar{\boldsymbol{\varepsilon}} \mathbf{C}_0 \boldsymbol{\varepsilon}'_r \, dv$$

$$+ \frac{1}{2} \sum_{r=1}^{N} \int_{V_r} \boldsymbol{\varepsilon}'_r (\mathbf{C}_r - \mathbf{C}_0) \boldsymbol{\varepsilon}'_r \, dv_r + \frac{1}{2} \sum_{r=1}^{N} \int_{V_r} \bar{\boldsymbol{\varepsilon}}_r (\mathbf{C}_r - \mathbf{C}_0) \boldsymbol{\varepsilon}'_r \, dv \tag{1.11}$$

where again (1.7) has been used.

The first term in (1.11) only involves averages; thus the integration can be performed directly. We are then left with

$$U \leqslant \frac{1}{2} \sum_{r=1}^{N} V_r \bar{\boldsymbol{\varepsilon}} \mathbf{C}_r \bar{\boldsymbol{\varepsilon}}_r + \frac{1}{2} \sum_{r=1}^{N} \int_{V_r} \boldsymbol{\varepsilon}'_r (\mathbf{C}_r - \mathbf{C}_0) \boldsymbol{\varepsilon}'_r \, dv$$

$$+ \frac{1}{2} \sum_{r=1}^{N} \int_{V_r} \left[\bar{\boldsymbol{\varepsilon}} \mathbf{C}_0 + \bar{\boldsymbol{\varepsilon}}_r (\mathbf{C}_r - \mathbf{C}_0) \right] \boldsymbol{\varepsilon}'_r \, dv \tag{1.12}$$

where the second and fourth terms in (1.11) have been combined to give the last term in (1.12). The last term in (1.12) vanishes because the variables involving averages can be extracted from the integral leaving

$$\int_{V_r} \boldsymbol{\varepsilon}'_r \, dv = 0$$

which follows from the definition of $\boldsymbol{\varepsilon}'_r$, Eq. (1.7). Finally, then, (1.12) is left as

$$U \leqslant \frac{1}{2} \sum_{r=1}^{N} V_r \bar{\boldsymbol{\varepsilon}} \mathbf{C}_r \bar{\boldsymbol{\varepsilon}}_r - \frac{1}{2} \sum_{r=1}^{N} \int_{V_r} \boldsymbol{\varepsilon}'_r (\mathbf{C}_0 - \mathbf{C}_r) \boldsymbol{\varepsilon}'_r \, dv \tag{1.13}$$

The lack of symmetry in the first term, involving both $\bar{\boldsymbol{\varepsilon}}$ and $\bar{\boldsymbol{\varepsilon}}_r$ rather than just either variable, causes us further problems, which we now remedy.

Define the tensor $\overline{\mathbf{A}}_r$ such that

$$\bar{\boldsymbol{\varepsilon}}_r = \overline{\mathbf{A}}_r \bar{\boldsymbol{\varepsilon}} \tag{1.14}$$

Note that because only the average variables are involved in (1.14)

$$\sum_{r=1}^{N} \mathbf{c}_r \overline{\mathbf{A}}_r = \mathbf{I} \tag{1.15}$$

where

$$\mathbf{I} = \tfrac{1}{2}(\delta_{ik}\delta_{jl} + \delta_{il}\delta_{jk})$$

Substitute (1.14) into (1.13) to obtain

$$U \leqslant \frac{1}{2} \sum_{r=1}^{N} V_r \bar{\boldsymbol{\varepsilon}} \mathbf{C}_r \overline{\mathbf{A}}_r \bar{\boldsymbol{\varepsilon}} - \frac{1}{2} \sum_{r=1}^{N} \int_{V_r} \boldsymbol{\varepsilon}'_r (\mathbf{C}_0 - \mathbf{C}_r) \boldsymbol{\varepsilon}'_r \, dv \tag{1.16}$$

The strain energy U is equated with the defining form involving the effective properties tensor $\mathbf{C}$, to give this relation as

$$\left(\tfrac{1}{2}\bar{\boldsymbol{\varepsilon}} \mathbf{C} \bar{\boldsymbol{\varepsilon}}\right) V \leqslant \frac{1}{2} \sum_{r=1}^{N} V_r \bar{\boldsymbol{\varepsilon}} \mathbf{C}_r \mathbf{A}_r \bar{\boldsymbol{\varepsilon}} - \frac{1}{2} \sum_{r=1}^{N} \int_{V_r} \boldsymbol{\varepsilon}'_r (\mathbf{C}_0 - \mathbf{C}_r) \boldsymbol{\varepsilon}'_r \, dv \tag{1.17}$$

Define $\overline{\mathbf{C}}$ as

$$\overline{\mathbf{C}} = \sum_{r=1}^{N} c_r \mathbf{C}_r \overline{A}_r \tag{1.18}$$

Then (1.17) can be written as

$$\bar{\boldsymbol{\varepsilon}}(\mathbf{C}-\overline{\mathbf{C}})\bar{\boldsymbol{\varepsilon}} \leqslant -\frac{1}{V}\sum_{r=1}^{N}\int_{V_r}\boldsymbol{\varepsilon}_r'(\mathbf{C}_0-\mathbf{C}_r)\boldsymbol{\varepsilon}_r'\,dv \tag{1.19}$$

This relation is finally beginning to look like a manageable form from which we can extract useful bounds information on $\mathbf{C}$. If relation (1.19) were not in tensor form, we see that by choosing C_0 sufficiently large, we are immediately led to the result $C \leqslant \overline{C}$, and the problem reduces to the determination of $\overline{C}$. Even in the full tensor form of (1.19) the problem still reduces to the determination of $\overline{\mathbf{C}}$ and we now assemble the considerable machinery needed to accomplish this task.

Strain Field

The problem now is to find $\overline{\mathbf{C}}$ as defined by (1.14) and (1.18); thus $\overline{\mathbf{C}}$ follows from $\boldsymbol{\varepsilon}$. We do not have complete freedom in picking $\boldsymbol{\varepsilon}$; it must be such that $\boldsymbol{\sigma}^*$ defined by (1.4) and (1.5) is self-equilibrated. Because variable $\boldsymbol{\tau}$, the polarization stress, is not self-equilibrated, it can be used to define a body force field. That is, in the isotropic comparison material, for a given field of $\boldsymbol{\tau}$ from (1.5) the equilibrium equations can be solved for the body forces to produce it. The strain state associated with $\boldsymbol{\tau}$ can also be found. Take $\hat{\boldsymbol{\varepsilon}}$ as being solved from $\boldsymbol{\tau}$ in the comparison material. Then a stress obtained from $\hat{\boldsymbol{\varepsilon}}$ and $\boldsymbol{\tau}$ in (1.4) will be self-equilibrated; proof of this is left as an exercise. Let us begin this latter procedure. The displacement due to stress $\boldsymbol{\tau}$ can be written as

$$\hat{u}_i(\mathbf{r}) = \int_S G_{ij}[\tau_{jk}]n_k\,ds + \int_V G_{ij}\tau_{jk,k}\,dv \tag{1.20}$$

where $[\tau_{jk}]$ is the discontinuity of stress $\boldsymbol{\tau}$ at the surface and where G_{ij} is the Green's function for the displacement u_i due to the force F_j in the three-dimensional isotropic, elastic medium; that is,

$$u_i = G_{ij}F_j$$

Alternatively, (1.20) can be written as

$$\hat{u}_i(\mathbf{r}) = -\int_V \frac{\partial G_{ij}}{\partial x_k'}\tau_{jk}\,dv \tag{1.21}$$

where F_j is at position $\mathbf{r}'$. From Love [4.4] the appropriate form of the

Green's function for the comparison material is

$$G_{ij} = \frac{1}{4\pi\mu_0} \frac{\delta_{ij}}{|\tilde{r} - \tilde{r}'|} - \frac{k_0 + \mu_0/3}{8\pi\mu_0\left(k_0 + \frac{4}{3}\mu_0\right)} \frac{\partial^2}{\partial x_i \partial x_j} |\mathbf{r} - \mathbf{r}'| \tag{1.22}$$

The strains $\hat{\varepsilon}_{ij}$ are obtained by differentiation of (1.21). Finally, the total strain $\boldsymbol{\varepsilon}$ is written as

$$\varepsilon_{ij} = \tfrac{1}{2}(\hat{u}_{i,j} + \hat{u}_{j,i}) + \varepsilon_{ij}^I \tag{1.23}$$

where the superimposed strain ε_{ij}^I is added to insure the satisfaction of the displacement boundary conditions of the problem. We later show that this strain ε_{ij}^I is uniform, which is needed to assure us that the stress obtained from using ε_{ij} in (1.4) still leads to a self-equilibrated stress.

Combining (1.21) into (1.23) gives

$$\varepsilon_{ij} = \frac{k_0 + \mu_0/3}{\mu_0\left(k_0 + \frac{4}{3}\mu_0\right)} \left[\sum_{r=1}^{N} \tau_{kl}^r \psi_{,ijkl}^r - \frac{1}{2\mu_0} \sum_{r=1}^{N} \left(\tau_{ik}^r \phi_{,kj}^r + \tau_{jk}^r \phi_{,ki}^r \right) \right] + \varepsilon_{ij}^I \tag{1.24}$$

where

$$\sigma^r(\mathbf{r}) = -\frac{1}{4\pi} \int_{V_r} \frac{dv}{|\mathbf{r} - \mathbf{r}'|} \tag{1.25}$$

and

$$\psi^r(\mathbf{r}) = -\frac{1}{8\pi} \int_{V_r} |\mathbf{r} - \mathbf{r}'| \, dv \tag{1.26}$$

It can be shown that the potential functions $\phi^r(\mathbf{r})$ and $\psi^r(\mathbf{r})$ satisfy

$$\nabla^4 \psi^r = \nabla^2 \phi^r = 1 \qquad \text{inside } V_r \tag{1.27}$$

Now we need the average values of $\phi_{,ij}^r$ and $\psi_{,ijkl}^r$ to obtain $\bar{\varepsilon}_r$. Consider first $\phi_{,ij}^r$. Potential ϕ^r is the ordinary Newtonian potential of attracting matter of unit density that fills V_r. This matter is distributed in a uniform and isotropic manner. Therefore the average value of $\phi_{,ij}^r$ must be an isotropic second order tensor. This requirement along with (1.27) gives

$$\phi_{,ij}^r \big|_{\text{AVG}} = \tfrac{1}{2} \delta_{ij} \tag{1.28}$$

Similarly, ψ is the biharmonic potential of matter and it can be shown that

$$\psi^r_{,ijkl}\big|_{\text{AVG}} = \tfrac{1}{15}(\delta_{ij}\delta_{kl} + \delta_{ik}\delta_{jl} + \delta_{il}\delta_{jk}) \tag{1.29}$$

Substitute (1.28) and (1.29) into (1.24) and use $\boldsymbol{\tau}$ from (1.5) to get

$$\bar{\boldsymbol{\varepsilon}}_r = -\mathbf{P}_0\boldsymbol{\tau}^r + \boldsymbol{\varepsilon}^I = \mathbf{P}_0(\mathbf{C}_0 - \mathbf{C}_r)\bar{\boldsymbol{\varepsilon}}_r + \boldsymbol{\varepsilon}^I \tag{1.30}$$

where

$$\mathbf{P}_0 = (\mathbf{C}_0^* + \mathbf{C}_0)^{-1} \tag{1.31}$$

with $\mathbf{C}_0^*$ defined by

$$C_0^* = k_0^*\delta_{ij}\delta_{kl} + \mu_0^*\left(\delta_{ik}\delta_{jl} + \delta_{il}\delta_{jk} - \tfrac{2}{3}\delta_{ij}\delta_{kl}\right) \tag{1.32}$$

$$k_0^* = \tfrac{4}{3}\mu_0$$

$$\mu_0^* = \frac{3}{2}\left(\frac{1}{\mu_0} + \frac{10}{9k_0 + 8\mu_0}\right)^{-1}$$

The inverse operation in (1.31) is defined by $\mathbf{P}_0\mathbf{P}_0^{-1} = \mathbf{I}$.

Write (1.30) as

$$\bar{\boldsymbol{\varepsilon}}_r = \mathbf{A}_r\boldsymbol{\varepsilon}^I \tag{1.33}$$

where

$$\mathbf{A}_r = \left[\mathbf{I} + \mathbf{P}_0(\mathbf{L}_r - \mathbf{L}_0)\right]^{-1} \tag{1.34}$$

Using relation (1.31), we can write (1.34) as

$$(\mathbf{C}_0^* + \mathbf{C}_r)\mathbf{A}_r = \mathbf{C}_0^* + \mathbf{C}_0 \tag{1.35}$$

Now multiply (1.33) by c_r and sum

$$\sum_{r=1}^{N} c_r\bar{\boldsymbol{\varepsilon}}_r = \sum_{r-1}^{N} c_r\mathbf{A}_r\boldsymbol{\varepsilon}^I \tag{1.36}$$

The first term is just $\bar{\boldsymbol{\varepsilon}}$; thus

$$\boldsymbol{\varepsilon}^I = \left(\sum_{r=1}^{N} c_r\mathbf{A}_r\right)^{-1}\bar{\boldsymbol{\varepsilon}} \tag{1.37}$$

At this point observe that the form of (1.37) assures us that $\boldsymbol{\varepsilon}^I$ is a uniform strain field since $\bar{\boldsymbol{\varepsilon}}$ is a volumetric average. Uniformity of $\boldsymbol{\varepsilon}^I$ was a condition we required in connection with (1.23).

Substitute (1.37) back into (1.33) to obtain

$$\bar{\boldsymbol{\varepsilon}}_r = \mathbf{A}_r\left(\sum_{r=1}^{N} c_r\mathbf{A}_r\right)^{-1}\bar{\boldsymbol{\varepsilon}} \tag{1.38}$$

This has exactly the same form as (1.14); thus we interrelate $\overline{\mathbf{A}}_r$ and $\mathbf{A}_r$ by

$$\overline{\mathbf{A}}_r = \mathbf{A}_r\left(\sum_{r=1}^{N} c_r\mathbf{A}_r\right)^{-1} \tag{1.39}$$

Substitute (1.39) into relation (1.18), giving

$$\overline{\mathbf{C}} = \sum_{r=1}^{N} c_r\mathbf{C}_r\mathbf{A}_r\left(\sum_{r=1}^{N} c_r\mathbf{A}_r\right)^{-1} \tag{1.40}$$

This relation can be written as

$$\overline{\mathbf{C}} = \sum_{r=1}^{N} c_r(\mathbf{C}_0^* + \mathbf{C}_r)\mathbf{A}_r\left(\sum_{r=1}^{N} c_r\mathbf{A}_r\right)^{-1} - \mathbf{C}_0^* \tag{1.41}$$

where $\mathbf{C}_0^*$ has just been added and subtracted in alternate positions. Using relation (1.35), Eq. (1.41) can be put into the form

$$\overline{\mathbf{C}} = \left[\sum_{r=1}^{N} c_r(\mathbf{C}_0^* + \mathbf{C}_r)^{-1}\right]^{-1} - \mathbf{C}_0^* \tag{1.42}$$

Explicit Bounds

Finally, we are in a position to establish the bounds on the effective moduli. From the basic result of the minimum energy theorem (1.19), we have

$$\gamma(\overline{\mathbf{C}} - \mathbf{C})\gamma \geqslant 0 \qquad \text{if} \quad \gamma(\mathbf{C}_0 - \mathbf{C}_r)\gamma \geqslant 0 \tag{1.43}$$

where γ is any symmetric second order tensor. Clearly, we must take $\mathbf{C}_0$ to be identified with the phase of the stiffest properties. In the case of an isotropic material we take

$$\begin{aligned} \mu_0 &= \mu_{\mathrm{MAX}} \\ k_0 &= k_{\mathrm{MAX}} \end{aligned} \tag{1.44}$$

where μ_{MAX} and k_{MAX} are the maximum properties from any of the N phases. Now, using $\overline{\mathbf{C}}$ from (1.42), in isotropic form, it follows from (1.43) that

$$k \leqslant \left[\sum_{r=1}^{N} c_r (k_g^* + k_r)^{-1} \right]^{-1} - k_g^*$$

$$\mu \leqslant \left[\sum_{r=1}^{N} c_r (\mu_g^* + \mu_r)^{-1} \right]^{-1} - \mu_g^* \tag{1.45}$$

where μ_g^* and k_g^* follow from (1.32) as

$$k_g^* = \tfrac{4}{3}\mu_{\mathrm{MAX}}$$

$$\mu_g^* = \frac{3}{2}\left(\frac{1}{\mu_{\mathrm{MAX}}} + \frac{10}{9k_{\mathrm{MAX}} + 8\mu_{\mathrm{MAX}}} \right)^{1} \tag{1.46}$$

For a two phase material we have from (1.18)

$$\overline{\mathbf{C}} = c_1 \mathbf{L}_1 \overline{\mathbf{A}}_1 + c_2 \mathbf{L}_2 \overline{\mathbf{A}}_2 \tag{1.47}$$

which can be written as

$$\overline{\mathbf{C}} = \mathbf{C}_2 + c_1(\mathbf{C}_1 - \mathbf{C}_2)\overline{\mathbf{A}}_1 \tag{1.48}$$

using (1.15). Using (1.35) and (1.39), then (1.48) becomes

$$\overline{\mathbf{C}} = \mathbf{C}_2 + c_1(\mathbf{C}_1 - \mathbf{C}_2)\left[\mathbf{I} + c_2(\mathbf{C}_0^* + \mathbf{C}_2)^{-1}(\mathbf{C}_1 - \mathbf{C}_2) \right]^{-1} \tag{1.49}$$

Then using $\overline{\mathbf{C}}$ from (1.49) in (1.43), with isotropic $\mathbf{C}_0$ properties taken as in (1.44), the bounds on properties result in a form corresponding to (1.45).

In an identical manner, the theorem of minimum complementary energy can be used to obtain lower bounds on the effective properties. The final results for a two phase media are

$$\frac{c_1}{1+[(1-c_1)(k_1-k_2)/(k_2+k_l)]} \leqslant \frac{k-k_2}{k_1-k_2} \leqslant \frac{c_1}{1+[(1-c_1)(k_1-k_2)/(k_2+k_u)]} \tag{1.50}$$

and

$$\frac{c_1}{1+[(1-c_1)(\mu_1-\mu_2)/(\mu_2+\mu_l)]} \leqslant \frac{\mu-\mu_2}{\mu_1-\mu_2} \leqslant \frac{c_1}{1+[(1-c_1)(\mu_1-\mu_2)/(\mu_2+\mu_u)]} \tag{1.51}$$

where for $(\mu_1-\mu_2)(k_1-k_2) \geqslant 0$

$$\begin{aligned} k_l &= \tfrac{4}{3}\mu_2 \\ k_u &= \tfrac{4}{3}\mu_1 \\ \mu_l &= \frac{3}{2}\left(\frac{1}{\mu_2}+\frac{10}{9k_2+8\mu_2}\right)^{-1} \\ \mu_u &= \frac{3}{2}\left(\frac{1}{\mu_1}+\frac{10}{9k_1+8\mu_1}\right)^{-1} \end{aligned} \tag{1.52}$$

and for $(\mu_1-\mu_2)(k_1-k_2) \leqslant 0$

$$\begin{aligned} k_l &= \tfrac{4}{3}\mu_1 \\ k_u &= \tfrac{4}{3}\mu_2 \\ \mu_l &= \frac{3}{2}\left(\frac{1}{\mu_2}+\frac{10}{9k_1+8\mu_2}\right)^{-1} \\ \mu_u &= \frac{3}{2}\left(\frac{1}{\mu_1}+\frac{10}{9k_2+8\mu_1}\right)^{-1} \end{aligned} \tag{1.53}$$

Comparison of these bounds with the Reuss and Voigt bounds (1.3) in particular examples shows that these bounds are much more definitive. Both sets of bounds (1.50), (1.51), and (1.3) are derived from the classical minimum principles of elasticity. Whereas uniform fields of stress and strain were used in deriving (1.3), the element of success that led to (1.50) and (1.51) was the incorporation of a variable field of admissible stress and strain. Even without specifying the geometry of phase combination, we were able to obtain variable fields of admissible stress and strain that gave us these restrictive bounds.

One of the main questions that should be posed at this point is as follows. Are the bounds (1.50) and (1.51) the most restrictive that can be obtained without specifying phase geometry? We will obtain at least a partial answer to this question. Before doing this, however, and before interpreting these bounds in general, we first give the corresponding bounds for transversely isotropic materials.

4.2 BOUNDS ON TRANSVERSELY ISOTROPIC EFFECTIVE MODULI

The minimum theorems of elasticity theory can be used to obtain bounds on the effective properties of heterogeneous media that are macroscopically transversely isotropic. We have seen in the preceding section how these results can be derived, so we here merely state the appropriate results for a two material transversely isotropic medium. These results were obtained by Hashin [4.5] and Hill [4.6]. The statement of these results presumes that $\mu_1>\mu_2$, $k_1>k_2$, and, as always, axis 1 is in the direction of the axis of symmetry. The bounds on the five independent properties are given by

$$K_2+\frac{c_1}{1/(K_1-K_2)+c_2/(K_2+\mu_2)}\leqslant K_{23}\leqslant K_1+\frac{c_2}{1/(K_2-K_1)+c_1/(K_1+\mu_1)} \tag{2.1}$$

$$\mu_2+\frac{c_1}{1/(\mu_1-\mu_2)+c_2(K_2+2\mu_2)/2\mu_2(K_2+\mu_2)}$$

$$\leqslant \mu_{23}\leqslant \mu_1+\frac{c_2}{1/(\mu_2-\mu_1)+c_1(K_1+2\mu_1)/2\mu_1(K_1+\mu_1)} \tag{2.2}$$

$$\mu_2+\frac{c_1}{1/(\mu_1-\mu_2)+c_2/2\mu_2}\leqslant \mu_{12}\leqslant \mu_1+\frac{c_2}{1/(\mu_2-\mu_1)+c_1/2\mu_1} \tag{2.3}$$

$$\frac{c_1c_2}{c_1/K_2+c_2/K_1+1/\mu_2}\leqslant\frac{E_{11}-c_1E_1-c_2E_2}{4(\nu_1-\nu_2)^2}\leqslant\frac{c_1c_2}{c_1/K_2+c_2/K_1+1/\mu_1} \tag{2.4}$$

and

$$\frac{c_1c_2}{c_1/K_2+c_2/K_1+1/\mu_2} \leqslant \frac{\nu_{12}-c_1\nu_1-c_2\nu_2}{(\nu_1-\nu_2)(1/K_2-1/K_1)} \leqslant \frac{c_1c_2}{c_1/K_2+c_2/K_1+1/\mu_1} \tag{2.5}$$

It is recalled that K is the plane strain bulk modulus defined by $K=k+\mu/3$ for an isotropic material.

The interpretation of these bounds is given next.

4.3 INTERPRETATION OF EFFECTIVE MODULI AND ASSOCIATED BOUNDS

There is much we can learn from the effective properties we have derived in the preceding two chapters, and from the present results on bounds. We jointly consider the two cases of macroscopically isotropic and transversely isotropic materials, since they have many common features.

First, note that when $\mu_1=\mu_2$ the bounds on the bulk moduli, (1.50), coincide, providing the exact solution in that case; the bounds on μ trivially coincide. Similarly, in the transversely isotropic case for $\mu_1=\mu_2$, the exact solutions are known for all five effective properties.

Next, observe that the expression for the lower bound of k (1.50) is identical with the result we derived for the composite spheres model (II-3.17) if we identify index 2 with the matrix phase and index 1 with the inclusion phase, in the case $(\mu_1-\mu_2)(k_1-k_2)\geqslant 0$. Alternatively, note that the upper bound for k (1.50) coincides with the composite spheres result if we identify index 1 with the matrix phase and index 2 with the inclusion phase, again in the case $(\mu_1-\mu_2)(k_1-k_2)\geqslant 0$. Thus because these bounds on k coincide with the exact results for a particular model, it follows that these are the best bounds that can be derived without specifying the geometry of the phase combinations.

One might expect a similar behavior for μ to that just found for k. However, this is not the case. The only model for which we have the exact solution for μ over the entire volume fraction range is the three phase model of Section 2.4. As we see shortly, the results of that model do not, in general, coincide with the bounds. Only in the case where the volume fraction of one phase is very small or very large, does this model give a prediction that coincides with the bounds (1.51). For example, the dilute result for μ, (II-2.23), coincides with the dilute forms of the lower and upper bounds of (1.51) when the proper identification of phases is made. This coincidence of model results for μ with the bounds also applies to the composite spheres model under very low or very high volume fraction.

An entirely similar situation exists in the transversely isotropic media case. Of the five bounds formulas (2.1)–(2.5) all but those of μ_{23} coincide with the proper material identification in the composites cylinders model, Section 3.2. Thus at least for these four properties the bounds results are the best that can be obtained without specifying geometry. Regarding the fifth property, μ_{23}, the three phase model result of Section 3.3 does not coincide with the bounds in (2.2) except at very low and very high volume fractions.

Two more bounds results should be mentioned. These are the explicit bounds found by Hashin [4.7] for the composite spheres model and Hashin and Rosen [4.8] for the composite cylinders model. These bounds should not be confused with those already stated in this chapter since the former are for particular geometric models, whereas the latter are applicable to all geometries, within their symmetry classes. It will be interesting to compare these different bounds results in particular examples of composite materials. Of greater importance will be the comparison of these bounds results with the solution of the three phase models of Sections 2.4 and 3.3, which are related to the composite spheres model and the composite cylinders model, respectively.

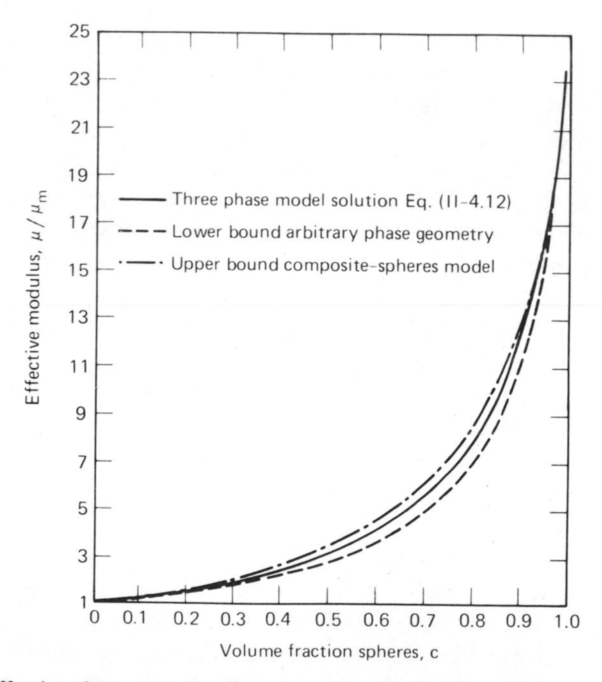

Fig. 4.1 Effective shear modulus for three phase spherical model, $\mu_i/\mu_m=23.46$, $\nu_i=0.22$, and $\nu_m=0.35$ (glass-epoxy composite).

In the macroscopically isotropic case, we take one example having properties appropriate to a suspension of glass spheres in an epoxy matrix. The results are as shown in Fig. 4.1. The upper bound from the arbitrary phase geometry case (1.51) and the lower bound from the composite spheres model fall outside the bounds shown in Fig. 4.1. Probably the primary thing to note in Fig. 4.1 is that for volume concentrations that are not too high, the lower bound formula from (1.51) gives a good approximation to the three phase model result (II-4.12). This feature is of some utility since the result (II-4.12) has an involved form, whereas the bound admits a very simple expression. However, any approximation of an exact result by a bound must be made with caution and with numerical justification for the system of interest. It is interesting to note that the result of Kerner, mentioned in Section 2.4, corresponds to the lower bound of (1.51). The error in the Kerner analysis apparently contributed a great simplification to the result.

In a manner parallel to that discussed previously, we now evaluate a system composed of glass fibers in an epoxy matrix. The results are shown in Fig. 4.2 for the transverse shear modulus, μ_{23}. The upper bound for the arbitrary phase geometry (2.2) and the lower bound for the composite cylinders model fall outside the bounds shown. The expression for the

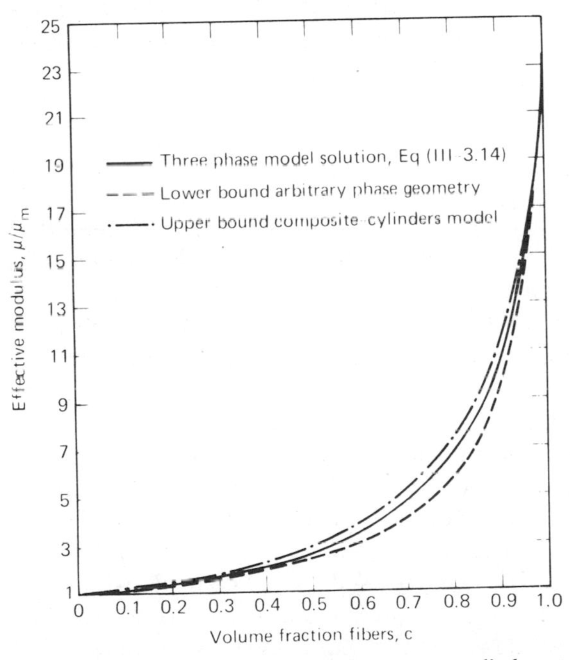

Fig. 4.2 Effective transverse shear modulus for three phase cylinder model with $\mu_f/\mu_m = 23.46$, $\nu_f = 0.22$, and $\nu_m = 0.35$ (glass epoxy composite).

upper bound of the composite cylinders model, as given by Hashin [4.9], is

$$\mu_{23}\Big|_{\substack{\text{COMP.}\\ \text{CYLS.}}} < \frac{(1+\alpha c^3)(\rho+\beta_m c)-3c(1-c)^2\beta_m^2}{(1+\alpha c^3)(\rho-c)-3c(1-c)^2\beta_m^2}\,\mu_m \tag{3.1}$$

where

$$\alpha=\frac{\beta_m-\gamma\beta_f}{1+\gamma\beta_f},\qquad \rho=\frac{\gamma+\beta_m}{\gamma-1},\qquad \gamma=\frac{\mu_f}{\mu_m},$$

$$\beta_f=\frac{1}{3-4\nu_f},\qquad \beta_m=\frac{1}{3-4\nu_m}$$

It can be noted that the lower bound for the arbitrary phase geometry case (2.2) gives a close approximation to the result (III-3.14) so long as the volume fraction is not too large. The result of Hermans, mentioned in Section 3.3, coincides with the lower bound of (2.2). Clearly the violated boundary condition in the Hermans work resulted in a great simplification of the final result. The evaluations of the status of the Kerner formula and the Hermans formula are interchangeable, one result being for the macroscopically isotropic case while the other is for the transversely isotropic case.

In Section 3.5 we developed the effective stiffness property for a dilute system of randomly oriented platelets in a continuous matrix phase. It is of interest to compare the properties predictions for that model with the general bounds obtained in this chapter. A comparison of the platelet results for k and μ (III-5.13, 5.18) with the upper bounds of (1.50) and (1.51), under dilute conditions, shows identity when phase 1 is identified with the stiff platelet material and phase 2 is identified with the more compliant matrix phase. This very interesting result shows the platelet form of reinforcement to provide the best possible stiffness type reinforcing results, under dilute conditions. It is important to observe, however, that the platelet geometry is not unique in this regard. We have already found that under dilute conditions the composite spheres model or, equivalently, the three phase model of Section 2.4, also gives the maximum reinforcing effect under dilute conditions when the more compliant phase is the inclusion phase. We already know that under nondilute conditions the three phase model of Section 2.4 does not coincide with the upper bound results of (1.51). It will be of great interest to us to see whether a nondilute platelet model produces results that coincide with the upper bounds. We develop a model of this type in the next section, and obtain an answer to the question posed.

As the last matter in this section, we consider the general applicability of the models developed here. We are only concerned here with the fiber case since the same matter was already dealt with in Chapter II in connection with spherical inclusion models. In the case of fiber systems the question is as follows. How well does the composite cylinders model of Section 3.2 and the corresponding model for transverse shear, Section 3.3, actually represent data for real fiber systems? After all, the models treated here idealize the system as having an infinite size gradation of cylinders, down to infinitesimal. Most of the actual fiber systems in use employ only single size fibers. The difference between the model idealization and the reality seems large. As is shown, however, the difference is not significant for typical systems of interest.

First, regarding the uniaxial modulus, E_{11}, formula (III-2.5) reveals the rule of mixtures to be the determining characteristic, and the acceptibility of this is widely recognized in practice. A similar situation exists with regard to the Poisson's ratio, ν_{12}. Of the remaining three independent properties, μ_{23}, K_{23}, and μ_{12}, probably the transverse shear modulus, μ_{23}, is the most sensitive to the geometry of the model. Alternatively, the transverse uniaxial modulus, E_{22}, would provide an equivalently critical test of the model. Experimental data can be found to support almost any result. We believe a more meaningful test of the prediction of the analytical models given here is to compare their predictions with those of numerical solutions for systems containing single size fibers. An extensive comparison of this type has been given by Behrens [4.10], based on the numerical solution of Chen and Cheng [4.11] for a hexagonal distribution of single size fibers. In particular, the transverse uniaxial modulus for the hexagonal array is shown to be very close or slightly above that calculated from the composite cylinders model for E_{11}, ν_{12}, K_{23}, ν_{12}, and the lower bound prediction for μ_{23} from (2.2). This is completely consistent with the observation from Fig. 4.2 that the lower bound result is very close to the three phase model prediction for μ_{23}.

We therefore conclude that for typical fiber systems of interest, the precise details of the geometry of the interface are of far less significance than merely accounting for which phase is the continuous phase and which phase is the inclusion phase, in the cross-section view. Of course, a conclusion of this type must be qualified; it depends on the ratio of stiffness of fiber phase to matrix phase. In the limiting case of single size infinitely stiff fibers, the composite cylinders model and the three phase model for μ_{23} cannot be expected to give meaningful results near the maximum volume fraction for the fiber phase. Despite these qualifications, as discussed earlier, the composite cylinders model and μ_{23} from the three phase model give realistic predictions for systems of practical interest. Equally important, these results from the composite cylinders model and the three phase model are the only rigorously derived analytical results in the field. The alternative to their use is the employment of one of the myriad empirical formulas in the field, one of which is mentioned in the problem set.

In Section 4.4 we investigate the effective properties for a distribution of randomly oriented fibers. In this work it is necessary to utilize the corresponding properties for an aligned fiber system. For properties E_{11}, ν_{12}, K_{23}, and μ_{12}, we use the results for the composite cylinders model. Logically, we should then also use the prediction (III-3.14) for μ_{23} of the three phase model. This latter result, however, takes such an involved analytical form that we approximate it by the lower bound from (2.2), which we recall from Fig. 4.2 is a realistic substitution. Next we proceed to the random fiber problem, after which we consider the random platelet problem. Fortunately, in the platelet case we have no problem in determining the five independent properties in the aligned platelet state. As is shown, all five of these platelet properties are directly derivable, as exact results.

4.4 SOME ISOTROPIC PROPERTIES OF FIBER SYSTEMS

We have given an extensive treatment to aligned fiber systems in the developments up to this point. Now we turn to the associated problem of a system involving a random orientation of fibers. We wish to determine the effective moduli for problems involving a three-dimensionally random orientation of fibers and also for the case of a two-dimensionally random orientation of fibers.

Three-Dimensional Case

The three-dimensional case is considered first. Obviously, the effective properties are isotropic. We certainly could proceed in the same manner as in Section 3.5, where we treated the case of a dilute suspension of randomly oriented platelets. That solution modeled the platelets as oblate ellipsoids; we could similarly model the fibers as prolate ellipsoids and then proceed as in Section 3.5. We do not take that approach, however, because we do not wish to have results that are limited to dilute conditions. The method we follow here leads to more general results. Let us begin by considering the geometry shown in Fig. 4.3. We envision the representative volume element as containing continuous fibers that terminate only on the surface. With the fibers taking random orientation, it would be unthinkable to try to solve the body shown in Fig. 4.3 as a boundary value

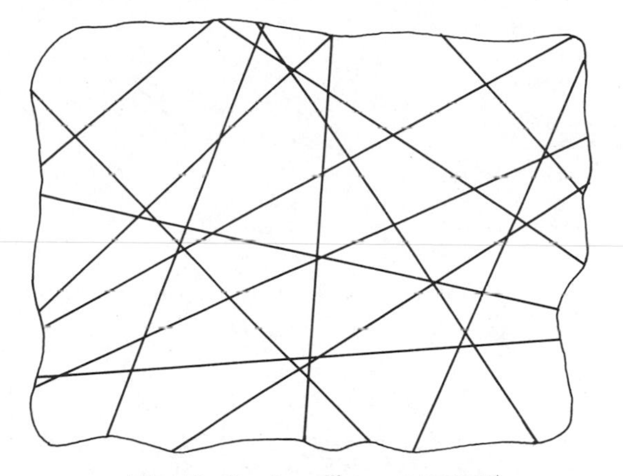

Fig. 4.3 Random fiber arrangement.

problem to determine the effective properties. We proceed otherwise.

First, note the distinguishing feature of the problem posed from the representative volume element of Fig. 4.3. This feature is simply the strain compatibility between the fibers and the deformation of the entire volume. Thus, imposed displacement boundary conditions on the surface of the representative volume element also cause compatible strain levels in the fibers. The key observation here then is the requirement of an imposed strain level. We approach the problem by taking a body of aligned fibers and imposing a given strain condition upon it. We then perform a geometric averaging procedure that effectively corresponds to taking a random direction of imposed strain in the aligned fiber system. We take the results of this analysis as being representative of the deformation of the actual system shown in Fig. 4.3. The procedure is now implemented.

The governing equations of the transversely isotropic medium (representative of an aligned fiber system) are already given as Eqs. (III-1.1). Following Christensen and Waals [4.12], we begin by imposing a strain in an arbitrary direction relative to the axis of symmetry of transversely isotropic media. The situation is as shown in Fig. 4.4, where axis 1 is in the direction of the symmetry axis and axis 3′ is in the direction of imposed strain. These conditions are stated by

$$
\begin{aligned}
&\varepsilon'_{33} \neq 0 \\
&\varepsilon'_{ij} = 0 \qquad \text{for} \quad i \neq 3, j \neq 3
\end{aligned}
\tag{4.1}
$$

Axis 3 can be taken to lie in the 1′2′ plane with no loss in generality. The appropriate tensor transformation law for strains is given by

$$
\varepsilon_{\alpha\beta} = l_{i\alpha} l_{j\beta} \varepsilon'_{ij} \tag{4.2}
$$

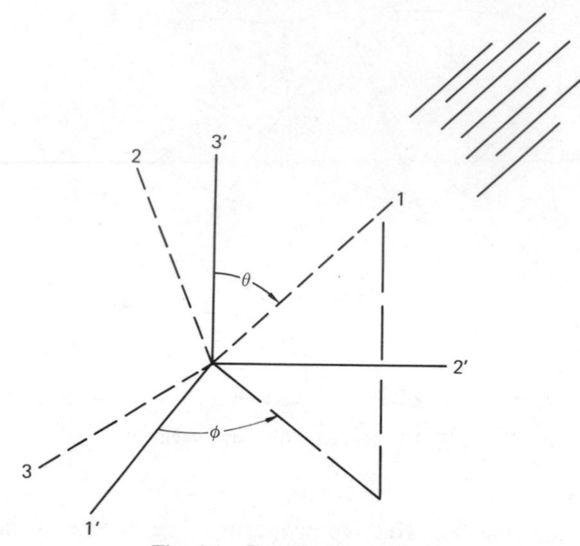

Fig. 4.4 Coordinate convention.

where $l_{i\alpha}$ is the direction cosine between the x_i' axis and the x_α axis. Combining equations (4.1) and (4.2) yields

$$\varepsilon_{\alpha\beta} = l_{3\alpha} l_{3\beta} \varepsilon_{33}' \tag{4.3}$$

Substituting (4.3) into the stress-strain relations (III-1.1) gives

$$\begin{aligned}
\frac{\sigma_{11}}{\varepsilon_{33}'} &= C_{11} l_{31}^2 + C_{12} l_{32}^2 \\
\frac{\sigma_{22}}{\varepsilon_{33}'} &= C_{12} l_{31}^2 + C_{22} l_{32}^2 \\
\frac{\sigma_{33}}{\varepsilon_{33}'} &= C_{12} l_{31}^2 + C_{23} l_{32}^2 \\
\frac{\sigma_{12}}{\varepsilon_{33}'} &= 2C_{66} l_{31} l_{32} \\
\frac{\sigma_{23}}{\varepsilon_{33}'} &= 0 \\
\frac{\sigma_{31}}{\varepsilon_{33}'} &= 0
\end{aligned} \tag{4.4}$$

where $l_{33} = 0$ has been used.

The stresses also have the tensor transformation law

$$\sigma'_{\alpha\beta} = l_{\alpha i} l_{\beta j} \sigma_{ij} \tag{4.5}$$

Substituting from (4.4) into (4.5) results in

$$\frac{\sigma'_{33}}{\varepsilon'_{33}} = C_{11} l_{31}^4 + (2C_{12} + 4C_{66}) l_{31}^2 l_{32}^2 + C_{22} l_{32}^4 \tag{4.6}$$

and

$$\frac{\sigma'_{22}}{\varepsilon'_{33}} = C_{11} l_{31}^2 l_{21}^2 + C_{12}\left(l_{32}^2 l_{21}^2 + l_{31}^2 l_{22}^2 + l_{31}^2 l_{23}^2\right)$$
$$+ C_{22} l_{32}^2 l_{22}^2 + 4C_{66} l_{31} l_{32} l_{21} l_{22} + C_{23} l_{32}^2 l_{23}^2 \tag{4.7}$$

The appropriate direction cosines from Fig. 4.4 are

$$l_{\alpha i} = \begin{bmatrix} \sin\theta\cos\phi & -\cos\theta\cos\phi & \sin\phi \\ \sin\theta\sin\phi & -\cos\theta\sin\phi & -\cos\phi \\ \cos\theta & \sin\theta & 0 \end{bmatrix} \tag{4.8}$$

As discussed earlier, it is postulated that the effect of a random orientation of fibers on determining the ratios of stress to strain $\sigma'_{ij}/\varepsilon'_{33}$ is analytically equivalent to finding the average value of $\sigma'_{ij}/\varepsilon'_{33}$ for the fiber direction, axis 1, taking all possible orientations relative to the fixed axes x'_i. That is,

$$\left.\frac{\sigma'_{ij}}{\varepsilon'_{33}}\right|_{\text{random}} = \frac{1}{2\pi}\int_0^{\pi}\int_0^{\pi} \frac{\sigma'_{ij}}{\varepsilon'_{33}} \sin\theta \, d\theta \, d\phi \tag{4.9}$$

Using (4.6) and (4.7) in (4.9) gives

$$\left.\frac{\sigma'_{33}}{\varepsilon'_{33}}\right|_{\text{random}} = \tfrac{1}{15}(3C_{11} + 4C_{12} + 8C_{22} + 8C_{66}) \tag{4.10}$$

and

$$\left.\frac{\sigma'_{22}}{\varepsilon'_{33}}\right|_{\text{random}} = \tfrac{1}{15}(C_{11} + 8C_{12} + C_{22} - 4C_{66} + 5C_{23}) \tag{4.11}$$

Now for an effectively isotropic medium the ratios of stress to strain shown in (4.10) and (4.11) relate, respectively, to the properties

$$k+\tfrac{4}{3}\mu$$

and

$$k-\tfrac{2}{3}\mu$$

Equating these respective forms to (4.10) and (4.11) gives

$$k=\tfrac{1}{9}\left[C_{11}+2(C_{22}+C_{23})+4C_{12}\right]$$

and

$$\mu=\tfrac{1}{30}\left[2C_{11}+7C_{22}-5C_{23}-4C_{12}+12C_{66}\right] \tag{4.12}$$

Equivalently, the E and ν properties can be related to the C values through their relation with μ and k.

We can also express μ and k in terms of the engineering properties, using relations (III-1.7). It is found that

$$k=\tfrac{1}{9}\left[E_{11}+4(1+\nu_{12})^2K_{23}\right] \tag{4.13}$$

$$\mu=\tfrac{1}{15}\left[E_{11}+(1-2\nu_{12})^2K_{23}+6(\mu_{12}+\mu_{23})\right] \tag{4.14}$$

Again E and ν may be found through their relationship with μ and k. There results

$$E=\frac{\left[E_{11}+4(1-\nu_{12})^2K_{23}\right]\left[E_{11}+(1-2\nu_{12})^2K_{23}+6(\mu_{12}+\mu_{23})\right]}{3\left[2E_{11}+\left(8\nu_{12}^2+12\nu_{12}+7\right)K_{23}+2(\mu_{12}+\mu_{23})\right]} \tag{4.15}$$

and

$$\nu=\frac{E_{11}+2\left(2\nu_{12}^2+8\nu_{12}+3\right)K_{23}-4(\mu_{12}+\mu_{23})}{2\left[2E_{11}+\left(8\nu_{12}^2+12\nu_{12}+7\right)K_{23}+2(\mu_{12}+\mu_{23})\right]} \tag{4.16}$$

With the properties for the aligned fiber reinforced media known, either by experiment or by theory, the values when substituted into (4.13)–(4.16) give the predictions for the properties of the randomly oriented fiber system. Before considering such predictions, however, we obtain the corresponding two-dimensional results.

Two-Dimensional Case

In the two-dimensional case practical considerations dictate that we consider plane stress conditions. The specification of the transversely isotropic stress-strain relations (III-1.1) to plane stress conditions gives

$$\sigma_{11} = Q_{11}\varepsilon_{11} + Q_{12}\varepsilon_{22}$$

$$\sigma_{22} = Q_{12}\varepsilon_{11} + Q_{22}\varepsilon_{22}$$

$$\sigma_{12} = 2Q_{66}\varepsilon_{12} \tag{4.17}$$

where 1,2 is the plane of interest and

$$Q_{11} = C_{11} - \frac{C_{12}^2}{C_{22}}$$

$$Q_{12} = C_{12} - \frac{C_{12}C_{23}}{C_{22}}$$

$$Q_{22} = C_{22} - \frac{C_{23}^2}{C_{22}}$$

$$Q_{66} = C_{66} \tag{4.18}$$

The procedure is similar to that outlined in the three-dimensional case. For the 1,2 axes 1 is taken to be the direction of symmetry of the transversely isotropic media. A coordinate system rotated at angle θ to the 1,2 system, is designated by 1′,2′. The only nonzero strain component is taken to be ε'_{11}. The ratios $\sigma'_{11}/\varepsilon'_{11}$ and $\sigma'_{22}/\varepsilon'_{11}$ are formed and substituted into the relations

$$\left.\frac{\sigma'_{ij}}{\varepsilon'_{11}}\right|_{\text{random}} = \frac{1}{\pi}\int_0^{\pi} \frac{\sigma'_{ij}}{\varepsilon'_{11}}\, d\theta \tag{4.19}$$

The resulting forms are

$$\left.\frac{\sigma'_{11}}{\varepsilon'_{11}}\right|_{\text{random}} = \tfrac{1}{8}(3Q_{11} + 3Q_{22} + 2Q_{12} + 4Q_{66}) \tag{4.20}$$

and

$$\left.\frac{\sigma'_{22}}{\varepsilon'_{11}}\right|_{\text{random}} = \tfrac{1}{8}(Q_{11} + Q_{22} + 6Q_{12} - 4Q_{66}) \tag{4.21}$$

Under plane stress conditions the ratios shown in (4.20) and (4.21) relate, respectively, to the planar isotropic properties

$$\frac{E}{1-\nu^2} \quad \text{and} \quad \frac{\nu E}{1-\nu^2}$$

Identifying these relations with (4.20) and (4.21), respectively, gives

$$\nu = \frac{Q_{11}+Q_{22}+6Q_{12}-4Q_{66}}{3Q_{11}+3Q_{22}+2Q_{12}+4Q_{66}} \tag{4.22}$$

and

$$E = \tfrac{1}{8}(3Q_{11}+3Q_{12}+2Q_{12}+4Q_{66})(1-\nu^2) \tag{4.23}$$

These results for planar isotropic E and ν can be related to the engineering properties of the aligned fiber system using the relations (4.18) and (III-1.7). It is found that

$$E = \frac{u_1^2 - u_2^2}{u_1}$$

and

$$\nu = \frac{u_2}{u_1} \tag{4.24}$$

where

$$u_1 = \tfrac{3}{8}E_{11} + \frac{\mu_{12}}{2} + \frac{(3+2\nu_{12}+3\nu_{12}^2)\mu_{23}K_{23}}{2(\mu_{23}+K_{23})}$$

$$u_2 = \tfrac{1}{8}E_{11} - \frac{\mu_{12}}{2} + \frac{(1+6\nu_{12}+\nu_{12}^2)\mu_{23}K_{23}}{2(\mu_{23}+K_{23})} \tag{4.25}$$

As in the three-dimensional case, the substitution of the engineering properties, from theory or experiment, into (4.25) and then into (4.24) gives the prediction for the planar isotropic properties in the case of two-dimensionally random orientation of fibers. Next we consider the implementation of this procedure.

Asymptotic Predictions

One may certainly use any effective properties values directly in relations (4.13)–(4.16) and (4.24) to give the predicted isotropic properties. As discussed extensively in Section 4.3, the models derived and considered here commend the use of the properties E_{11}, ν_{12}, K_{23} and μ_{12}, (III-2.5), and (III-2.7)–(III-2.9) from the composite cylinders model and the transverse modulus from the three phase model (III-3.14). A scheme such as this would best be implemented with a small computer program, and it would be simple to program and use. However, for our purpose of interpretation we now seek simple analytical forms for the procedure just described. We do not wish to use empirical methods, of uncertain meaning; therefore we proceed by asymptotic methods appropriate to the case of stiff fiber systems.

The properties to be used in this asymptotic method are those mentioned earlier, with one exception. The transverse shear modulus property (III-3.14) has a very complicated form, and we therefore approximate it by the lower bound from (2.2), which as demonstrated and discussed in Section 4.3, is a realistic procedure. We collect the properties here for use. These are

$$E = cE_f + (1-c)E_m + 4c(1-c)\mu_m \left[\frac{(\nu_f - \nu_m)^2}{\dfrac{[(1-c)\mu_m]}{(k_f + \mu_m/3)} + \dfrac{c\mu_m}{(k_m + \mu_m/3)} + 1} \right]$$

$$\nu_{12} = \nu_{13} = c\nu_f + (1-c)\nu_m + \frac{c(1-c)(\nu_f - \nu_m)\left[\dfrac{\mu_m}{(k_m + \mu_m/3)} - \dfrac{\mu_m}{(k_f + \mu_f/3)} \right]}{\dfrac{(1-c)\mu_m}{(k_f + \mu_f/3)} + \dfrac{c\mu_m}{(k_m + \mu_m/3)} + 1}$$

$$K_{23} = k_m + \frac{\mu_m}{3} + \frac{c}{\dfrac{1}{[k_f - k_m + (\mu_f - \mu_m)/3]} + \dfrac{(1-c)}{[k_m + (4\mu_m/3)]}}$$

$$\frac{\mu_{12}}{\mu_m} = \frac{\mu_f(1+c) + \mu_m(1-c)}{\mu_f(1-c) + \mu_m(1+c)} \tag{4.26}$$

and

$$\frac{\mu_{23}}{\mu_m} = 1 + \frac{c}{\dfrac{\mu_m}{(\mu_f - \mu_m)} + \dfrac{[(k_m + 7\mu_m/3)(1-c)]}{[2(k_m + 4\mu_m/3)]}}$$

The direct substitution of (4.26) into (4.13)–(4.16) or (4.24) and (4.25) then gives the final prediction.

Before seeking asymptotic results, a closed form expression for the preceding procedure can be found in one case. Take both phases as being incompressible. Then (4.14) can be shown to reduce to

$$\mu_{3D}|_{\nu_f=\nu_m=1/2}=\tfrac{1}{15}\left[E_{11}+6(\mu_{12}+\mu_{23})\right] \tag{4.27}$$

Substituting into (4.27) from (4.26) gives

$$\mu_{3D}|_{\nu_f=\nu_m=1/2}=\frac{c}{5}\mu_f+\frac{1}{5}\left[\frac{(5+2c+c^2)\mu_f+(5+c)(1-c)\mu_m}{(1-c)\mu_f+(1+c)\mu_m}\right]\mu_m \tag{4.28}$$

With incompressibility, $E_{3D}=3\mu_{3D}$, and from (4.28) there results

$$E_{3D}|_{\nu_f=\nu_m=1/2}=\frac{c}{5}E_f+\frac{1}{5}\left[\frac{(5+2c+c^2)E_f+(5+c)(1-c)E_m}{(1-c)E_f+(1+c)E_m}\right]E_m \tag{4.29}$$

To obtain results for other cases we must proceed by asymptotic methods, taken from Christensen [4.13]. Proceed first with the three-dimensional case.

Rewrite expression (4.15) as

$$E_{3D}=\frac{cE_f+\left[2\hat{E}_{11}+(5+4\nu_{12}+8\nu_{12}^2)K_{23}+6(\mu_{12}+\mu_{23})\right]+O(1/cE_f)}{6\{1+\left[2\hat{E}_{11}+(7+12\nu_{12}+8\nu_{12}^2)K_{23}+2(\mu_{12}+\mu_{23})\right]/2cE_f\}} \tag{4.30}$$

where

$$\hat{E}_{11}+E_{11}-cE_f \tag{4.31}$$

Expand the quotient in (4.30) as a power series in (E_m/cE_f) to obtain

$$\frac{E_{3D}}{E_m}=\frac{1}{6}\frac{cE_f}{E_m}+\frac{\hat{E}_{11}}{6E_m}+\frac{(3-4\nu_{12}+8\nu_{12}^2)K_{23}}{12E_m}+\frac{5}{6}\frac{(\mu_{12}+\mu_{23})}{E_m}+O\left(\frac{E_m}{cE_f}\right) \tag{4.32}$$

Now under conditions $cE_f \gg E_m$ and $cE_f \gg k_m$ relations (4.26) along with

(4.31) give

$$\begin{aligned}
\hat{E}_{11} &= (1-c)E_m + 4c(1-c)\mu_m\left[\frac{(\nu_f-\nu_m)^2}{c\mu_m/(k_m+\mu_m/3)+1}\right] \\
\nu_{12} &= c\nu_f + (1-c)\nu_m + \frac{c(1-c)(\nu_f-\nu_m)\left[\mu_m/(k_m+\mu_m/3)\right]}{1+\left[c\mu_m/(k_m+\mu_m/3)\right]} \\
K_{23} &= \frac{k_m}{1-c} + \frac{(1+3c)}{(1-c)}\frac{\mu_m}{3} \\
\mu_{12} &= \frac{(1+c)}{(1-c)}\mu_m
\end{aligned} \tag{4.33}$$

and

$$\mu_{23} = \mu_m + \frac{2c\left(k_m+\frac{4}{3}\mu_m\right)}{(1-c)\left(k_m+\frac{7}{3}\mu_m\right)}\mu_m$$

although it must be noted that these formulas are not valid at $c=1$.

The substitution of relations (4.33) into (4.32) completes the derivation. However, the results are still somewhat complicated unless specific values of Poisson's ratio are assigned. The case of $\nu_f=\nu_m=\frac{1}{4}$ has been computed, with the result from (4.33) that

$$\left.\begin{aligned}
\hat{E}_{11} &= (1-c)E_m \\
\nu_{12} &= \tfrac{1}{4} \\
K_{23} &= \frac{2}{5}\frac{(2+c)}{(1-c)}E_m \\
\mu_{12} &= \frac{2}{5}\frac{(1+c)}{(1-c)}E_m \\
\mu_{23} &= \frac{1}{5}\frac{(2+c)}{(1-c)}E_m
\end{aligned}\right\}
\begin{aligned}
&\nu_f=\nu_m=\tfrac{1}{4} \\
&E_f \gg E_m \\
&E_f \gg k_m \\
&c \neq 1
\end{aligned} \tag{4.34}$$

Finally, the substitution of (4.34) into (4.32) gives

$$\left.\frac{E_{3D}}{E_m}\right|_{\nu_f=\nu_m=1/4} = \frac{1}{6}\frac{cE_f}{E_m} + \left(\frac{1+c/4+c^2/6}{1-c}\right) + O\left(\frac{E_m}{cE_f}\right) \tag{4.35}$$

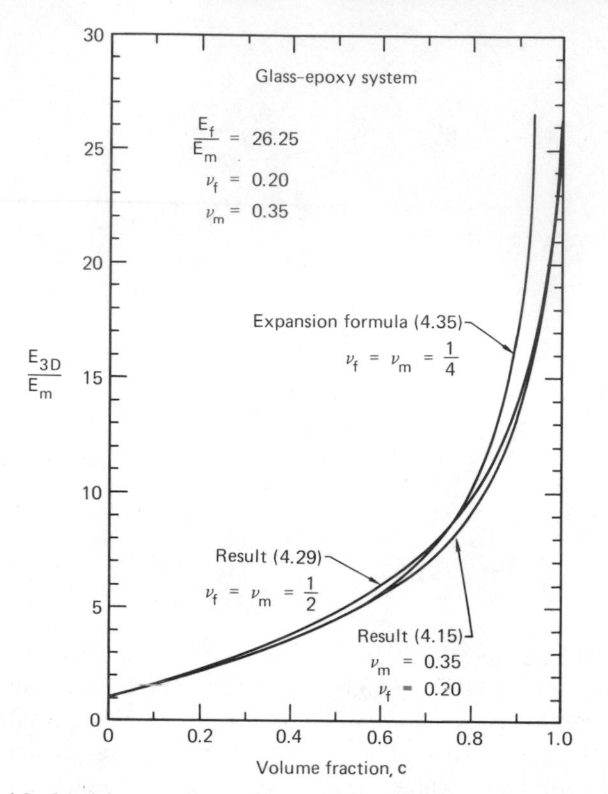

Fig. 4.5 Modulus prediction, three dimensionally random fiber orientation case.

where it is recalled that $c \neq 1$, a restriction required by the approximation (4.33).

A comparison of the predictions of formulas (4.29) and (4.35) with the results obtained from the use of the full forms (4.26) in (4.15) is shown in Fig. 4.5. The simple formulas (4.29) and (4.35) are seen to give very reliable predictions.

Consider the limiting case behavior when $E_m \rightarrow 0$. The results from formula (4.29) are different from that from (4.35). The difference in these relations is reconciled when one recalls the limit processes involved in reaching the result using (4.29). First there was the incompressibility process $\nu_m \rightarrow \frac{1}{2}$, followed by the vanishing modulus process $E_m \rightarrow 0$. An examination of the procedure involved reveals that if the order of the processes is reversed, then the factor $\frac{1}{6}$ rather than $\frac{1}{5}$ results. In light of these remarks the proper result for the vanishing matrix case is

$$E_{3D}|_{E_m=0} = \frac{c}{6} E_f$$

This special result was first obtained by Cox [4.14], as was the corresponding two-dimensional form.

In the two-dimensional case, we write (4.24) as

$$E_{2D} = \frac{1}{u_1}\left(u_1^2 - u_2^2\right) \tag{4.36}$$

where

$$\begin{aligned} u_1 &= \tfrac{3}{8}cE_f + \hat{u}_1 \\ u_2 &= \tfrac{1}{8}cE_f + \hat{u}_2 \end{aligned} \tag{4.37}$$

with

$$\begin{aligned} \hat{u}_1 &= \tfrac{3}{8}\hat{E}_{11} + \frac{\mu_{12}}{2} + \frac{(3+2\nu_{12}+3\nu_{12}^2)\mu_{23}K_{23}}{2(\mu_{23}+K_{23})} \\ \hat{u}_2 &= \tfrac{1}{8}\hat{E}_{11} - \frac{\mu_{12}}{2} + \frac{(1+6\nu_{12}+\nu_{12}^2)\mu_{23}K_{23}}{2(\mu_{23}+K_{23})} \end{aligned} \tag{4.38}$$

Substituting (4.37) into (4.36) and using (4.38) the result can be expressed as the expansion

$$\frac{E_{2D}}{E_m} = \frac{1}{3}\frac{cE_f}{E_m} + \frac{1}{E_m}\left[\frac{\hat{E}_{11}}{3} + \frac{8}{9}\mu_{12} + \frac{4}{9}\frac{(3-2\nu_{12}+3\nu_{12}^2)}{(\mu_{12}+K_{23})}\mu_{23}K_{23}\right] + O\left(\frac{E_m}{cE_f}\right) \tag{4.39}$$

Relation (4.39) is still somewhat complicated, and it is evaluated here only for specific values of the Poisson's ratios.

For the incompressible case $\nu_f = \nu_m = \frac{1}{2}$, it can be shown that

$$\begin{aligned} &\nu_{12} = \tfrac{1}{2} \\ &K_{23} \to \infty \\ &\hat{E}_{11} = (1-c)E_m \\ &\mu_{23} = \mu_{12} \end{aligned}$$

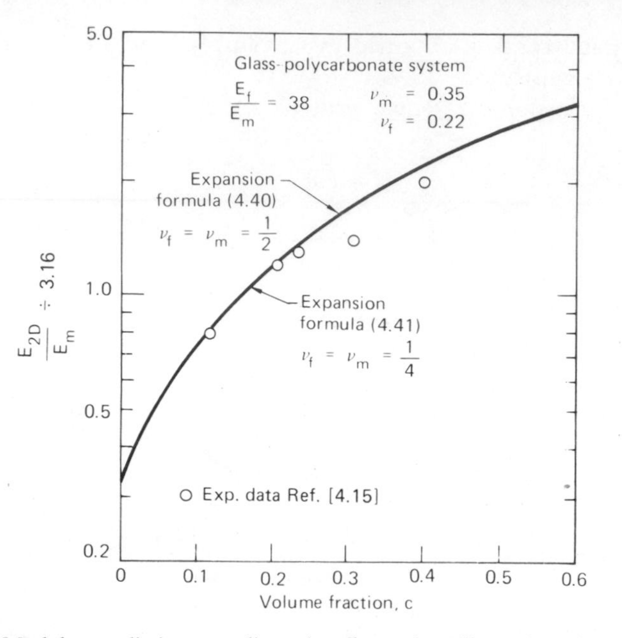

Fig. 4.6 Modulus prediction, two dimensionally random fiber orientation case, comparison with experiment.

with μ_{12} given by (4.26). Combining these relations into (4.39) gives

$$\left.\frac{E_{2D}}{E_m}\right|_{\nu_f=\nu_m=1/2}=\frac{1}{3}\frac{cE_f}{E_m}+\frac{(1-c)}{3}+\frac{19}{27}\left[\frac{E_f(1+c)+E_m(1-c)}{E_f(1-c)+E_m(1+c)}\right]+O\left(\frac{E_m}{cE_f}\right) \tag{4.40}$$

For the case of $\nu_f=\nu_m=\frac{1}{4}$, relations (4.34) for stiff fibers, when substituted into (4.39), give

$$\left.\frac{E_{2D}}{E_m}\right|_{\nu_f=\nu_m=1/4}=\frac{1}{3}\frac{cE_f}{E_m}+\left[\frac{272-41c+90c^2}{270(1-c)}\right]+O\left(\frac{E_m}{cE_f}\right) \tag{4.41}$$

The difference of the zero order term in (E_m/cE_f) in (4.40) and (4.41) should be noted. In arriving at (4.40) it was not necessary to write explicitly μ_{12} and μ_{23} in the form appropriate to stiff fibers; thus both E_f and E_m appear in the zero order term in (4.40). However, in arriving at (4.41) the formulas (4.34) are those appropriate to stiff fibers, $E_f \gg E_m$ and $E_f \gg k_m$; thus E_f does not appear in the zero order term in (4.41). Neither formula (4.40) or (4.41) is valid at $c=0$ or $c=1$, as follows from the method of derivation.

The two formulas (4.40) and (4.41) are compared with experimental data in Fig. 4.6. The experimental data are from Halpin, Jerina, and Whitney [4.15]. As expected for a state of plane stress, the results are not sensitive to the Poisson's ratios of the two phases.

In both the two-dimensional and three-dimensional random fiber orientation case, we have proceeded by a method suggested by the interpenetrating network of Fig. 4.3. As a practical matter, there is, of course, some volume fraction at which interference between fibers becomes a problem. There is one exception, however, to this practical difficulty. The exception occurs in the case where lamina of aligned fibers are stacked at various orientations to obtain an isotropic laminate. The present results apply equally well to that situation. The general subject of lamination theory is taken up in the next chapter. Finally, there is one more practical matter to be remembered. In using these properties results to model randomly oriented chopped fiber systems, end effects are neglected. As we recall from Section 3.4, fibers must have a very large aspect ratio for the end effects to be negligible.

4.5 SOME ISOTROPIC PROPERTIES OF PLATELET SYSTEMS

Having just obtained the properties for a randomly oriented fiber system, we wish to use the same method to determine the effective properties for a randomly oriented platelet system. Of course, we carefully check the results found to make sure they reduce to the results found in Section 3.5 for a dilute system of platelets.

Two-Dimensional Case

First we examine the two-dimensional case. The geometry under consideration is that of a lamination sequence of two alternating materials, of thicknesses h_1 and h_2 and corresponding isotropic properties E_1, ν_1 and E_2, ν_2. Obviously, by considering infinite lamina, we are neglecting any edge effects in the lamina representative of a suspension of platelets of finite planar dimensions.

Now as a practical matter the two-dimensional conditions of interest are those relevant to a state of plane stress. Thus it suffices to consider the planar deformation of a two material combination $h_1 + h_2 = h$, with no surface tractions on the lateral surfaces. The effective planar shear modulus is given by

$$\mu = \frac{1}{h}(h_1 \mu_1 + h_2 \mu_2) \tag{5.1}$$

since the lamina are in a parallel arrangement. For plane stress, relative to x,y coordinates, the general form of the stress-strain relations is

$$\sigma_{xx} = \frac{E}{1-\nu^2}(\varepsilon_{xx} + \nu\varepsilon_{yy})$$

$$\sigma_{yy} = \frac{E}{1-\nu^2}(\varepsilon_{yy} + \nu\varepsilon_{xx}) \tag{5.2}$$

Take $\varepsilon_{yy} = 0$; then the average stress across the two layer combination is given by

$$\sigma_{xx} = \frac{1}{h}\left(h_1\frac{E_1}{1-\nu_1^2} + h_2\frac{E_2}{1-\nu_2^2}\right)\varepsilon_{xx} \tag{5.3}$$

The appropriate statement of the effective properties related to average stress and strain in (5.3) is

$$\frac{\sigma_{xx}}{\varepsilon_{xx}} = \frac{E}{1-\nu^2}$$

thus

$$\frac{E}{1-\nu^2} = \frac{1}{h}\left(h_1\frac{E_1}{1-\nu_1^2} + h_2\frac{E_2}{1-\nu_2^2}\right) \tag{5.4}$$

Relations (5.1), (5.4), and the elasticity relation

$$\mu = \frac{E}{2(1+\nu)} \tag{5.5}$$

suffice to determine the effective properties E and ν. It must be remembered that these are the in-plane isotropic effective properties. The deformation with regard to the coordinate in the direction normal to the x,y plane would be governed by other properties.

It is found that

$$E = c_1E_1 + c_2E_2 + \frac{c_1c_2E_1E_2(\nu_1-\nu_2)^2}{\left[c_1E_1(1-\nu_2^2) + c_2E_2(1-\nu_1^2)\right]} \tag{5.6}$$

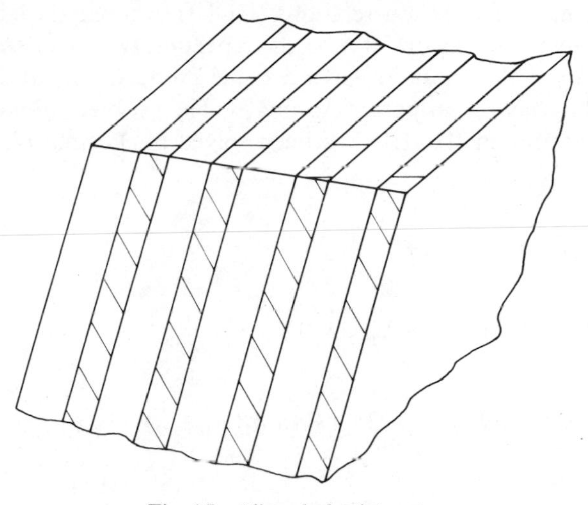

Fig. 4.7 Aligned platelet system.

and

$$\nu = \frac{c_1\nu_1 E_1(1-\nu_2^2) + c_2\nu_2 E_2(1-\nu_1^2)}{c_1 E_1(1-\nu_2^2) + c_2 E_2(1-\nu_1^2)} \tag{5.7}$$

where

$$c_1 = \frac{h_1}{h}, \qquad c_2 = \frac{h_2}{h}$$

are the volume fractions. Clearly, the rule of mixtures is a very good approximation for E in (5.6), but not for ν in (5.7). The planar case with alternating lamina is thus a very simple and direct problem insofar as determining the effective properties.

Three-Dimensional Case

Now the three-dimensional case of randomly oriented platelets is formulated and solved. The solution follows that of Christensen [4.16]. Following the method of the previous section, the first step is to characterize the properties in a state of complete alignment of lamina. We idealize the aligned platelet system as shown in Fig. 4.7, where alternating layers are the respective platelet and matrix phases. The system is transversely isotropic, and the effective stiffness properties are specified through their presence in stress-strain relations (III-1.1), where axis 1 is the axis of transverse symmetry. Again let h be the repeating two layer total thickness with the layer of thickness h_1, having Lamé properties λ_1 and μ_1 and that of thickness h_2 having properties λ_2 and μ_2. The problem of solving for the effective constants in (III-1.1) has been solved by Postma [4.17], with the result

$$C_{11}=\frac{1}{D}h^2(\lambda_1+2\mu_1)(\lambda_2+2\mu_2)$$

$$C_{12}=\frac{h}{D}[\lambda_1 h_1(\lambda_2+2\mu_2)+\lambda_2 h_2(\lambda_1+2\mu_1)]$$

$$C_{22}=\frac{1}{D}\{h^2(\lambda_1+2\mu_1)(\lambda_2+2\mu_2)+4h_1h_2(\mu_1-\mu_2)[(\lambda_1+\mu_1)-(\lambda_2+\mu_2)]\}$$

$$C_{23}=\frac{1}{D}\{h^2\lambda_1\lambda_2+2(\lambda_1h_1+\lambda_2h_2)(\mu_2h_1+\mu_1h_2)\}$$

$$C_{66}=\frac{h\mu_1\mu_2}{h_1\mu_2+h_2\mu_1} \tag{5.8}$$

where

$$D=h[h_1(\lambda_2+2\mu_2)+h_2(\lambda_1+2\mu_1)] \tag{5.9}$$

The geometric averaging process used to find the isotropic effective properties is identical to that of the preceding section. The effective properties k and μ are given by (4.12).

Substituting the C expressions from (5.8) into (4.12) gives

$$\begin{aligned}9Dk=&3(\lambda_1+2\mu_1)(\lambda_2+2\mu_2)+2\lambda_1\lambda_2\\&+4\lambda_1c_1(\lambda_2+2\mu_2)+4\lambda_2c_2(\lambda_1+2\mu_1)\\&+4(\lambda_1c_1+\lambda_2c_2)(\mu_2c_1+\mu_1c_2)\\&+8c_1c_2(\mu_1-\mu_2)^2[(\lambda_1+\mu_1)-(\lambda_2+\mu_2)]\end{aligned} \tag{5.10}$$

where c_1 and c_2 are the volume fractions, h_1/h and h_2/h, respectively, and where D is given by (5.9).

The isotropic material upper bound from (1.50) is repeated here for convenience. It is

$$\frac{k-k_2}{k_1-k_2} \leqslant \frac{c_1}{1+\left[(k_1-k_2)c_2\right]/\left(k_2+\frac{4}{3}\mu_1\right)} \tag{5.11}$$

where $(\mu_1-\mu_2)(k_1-k_2) \geqslant 0$. It is easy to demonstrate that the prediction (5.10) for k is less than the bound given by (5.11).

It is of interest to examine the dilute suspension case. Let

$$c_1 \ll 1$$

Retaining only zero and first order terms in c_1 gives (5.10) as

$$k = k_2 + c_1\left[\frac{(k_1-k_2)\left(k_2+\frac{4}{3}\mu_1\right)}{k_1+\frac{4}{3}\mu_1}\right] \tag{5.12}$$

which is identical with the bound (5.11) under dilute suspension conditions. Thus although the present platelet solution does not coincide with the upper bound under general conditions, they do coincide under dilute conditions. This result, then, is completely compatible with the dilute platelet solution given in Section 3.5.

Following the same procedure as that just outlined for the bulk modulus, we substitute relations (5.8) into the second part of (4.12) to obtain the expression for the effective shear modulus μ in terms of the individual phase properties. There results

$$\begin{aligned}
\frac{30D}{h^2}\mu = {} & 9(\lambda_1+2\mu_1)(\lambda_2+2\mu_2) \\
& + 28c_1c_2(\mu_1-\mu_2)\left[(\lambda_1+\mu_1)-(\lambda_2+\mu_2)\right] \\
& - 5\lambda_1\lambda_2 - 10(\lambda_1c_1+\lambda_2c_2)(\mu_2c_1+\mu_1c_2) \\
& - 4\lambda_1c_1(\lambda_2+2\mu_2) - 4\lambda_2c_2(\lambda_1+2\mu_1) \\
& + \frac{12\mu_1\mu_2}{c_1\mu_2+c_2\mu_1}\left(\frac{D}{h^2}\right)
\end{aligned} \tag{5.13}$$

where D is given by (5.9).

First we examine the dilute suspension case specified by

$$c_1 \ll 1$$

and using this result in (5.13) gives, after lengthy algebra,

$$\mu = \mu_2 + \frac{c_1(\mu_1 - \mu_2)}{15}\left[\frac{9k_1 + 4(2\mu_1 + \mu_2)}{k_1 + \frac{4}{3}\mu_1} + 6\frac{\mu_2}{\mu_1}\right] \tag{5.14}$$

From (1.51) the corresponding isotropic material upper bound is given by

$$\frac{\mu - \mu_2}{\mu_1 - \mu_2} \leqslant \frac{c_1}{1 + \dfrac{(\mu_1 - \mu_2)}{\mu_2 + \dfrac{3}{2}\left(\dfrac{1}{\mu_1} + \dfrac{10}{9k_1 + 8\mu_1}\right)^{-1}}} \tag{5.15}$$

for $(\mu_1 - \mu_2)(k_1 - k_2) \geqslant 0$. It can be shown that the result (5.14) and the bound (5.15) coincide, which again agrees with the results of Section 3.5. In the nondilute case it is easy to show that the general result (5.13) is less than the corresponding upper bound.

For the convenience of engineering measurements, we next obtain useful formulas for the effective uniaxial modulus E, under dilute conditions. The appropriate relation between E, k, and μ is given by

$$E = \frac{9k\mu}{3k + \mu} \tag{5.16}$$

Under dilute conditions, and assuming the platelet phase is very stiff compared with the matrix phase, relations (5.12) and (5.14), respectively, reduce to

$$k = k_m + \frac{2cE_p}{9(1 - \nu_p)} \tag{5.17}$$

and

$$\mu = \mu_m + \frac{c}{30}\left(\frac{7 - 5\nu_p}{1 - \nu_p^2}\right)E_p \tag{5.18}$$

where a more direct phase designation has been adopted and c is the volume fraction of the platelet phase.

Substituting the dilute results (5.17) and (5.18) into (5.16) gives

$$E=\frac{9\left\{\frac{E_m}{[3(1-2\nu_m)]}+\frac{2cE_p}{[9(1-\nu_p)]}\right\}\left\{\frac{E_m}{[2(1+\nu_m)]}+\frac{(7-5\nu_p)}{30(1-\nu_p^2)}cE_p\right\}}{\frac{3E_m}{[2(1-2\nu_m)(1+\nu_m)]}+\frac{(9+5\nu_p)cE_p}{[10(1-\nu_p^2)]}} \tag{5.19}$$

We wish to proceed further to obtain a simple result that is amenable to direct interpretation. To this end we must consider the separate cases of $cE_p \ll E_m$ and $cE_p \gg E_m$.

$cE_p \ll E_m$ Case

In the case of $cE_p \ll E_m$ relation (5.19) is arranged as a power series in cE_p/E_m with the result

$$\frac{E}{E_m}=1+\frac{1}{(1-\nu_p^2)}\left(\frac{cE_p}{E_m}\right)[(1+\nu_m)(7-5\nu_p)+10(1-2\nu_m)(1+\nu_p)$$
$$-(1-2\nu_m)(1+\nu_m)(9+5\nu_p)]+O\left(\frac{c^2E_p^2}{E_m^2}\right) \tag{5.20}$$

Next, the alternate case is treated.

$cE_p \gg E_m, cE_p \gg k_m$ Case

Now, expanding (5.19) as a power series in E_m/cE_p gives

$$\frac{E}{E_m}=\frac{2(7-5\nu_p)}{3(1-\nu_p)(9+5\nu_p)}\left(\frac{cE_p}{E_m}\right)+\frac{10(1+\nu_p)}{9+5\nu_p}$$
$$\left[\frac{1}{1+\nu_m}+\frac{7-5\nu_p}{10(1+\nu_p)(1-2\nu_m)}-\frac{7-5\nu_p}{(1+\nu_m)(1-2\nu_m)(9+5\nu_p)}\right]+O\left(\frac{E_m}{cE_p}\right) \tag{5.21}$$

The evaluation of the coefficients of the terms in (5.20) and (5.21), for

$\nu_p = \nu_m = \frac{1}{4}$, gives

$$\frac{E}{E_m} = 1 + \frac{1}{2}\left(\frac{cE_p}{E_m}\right) + O\left(\frac{c^2E_p^2}{E_m^2}\right) \tag{5.22}$$

and

$$\frac{E}{E_m} = \frac{184}{369}\left(\frac{cE_p}{E_m}\right) + \frac{1686}{1681} + O\left(\frac{E_m}{cE_p}\right) \tag{5.23}$$

Other values of Poisson's ratios give similar results, with the conclusion that

$$E \simeq \frac{cE_p}{2} + E_m \tag{5.24}$$

under practical conditions.

Under dilute conditions the corresponding result in the fiber case from (4.35) is

$$\left.\frac{E}{E_m}\right|_{\nu_f = \nu_m = 1/4} = \frac{cE_f}{6} + E_m \tag{5.25}$$

Thus the platelet stiffness reinforcing mechanism is about three times as effective as the fiber form mechanism.

Incompressible Case

It was necessary to use asymptotic methods to obtain the simple result (5.24). Now we show that we can obtain exact results of very simple form in the special case where both phases are incompressible.

Letting

$$\nu_1 = \nu_2 = \tfrac{1}{2}$$

in the general result (5.13), it can be shown to reduce to the result

$$\mu = \tfrac{3}{5}(c_1\mu_1 + c_2\mu_2) + \frac{2\mu_1\mu_2}{5(c_1\mu_2 + c_2\mu_1)} \tag{5.26}$$

An alternate form of this result is

$$\mu = \tfrac{3}{5} c_1 \mu_1 + \left[\frac{(5 - 3c_1^2 - 6c_1c_2)\mu_1 + 3c_1c_2\mu_2}{5(c_1\mu_2 + c_2\mu_1)} \right] \mu_2 \tag{5.27}$$

To interpret these results first observe from (5.26) that the rule of mixtures gives a poor prediction of the properties, unless the shear moduli of the individual phases are very close to each other. To compare with fiber form predictions, it is convenient to convert (5.27) to the appropriate form employing the uniaxial modulus. Using the incompressibility condition, (5.27) can be converted to

$$E = \tfrac{3}{5} c_1 E_1 + \left[\frac{(5 - 3c_1^2 - 6c_1c_2)E_1 + 3c_1c_2E_2}{5(c_1E_2 + c_2E_1)} \right] E_2 \tag{5.28}$$

This form should be compared with the corresponding fiber result for incompressible phases (4.29). Again, it is seen by this comparison that the platelet form of reinforcing is three times as effective as the fiber form.

Perhaps the character of the present model for lamellar or platelet type media is best understood by noting from all the formulas derived in this section that a phase inversion leaves the effective properties unchanged. For example, in the formula (5.26) for μ, interchange indices 1 and 2. The result for μ is unaltered. The formulas for particulate or fiber reinforcement undergo a drastic change in prediction under phase inversion. One can thus rationalize the present results as applying to a type of heterogeneous material with interpenetrating phases. In this type of medium, both phases are continuous, and there is no geometric characteristic to the interface that provides a code by which a material can be designated as being on a certain side of the interface.

Alternatively, if we are trying to model the properties of an equivalent isotropic medium made up of discrete, randomly oriented grains composed of alternating lamina, then the present results must be viewed as upper bounds on the properties. The upper bound character follows from the fact that we would be neglecting interface conditions at grain surfaces, and the present approach using imposed strains corresponds to the use of an admissible displacement field in the theorem of minimum potential energy. This same conclusion also applies to the fiber results of the preceding section. If the fiber structure were considered to be composed of randomly oriented, fiber containing grains, rather than the interpenetrating fiber model of Fig. 4.3, then the results of Section 4.4 must be interpreted as having an upper bound character.

4.6 SUMMARY OF EFFECTIVE STIFFNESS PROPERTIES AND CONCLUSIONS

The preceding three chapters have been spent covering the subject of the determination of the effective linear elastic properties for various types of heterogeneous media. More emphasis has been placed on this aspect of the general subject than on any other topic to be covered in the subsequent chapters. Why do we give the place of first importance to the determination of effective properties? For one thing, it is the most highly developed aspect of the subject of heterogeneous media behavior. These developments did not just happen by accident; obviously, there has been a driving force for such developments. An understanding of the role individual phases play in the overall macroscopic behavior of a composite material provides the key by which materials can be judiciously selected for optimal combinations. Without such knowledge, the field of heterogeneous material behavior primarily would be just an offshoot of anisotropic elasticity. Problems would be posed in a design context and stress analysis would be used to seek final designs. By going one step back, to relate composite behavior to the properties of individual phases, one gains great flexibility in designing the composite materials to meet the objectives of the end product use. This study of individual phase properties as related to composite behavior, sometimes called micromechanics, has provided an "extra degree of freedom" in the selection of composite materials that has greatly contributed to their success. Of course, it must also be remembered that the rationale for composite materials relates to the fact that the combination of materials gives "composite properties" that otherwise may not be attainable. Without a knowledge of how the individual phase properties contribute to the composite properties, the field would be hopelessly empirical, at the level of selecting the constituent phases. Fortunately, the field is now advancing rapidly beyond the empirical stage.

We do not here summarize all the results of the preceding developments. We only wish to reinforce the main conclusions. Considered as a whole, we have covered only a very limited number of models and analyses. However, our models cover a very wide range of geometric types, including spherical, cylindrical, and lamellar interface geometry. We have seen that these different geometric forms give drastically different reinforcing effects, so far as the effective moduli are concerned. Consider first the case of the stiffer material being the inclusion phase in a more compliant matrix phase. When inclusion orientations are random, such that the effective properties are isotropic, we have seen that the least reinforcing effect comes from the spherical inclusion geometry, and the greatest effect is obtained from the lamellar geometry. The cylindrical inclusion geometry is intermediate between these two cases. Conversely, when the more compliant phase is the inclusion phase, then the spherical inclusion form has the least degrading effect, and the lamellar form (suggestive of material cuts in the extreme case) has the greatest effect.

The bounds that have been found are impressively definitive. Again, in the case of macroscopically isotropic media, the bounds for the effective bulk modulus coincide with the results from the composite spheres model. Thus the bounds on k are the best that can be obtained without specifying the phase geometry. Furthermore, the model thus reveals the phase geometry for optimal material combinations. The same situation is not true for the effective shear modulus μ. Except under dilute conditions, none of the model predictions for μ coincide with the bounds. However, some of the models are at least quite close to the bound, suggesting that, even at best, there is very little latitude for possible tightening of the bounds. The situation for transversely isotropic media was entirely similar. The predictions from the composite cylinders model have values that coincide with the bounds for four of the five independent effective properties, E_{11}, ν_{12}, K_{23}, and μ_{12}. The bounds on the transverse shear modulus μ_{23} do not in general coincide with the predictions from any geometric model.

We placed strong emphasis on the composite spheres model and the composite cylinders model. The other prominent model in widespread use is that of the self-consistent scheme. The resulting predictions from the self-consistent scheme are not actually relevant to a particular geometric model, except in the case of a single phase polycrystalline media. For multiphase media the method makes alternate geometric arrangements of the phases in order to estimate the average strain in the phase for use in calculating the effective properties. It makes no difference whether or not the phase is continuous; the geometry is still rearranged to view the particular phase under consideration as an inclusion in an infinite media. It is not surprising that the self-consistent scheme gives unrealistic results in the case of multiphase media. This situation was discussed in Section 2.5. Because of its abnormal predictions there will be no further involvement here with applications of the self-consistent scheme to multiphase media.

With regard to the composite spheres and cylinders models, we should state some observations about their range of applicability. The geometry of these models inherently allows the full range of volume fraction of inclusions, $0 \leqslant c \leqslant 1$. Thus for composite materials with inclusions having a broad gradation in sizes, the composite spheres or cylinders models are ideally suited for application. The most extreme test of these models, however, is provided by composites having single size inclusions. This latter situation is the realistic case with fiber reinforced materials. First, consider the spherical inclusion case. In Section 2.4 experimental data for the effective modulus were compared with the prediction from the composite spheres model and the associated three phase model of Section 2.4. The composite involved a very narrow size distribution of glass spheres in a polyester matrix. As seen from Fig. 2.4, the comparison between the theoretical results and the experimental results was very close, at least up to a volume fraction of $c=0.45$.

Considering that the "loose" (cubical) packing of spheres has a maximum volume fraction of $c=\pi/6=0.52$, the theoretical prediction from the composite spheres model is good until an amazingly large volume fraction. In the case of single size packing of fibers, the maximum "loose" (cubical) packing volume fraction is $c=\pi/4=0.79$. Typical maximum volume fractions for filament winding are about $c=\frac{2}{3}$. Comparison of the fiber case with the results just discussed for spherical inclusions suggests that the composite cylinders model is expected to give a close estimate of actual properties at normal volume fractions in typical applications. This observation is consistent with the results cited in Section 4.3, which involved a comparison of composite cylinders models and numerical solutions for the case of hexagonal packing.

Of course, comparisons such as those given earlier must depend on the ratio of the inclusion to matrix moduli. The stiffer the inclusions, in a relative sense, the more restrictive is the reasonable range of applicability of the composite spheres or cylinders models to single size inclusion type composites. Reliable experimental data for all five independent properties of fiber composites are difficult to obtain. However, it is sometimes stated that the experimental results are found to be larger than those predicted by the composite cylinders model. This possible effect is probably due to the irregular geometry in actual systems. The agglomeration of fibers into a bunch can cause an excluded volume effect whereby for transverse properties there is an effective volume fraction of fiber inclusions larger than the actual volume fraction. This effect can also occur with particulate systems. Despite these complications, the composite spheres and cylinders models and the associated models of Sections 2.4 and 3.3 remain the only soundly based theoretical models that give a realistic prediction of the behavior of actual composite systems.

In the literature, one finds copious use of series and parallel models, corresponding to the Reuss and Voigt bounds derived in Section 4.1. The present developments show that series and parallel models are practically meaningless. Only in a few cases have we found such models to give reasonable approximations to the effective properties. The shortcomings of the individual series and parallel models are often remedied by taking complicated combinations of the two. This, of course, corresponds to curve fitting, and it is no substitute for a rational derivation of the appropriate property.

Our intention here has not been to survey the entire field in detail. Indeed, we have not even mentioned, until now, the great amount of work that has gone into the statistical characterization of composite materials. Such aspects of the subject are beyond the scope intended here, but, as typical efforts along these lines, we mention the book by Beran [4.18] and the bounds developments by Kröner [4.19] in which results are given involving correlation functions of third order. Effective properties can also be defined in the dynamic context, for long wave length conditions. Work along these lines, also including a statistical characterization of the media, has been presented by Bose and Mal [4.20]. Also, on the deterministic side there is a vast literature. For example, Walpole [4.21] has obtained bounds results for materials involving phases that are anisotropic; applications to polycrystalline aggregates are given. Although we have not provided a literature survey, that is not our objective. Rather we have taken those elements of research work that we need to construct a reasonably comprehensive treatment of the subject from a deterministic theory point of view.

There have been many idealizations involved in the present developments. For example, most of the results have been for two material combinations, although the bounds were derived in general form. Composite materials are often employed in forms that involve the combination of three or more material phases. How does one treat these problems when determining effective properties? There is no easy and simple answer to this question. With a strong reliance on judgment, one can treat the overall problem as a sequence of two phase problems. Such directions of inquiry are relatively unexplored. Another idealization is involved in most of the work on the effective properties of fiber and platelet systems. This idealization is that of the neglect of end and edge effects. However, we did obtain explicit results on finite length fiber effects in Section 3.4, for a dilute suspension of aligned fibers. Effects such as these are of great practical importance.

It is important to observe that strength is a property of equal importance to that of stiffness. Our emphasis thus far on stiffness is not meant to convey a contrary impression. We treat strength as a topic in Section 6.2, but as we see there, the subject does not have the same high state of development that we have just given for the effective stiffness properties of heterogeneous media. Also, it should be noted that the mathematical problems posed in determining effective moduli have their counterparts in many other fields, such as dielectric, electromagnetic, and thermal behavior of materials. In Chapter IX we treat the subject of effective thermal properties of heterogeneous media because of its close relationship to mechanical properties.

PROBLEMS

1. Explicitly prove that $\boldsymbol{\sigma}^*$ in (1.4) is self-equilibrated. In effecting this proof, explain the relationship between the polarization stress $\boldsymbol{\tau}$, the comparison material stiffness tensor, $\mathbf{C}_0$, and the strain $\boldsymbol{\varepsilon}$.
2. In the macroscopically isotropic case, verify that the bounds expressions for k in (1.50) are interchangeable upon phase inversion. That is, for index $1 \rightarrow 2$ and index $2 \rightarrow 1$, the lower bound takes the form of the upper bound after phase inversion, and so on. Is the same observation true for the bounds on the effective shear modulus μ?
3. Explain how one would approach the problem of determining bounds on stiffness properties for the hexagonal packing of single size fibers in a separate matrix phase. Consult literature references.
4. Obtain the formulas for the maximum volume fractions of single size spheres in square and hexagonal packing arrangements. Obtain corresponding results for the packing of single size cylinders.
5. There is an empirical result known as the Halpin-Tsai equation for predicting effective composite properties. Consult the composites material literature for this form and discuss its limitations and significance compared with the results derived here.
6. Obtain the special approximate forms for the effective properties of the composite cylinders model that result from the use of $cE_f/E_m \gg 1$ and $cE_f/k_m \gg 1$.
7. Derive the formulas (5.6) and (5.7) for the plane stress properties of two material lamellar systems. Examine a practical example to deduce the deviation of the result (5.7) for ν, from a rule of mixtures prediction.
8. Discuss the practical limitations of using fibers and platelets in three-dimensional random orientation configurations.
9. Discuss the means by which one could obtain the effective moduli for finite length randomly oriented fibers. Consider combining the methods given in Section 3.4 on finite length, aligned fibers and Section 4.4 on infinitely long, randomly oriented fibers.

REFERENCES

4.1 B. Paul, "Prediction of elastic constants of multiphase materials," *Trans. ASME*, vol. 218, 36 (1960).

4.2 Z. Hashin and S. Shtrikman, "A variational approach to the theory of the elastic behavior of multiphase materials," *J. Mech. Phys. Solids*, vol. 11, 127 (1963).

4.3 L. J. Walpole, "On bounds for the overall elastic moduli of inhomogeneous systems—I," *J. Mech. Phys. Solids*, vol. 14, 151 (1966).

4.4 A. E. H. Love, *A Treatise on the Mathematical Theory of Elasticity*, Dover, New York, 1944.

4.5 Z. Hashin, "On elastic behavior of fibre reinforced materials of arbitrary transverse phase geometry," *J. Mech. Phys. Solids*, vol. 13, 119 (1965).

4.6 R. Hill, "Theory of mechanical properties of fiber-strengthened materials: I. Elastic behavior," *J. Mech. Phys. Solids*, vol. 12, 199 (1964).

4.7 Z. Hashin, "The elastic moduli of heterogeneous materials," *J. Appl. Mech.*, vol. 29, 143 (1962).

4.8 Z. Hashin and W. Rosen, "The elastic moduli of fiber reinforced materials," *J. Appl. Mech.*, vol. 31, 223 (1964).

4.9 Z. Hashin, "Theory of fiber reinforced materials," NASA CR-1974, 1972.

4.10 E. Behrens, "Elastic constants of fiber-reinforced composites with transversely isotropic constituents," *J. Appl. Mech.*, vol. 38, 1062 (1971).

4.11 C. H. Chen and S. Cheng, "Mechanical properties of anisotropic fiber-reinforced composites," *J. Appl. Mech.*, vol. 37, 186 (1970).

4.12 R. M. Christensen and F. M. Waals, "Effective stiffness of randomly oriented fiber composites," *J. Comp. Mater.*, vol. 6, 518 (1972).

4.13 R. M. Christensen, "Asymptotic modulus results for composites containing randomly oriented fibers," *Int. J. Solids Structures*, vol. 12, 537 (1976).

4.14 H. L. Cox, "The elasticity and strength of paper and other fibrous materials," *Brit. J. Appl. Phys.*, vol. 3, 72 (1952).

4.15 J. C. Halpin, K. Jerina, and J. M. Whitney, "The laminate analogy for 2- and 3-dimensional composite materials," *J. Comp. Mater.*, vol. 5, 36 (1971).

4.16 R. M. Christensen, "Isotropic properties of platelet reinforced media," *J. Engrg. Mater. Tech.*, vol. 101 (1979).

4.17 G. W. Postma, "Wave propagation in a stratified medium," *Geophysics*, vol. 20, 780 (1955).

4.18 M. J. Beran, *Statistical Continuum Theories*, Wiley, New York, 1968.

4.19 E. Kröner, "Bounds for effective elastic moduli of disordered materials," *J. Mech. Phys. Solids*, vol. 25, 137 (1977).

4.20 S. K. Bose and A. K. Mal, "Elastic waves in a fiber-reinforced composite," *J. Mech. Phys. Solids*, vol. 22, 217 (1974).

4.21 L. J. Walpole, "On bounds for the overall elastic moduli of inhomogeneous systems—II," *J. Mech. Phys. Solids*, vol. 14, 289 (1966).

CHAPTER V
LAMINATES

We have seen how one can use the properties of individual phases to predict the properties of a composite combination of materials. This certainly suggests that different materials can be combined to give special properties to the combination. However, it does not tell us much about the form in which composite materials are best made for practical use.

To be more specific, let us consider the case of fiber reinforced materials. One way to prepare a material of this type would be to take a prepared mat of randomly oriented chopped fibers and impregnate it with a resin. The analysis of Section 4.4 could be used to predict the stiffness properties of the combination. This method is commonly used, but there is an even more widely used fabrication technique, involving the stacking of lamina to obtain a laminate. With the fibers being aligned in each lamina, a higher loading of fibers in the composite can be obtained with laminates than with the random orientation fiber mat.

There are two main methods by which laminates can be constructed. One method uses "wet" filament winding, whereas the other method involves the pressing and curing of preimpregnated fiber lamina. The laminate can have the various lamina arranged at predetermined angles to give in-plane isotropic properties to the laminate, or the fiber lamina can be given preferential orientation to produce anisotropic properties. It is one of the main advantages of fiber reinforced materials that they can be made to produce the optimal degree of anisotropy for a particular application. We delve further into this matter in Chapter VI. At this point we merely want to observe the high degree of flexibility we have in designing and using composite materials. One of the first problems we face in this chapter is exactly how one should arrange the lamina to produce isotropic in-plane properties for the laminate. Common sense dictates that the lamina should be stacked at angles that vary by a fixed amount with each successive lamina. Common sense, does not, however, reveal the minimum number of lamina needed to obtain isotropic in-plane properties for the laminate. We obtain a somewhat surprising answer to this question.

To use laminate construction in engineering design, we need to know the properties of the laminate. It is not sufficient just to know the properties of the individual lamina. Since laminates are primarily used in thin plate and thin shell construction, we seek the properties of engineering laminates appropriate to the small thickness conditions. We first approach the problem at a level consistent with the classical theory of thin homogeneous plates. Some special effects are shown to occur that do not even exist with homogeneous materials. We also consider a high-order theory of laminate behavior that relaxes some of the restrictive assumptions made at the level of the classical theory. It is beyond the scope intended here to study laminated shells.

We begin with the consideration of basic properties of a lamina, and their relationship to certain laminate properties.

5.1 TRANSFORMATION RELATIONS

It is very useful for us to have easy access to the rules that allow us to specify material properties in rotated coordinate systems. The obvious motivation for this relates to the use of laminates that involve lamina stacked at various angles relative to each other.

Let us begin by recalling from Section 1.1 the pertinent form of the elastic stress-strain relations,

$$\sigma_{ij} - C_{ijkl}\varepsilon_{kl} \tag{1.1}$$

In Section 1.1 we introduced a contracted notation, as

$$\sigma_i = C_{ij}\varepsilon_j \tag{1.2}$$

where σ_i and ε_i are defined in terms of σ_{ij} and ε_{ij} in Section 1.1.

The various types of material symmetry are discussed in Section 1.1. Our interest here relates to the forms having symmetry with respect to a plane and the case of orthotropy. For coordinate x_3 normal to a plane of material symmetry, we have a material with 13 independent constants, as

$$C_{ij} = \begin{bmatrix} C_{11} & C_{12} & C_{13} & 0 & 0 & C_{16} \\ & C_{22} & C_{23} & 0 & 0 & C_{26} \\ & & C_{33} & 0 & 0 & C_{36} \\ & & & C_{44} & C_{45} & 0 \\ & & & & C_{55} & 0 \\ & & & & & C_{66} \end{bmatrix} \tag{1.3}$$

For the case of three mutually perpendicular planes of symmetry, orthotropy, there are nine independent constants, as

$$C_{ij} = \begin{bmatrix} C_{11} & C_{12} & C_{13} & 0 & 0 & 0 \\ & C_{22} & C_{23} & 0 & 0 & 0 \\ & & C_{33} & 0 & 0 & 0 \\ & & & C_{44} & 0 & 0 \\ & & & & C_{55} & 0 \\ & & & & & C_{66} \end{bmatrix} \tag{1.4}$$

Inverting relations (1.1) allows us to write

$$\varepsilon_i = S_{ij}\sigma_j \tag{1.5}$$

For orthotropy the form of S_{ij} is identical to the form of C_{ij} (1.4). In a manner similar to that which we discussed in Section 3.1 for transversely isotropic materials, we write S_{ij} for orthotropic materials as

$$S_{ij} = \begin{bmatrix} \frac{1}{E_{11}} & -\frac{\nu_{12}}{E_{11}} & -\frac{\nu_{13}}{E_{11}} & 0 & 0 & 0 \\ -\frac{\nu_{21}}{E_{22}} & \frac{1}{E_{22}} & -\frac{\nu_{23}}{E_{22}} & 0 & 0 & 0 \\ -\frac{\nu_{31}}{E_{33}} & -\frac{\nu_{32}}{E_{33}} & \frac{1}{E_{33}} & 0 & 0 & 0 \\ 0 & 0 & 0 & \frac{1}{2\mu_{23}} & 0 & 0 \\ 0 & 0 & 0 & 0 & \frac{1}{2\mu_{13}} & 0 \\ 0 & 0 & 0 & 0 & 0 & \frac{1}{2\mu_{12}} \end{bmatrix} \tag{1.6}$$

From symmetry we have

$$\nu_{ij}E_j=\nu_{ji}E_i \qquad (i,j \text{ not summed}) \tag{1.7}$$

where we recall that in the notation for Poisson's ratios, ν_{ij}, the first index refers to the direction of imposed strain and the second index refers to the response direction.

For a state of plane stress we have

$$\sigma_3=\sigma_{33}=0$$

$$\sigma_4=\sigma_{23}=0$$

$$\sigma_5=\sigma_{31}=0$$

leaving the stress-strain relations as

$$\begin{bmatrix}\varepsilon_1\\ \varepsilon_2\\ \varepsilon_6\end{bmatrix}=\begin{bmatrix}\dfrac{1}{E_{11}} & -\dfrac{\nu_{12}}{E_{11}} & 0\\ -\dfrac{\nu_{21}}{E_{22}} & \dfrac{1}{E_{22}} & 0\\ 0 & 0 & \dfrac{1}{2\mu_{12}}\end{bmatrix}\begin{bmatrix}\sigma_1\\ \sigma_2\\ \sigma_6\end{bmatrix} \tag{1.8}$$

which involves four independent constants.

These equations can be inverted to give

$$\begin{bmatrix}\sigma_1\\ \sigma_2\\ \sigma_6\end{bmatrix}=\begin{bmatrix}\dfrac{E_{11}}{1-\nu_{12}\nu_{21}} & \dfrac{\nu_{21}E_{11}}{1-\nu_{12}\nu_{21}} & 0\\ \dfrac{\nu_{12}E_{22}}{1-\nu_{12}\nu_{21}} & \dfrac{E_{22}}{1-\nu_{12}\nu_{21}} & 0\\ 0 & 0 & 2\mu_{12}\end{bmatrix}\begin{bmatrix}\varepsilon_1\\ \varepsilon_2\\ \varepsilon_6\end{bmatrix} \tag{1.9}$$

The relationship of the terms in (1.9) to the C_{ij} constants of (1.2) are easily obtainable by using $\sigma_3=0$ in (1.2). There results

$$\begin{bmatrix}\sigma_1\\ \sigma_2\\ \sigma_6\end{bmatrix}=\begin{bmatrix}Q_{11} & Q_{12} & 0\\ Q_{21} & Q_{22} & 0\\ 0 & 0 & Q_{66}\end{bmatrix}\begin{bmatrix}\varepsilon_1\\ \varepsilon_2\\ \varepsilon_6\end{bmatrix} \tag{1.10}$$

where

$$Q_{11}=C_{11}-\frac{C_{13}^2}{C_{33}}$$

$$Q_{12}=Q_{21}=C_{12}-\frac{C_{13}C_{23}}{C_{33}}$$

$$Q_{22}=C_{22}-\frac{C_{23}^2}{C_{33}}$$

$$Q_{66}=C_{66} \tag{1.11}$$

Tensor Transformations

Our general interest here centers on the case of an orthotropic material. However, when the material is rotated about one of the axes perpendicular to a plane of symmetry, then recourse must be made to the properties description for a material with but one plane of symmetry. The tensor transformation law for (1.1) is given by

$$C'_{ijkl}=a_{im}a_{jn}a_{ko}a_{lp}C_{mnop} \tag{1.12}$$

Taking the 1′, 2′, 3′ axes as being rotated about the 3 axis of the 1, 2, 3 system gives

$$C'_{11}=m^4C_{11}+2m^2n^2(C_{12}+2C_{66})+4mn(m^2C_{16}+n^2C_{26})+n^4C_{22}$$

$$C'_{12}=m^2n^2(C_{11}+C_{22}-4C_{66})-2mn(m^2-n^2)(C_{16}-C_{26})+(m^4+n^4)C_{12}$$

$$C'_{13}=m^2C_{13}+n^2C_{23}+2mnC_{36}$$

$$C'_{16}=m^2(m^2-3n^2)C_{16}-mn[m^2C_{11}-n^2C_{22}-(m^2-n^2)(C_{12}+2C_{66})]$$
$$+n^2(3m^2-n^2)C_{26}$$

$$C'_{22}=n^4C_{11}+2m^2n^2(C_{12}+2C_{66})-4mn(m^2C_{26}+n^2C_{16})+m^4C_{22}$$

$$C'_{23}=n^2C_{13}+m^2C_{23}-2mnC_{36}$$

$$C'_{26}=m^2(m^2-3n^2)C_{26}-mn[n^2C_{11}-m^2C_{22}+(m^2-n^2)(C_{12}+2C_{66})]$$
$$+n^2(3m^2-n^2)C_{16} \tag{1.13}$$

$$C'_{33} = C_{33}$$

$$C'_{36} = (m^2 - n^2)C_{36} + mn(C_{23} - C_{13})$$

$$C'_{44} = m^2C_{44} - 2mnC_{45} + n^2C_{55}$$

$$C'_{45} = (m^2 - n^2)C_{45} + mn(C_{44} - C_{55})$$

$$C'_{55} = m^2C_{55} + 2mnC_{45} + n^2C_{44}$$

$$C'_{66} = m^2n^2(C_{11} + C_{22} - 2C_{12}) + 2mn(m^2 - n^2)(C_{22} - C_{16}) + (m^2 - n^2)^2C_{66}$$

where $m = \cos\theta$, $n = \sin\theta$, and θ is the angle of rotation.

We now wish to obtain the corresponding transformation relations for the plane stress form of the stress-strain relations. For an orthotropic material that has only one symmetry plane coincident with the coordinate plane, 1-2, the stress-strain relations (1.10) must be generalized as

$$\begin{bmatrix} \sigma_1 \\ \sigma_2 \\ \sigma_6 \end{bmatrix} = \begin{bmatrix} Q_{11} & Q_{12} & Q_{16} \\ Q_{21} & Q_{22} & Q_{26} \\ Q_{61} & Q_{62} & Q_{66} \end{bmatrix} \begin{bmatrix} \varepsilon_1 \\ \varepsilon_2 \\ \varepsilon_6 \end{bmatrix} \tag{1.14}$$

where it can be shown that

$$Q_{ij} = C_{ij} - \frac{C_{i3}C_{j3}}{C_{33}} \tag{1.15}$$

The presence of the $Q_{16} = Q_{61}$, $Q_{26} = Q_{62}$ terms in (1.14) shows a coupling between shear stress/strain and normal stress/strain. The transformation of Q_{ij} follows exactly the same form as that of C_{ij} (1.13), since the same tensor transformation law must apply. Following Tsai and Pagano [5.1], the transformations (1.13) can be written in a very compact form, as is now done for the Q_{ij}'s. Noting the identities

$$m^4 = \tfrac{1}{8}(3 + 4\cos 2\theta + \cos 4\theta)$$

$$m^3n = \tfrac{1}{8}(2\sin 2\theta + \sin 4\theta)$$

$$m^2n^2 = \tfrac{1}{8}(1 - \cos 4\theta)$$

$$mn^3 = \tfrac{1}{8}(2\sin 2\theta - \sin 4\theta)$$

$$n^4 = \tfrac{1}{8}(3 - 4\cos 2\theta + \cos 4\theta) \tag{1.16}$$

the Q's have the following transformation

$$\begin{bmatrix} Q'_{11} \\ Q'_{22} \\ Q'_{12} \\ Q'_{66} \\ 2Q'_{16} \\ 2Q'_{26} \end{bmatrix} = \begin{bmatrix} U_1 & U_2 & 2U_6 & U_3 & U_7 \\ U_1 & -U_2 & -2U_6 & U_3 & U_7 \\ U_4 & 0 & 0 & -U_3 & -U_7 \\ U_5 & 0 & 0 & -U_3 & -U_7 \\ 0 & 2U_6 & -U_2 & 2U_7 & -2U_3 \\ 0 & 2U_6 & -U_2 & -2U_7 & 2U_3 \end{bmatrix} \begin{bmatrix} 1 \\ \cos 2\theta \\ \sin 2\theta \\ \cos 4\theta \\ \sin 4\theta \end{bmatrix} \tag{1.17}$$

where

$$\begin{aligned} U_1 &= \tfrac{1}{8}(3Q_{11}+3Q_{22}+2Q_{12}+4Q_{66}) \\ U_2 &= \tfrac{1}{2}(Q_{11}-Q_{22}) \\ U_3 &= \tfrac{1}{8}(Q_{11}+Q_{22}-2Q_{12}-4Q_{66}) \\ U_4 &= \tfrac{1}{8}(Q_{11}+Q_{22}+6Q_{12}-4Q_{66}) \\ U_5 &= \tfrac{1}{8}(Q_{11}+Q_{22}-2Q_{12}+4Q_{66}) \\ U_6 &= \tfrac{1}{2}(Q_{16}+Q_{26}) \\ U_7 &= \tfrac{1}{2}(Q_{16}-Q_{26}) \end{aligned} \tag{1.18}$$

There are some invariant properties that should be noted. By direct observation from (1.17).

$$Q'_{11}+Q'_{22}+2Q'_{12}=Q_{11}+Q_{22}+2Q_{12}$$

and

$$Q'_{66}-Q'_{12}=Q_{66}-Q_{12} \tag{1.19}$$

These invariants are the same as those given by Hearmon [5.2].

Quasi-Isotropic Case

Consider now the regular stacking of orthotropic lamina, to form a laminate. At this point we only consider in-plane stiffness properties of the laminate. Bending effects are considered in the next section. Suppose we stack three or more identical laminae, at equal angles between successive lamina. The stress resultants for the laminate have the form of (1.14), and are explicitly given by

$$\begin{bmatrix} N_1 \\ N_2 \\ N_6 \end{bmatrix} = \begin{bmatrix} A_{11} & A_{12} & A_{16} \\ A_{21} & A_{22} & A_{26} \\ A_{61} & A_{62} & A_{66} \end{bmatrix} \begin{bmatrix} \varepsilon_1 \\ \varepsilon_2 \\ \varepsilon_6 \end{bmatrix} \tag{1.20}$$

where

$$A_{ij} = \sum_{k=1}^{N} (Q_{ij})_k h_k \tag{1.21}$$

with h_k being the thickness of the kth layer. Consider for example the term A_{11}. Write (1.21) as

$$A'_{11} = \sum_{k=1}^{N} (Q'_{11})_k h_k \tag{1.22}$$

where now we seek the property A'_{11} in the i direction. The 1,2 coordinates are taken to be aligned with the principal directions of the orthotropic lamina of interest. Under this condition it follows from (1.17) that

$$(Q'_{11})_k = U_1 + U_2 \cos 2\theta_k + U_3 \cos 4\theta_k \tag{1.23}$$

where θ_k is the angle between the 1 axis of the k lamina and the 1′ direction of the laminate. Let the total thickness of the laminate be h; then $h_k = h/N$, for equal thickness laminae, and (1.22) and (1.23) form

$$A'_{11} = \frac{h}{N}\left\{ U_1 + U_2 \sum_{k=1}^{N} \cos 2\theta_k + U_3 \sum_{k=1}^{N} \cos 4\theta_k \right\}$$

Let any direction relative to the 1′ direction be given by angle ϕ. Then the laminate property in the direction at angle ϕ from 1′ is given by

$$A''_{11} = \frac{h}{N}\left\{ U_1 + U_2 \sum_{k=1}^{N} \cos 2(\theta_k - \phi) + U_3 \sum_{k=1}^{N} \cos 4(\theta_k - \phi) \right\} \tag{1.24}$$

Using trigonometric identities this can be written as

$$A''_{11} = \frac{h}{N}\left\{ U_1 + U_2 \cos 2\phi \sum_{k=1}^{N} \cos 2\theta_k + U_2 \sin 2\phi \sum_{k=1}^{N} \sin 2\theta_k \right.$$
$$\left. + U_3 \cos 4\phi \sum_{k=1}^{N} \cos 4\theta_k + U_3 \sin 4\phi \sum_{k=1}^{N} \sin 4\theta_k \right\} \tag{1.25}$$

Now the angle separating the N laminae is π/N and the typical summation in (1.25) has the form

$$\sum_{k=1}^{N} \cos 2\theta_k = \cos\frac{2\pi}{N} + \cos\frac{4\pi}{N} + \cdots + \cos 2\pi$$

But the form above can be summed, as in

$$\cos x + \cos 2x + \cdots + \cos Nx = \frac{\sin\left(N+\frac{1}{2}\right)x}{2\sin\frac{x}{2}} - \frac{1}{2} \tag{1.26}$$

Using $x = 2\pi/N$ in (1.26) to give the result

$$\sum_{k=1}^{N} \cos 2\theta_k = 0 \tag{1.27}$$

Similarly, all other summations in (1.25) vanish for $N \geqslant 3$, leaving (1.25) as

$$A''_{11} = \frac{h}{N} U_1 \tag{1.28}$$

which is independent of angle ϕ. Similarly,

$$A'_{ij} = \text{constant}$$

independent of the orientation of the 1′, 2′ axes. It is easy to show that this behavior is not true for $N=2$. Thus we see that when three or more identical laminae are stacked at equal angles, the resulting in-plane properties are isotropic. Actually a laminate of this type is said to be quasi-isotropic, since although the in-plane properties are isotropic, as we see in the next section, this does not necessarily mean that the properties that govern bending are isotropic. We recall from our results on randomly oriented fibers in Section 4.4 that the isotropic in-plane properties are given by (4.22) and (4.23). These formulas are also applicable here for a quasi-isotropic laminate composed of three or more laminae at equal angles.

5.2 CLASSICAL THEORY OF LAMINATED PLATES

Having seen how to transform the lamina properties between various coordinate systems, we now are ready to deal with laminates. In fact, we now develop a full bending and stretching theory for the behavior of laminated plates. The theory is at the comparable level of the classical theory for homogeneous plates and admits the homogeneous plate theory as a special case.

Governing Relations

The kth lamina in the laminate has the plane stress form of the stress-strain relations designated by

$$[\sigma]_k=[Q]_k[\varepsilon]_k \tag{2.1}$$

where the full form is given by (1.14). The Kirchhoff-Love hypothesis is invoked to write the displacement forms as

$$\begin{aligned} u &= u_0(x,y) - z\frac{\partial w_0(x,y)}{\partial x} \\ v &= v_0(x,y) - z\frac{\partial w_0(x,y)}{\partial y} \\ w &= w_0(x,y) \end{aligned} \tag{2.2}$$

where the z-coordinate is normal to the undeformed center plane of the laminate. Relations (2.2) are nothing more than the usual requirement that plane sections remain plane during bending. The strain displacement relations give

$$\begin{aligned} \varepsilon_{xx} &= \frac{\partial u_0}{\partial x} - z\frac{\partial^2 w_0}{\partial x^2} \\ \varepsilon_{yy} &= \frac{\partial v_0}{\partial y} - z\frac{\partial^2 w_0}{\partial y^2} \\ \varepsilon_{xy} &= \frac{1}{2}\left(\frac{\partial u_0}{\partial y} + \frac{\partial v_0}{\partial x}\right) - z\frac{\partial^2 w_0}{\partial x\,\partial y} \end{aligned} \tag{2.3}$$

The strains are written as

$$\begin{bmatrix} \varepsilon_{xx} \\ \varepsilon_{yy} \\ \varepsilon_{xy} \end{bmatrix} = \begin{bmatrix} \varepsilon_{xx}^0 \\ \varepsilon_{yy}^0 \\ \varepsilon_{xy}^0 \end{bmatrix} + z\begin{bmatrix} \kappa_x \\ \kappa_y \\ \kappa_{xy} \end{bmatrix} \tag{2.4}$$

where

$$\varepsilon_{xx}^0 = \frac{\partial u_0}{\partial x} \qquad \kappa_x = -\frac{\partial^2 w_0}{\partial x^2}$$

$$\varepsilon_{yy}^0 = \frac{\partial v_0}{\partial y} \qquad \kappa_y = -\frac{\partial^2 w_0}{\partial y^2}$$

$$\varepsilon_{xy}^0 = \frac{1}{2}\left(\frac{\partial u_0}{\partial y} + \frac{\partial v_0}{\partial x}\right) \qquad \kappa_{xy} = -\frac{\partial^2 w_0}{\partial x\, \partial y} \tag{2.5}$$

Define the normal and shear stress resultants as

$$(N_x, N_y, N_{xy}) = \int_{-h/2}^{h/2} \left[\sigma_{xx}^{(k)}, \sigma_{yy}^{(k)}, \sigma_{xy}^{(k)}\right] dz \tag{2.6}$$

and

$$(Q_x, Q_y) = \int_{-h/2}^{h/2} \left[\sigma_{zx}^{(k)}, \sigma_{yz}^{(k)}\right] dz \tag{2.7}$$

where $z=0$ is taken as the middle plane of the plate and h is the total thickness. The bending moments are defined as

$$(M_x, M_y, M_{xy}) = \int_{-h/2}^{h/2} \left[\sigma_{xx}^{(k)}, \sigma_{yy}^{(k)}, \sigma_{xy}^{(k)}\right] dz \tag{2.8}$$

The equations of equilibrium, when written out, have the forms

$$\frac{\partial \sigma_{xx}}{\partial x} + \frac{\partial \sigma_{xy}}{\partial y} + \frac{\partial \sigma_{zx}}{\partial z} = 0 \tag{2.9}$$

$$\frac{\partial \sigma_{yy}}{\partial y} + \frac{\partial \sigma_{xy}}{\partial x} + \frac{\partial \sigma_{yz}}{\partial z} = 0 \tag{2.10}$$

$$\frac{\partial \sigma_{zz}}{\partial_z} + \frac{\partial \sigma_{zx}}{\partial x} + \frac{\partial \sigma_{yz}}{\partial y} = 0 \tag{2.11}$$

When integrated with respect to z, these equations give

$$\frac{\partial N_x}{\partial x} + \frac{\partial N_{xy}}{\partial y} = 0 \tag{2.12}$$

$$\frac{\partial N_y}{\partial y} + \frac{\partial N_{xy}}{\partial x} = 0 \tag{2.13}$$

and

$$\frac{\partial Q_x}{\partial x}+\frac{\partial Q_y}{\partial y}+q=0 \tag{2.14}$$

where

$$q=\sigma_{zz}\left(\frac{h}{2}\right)-\sigma_{zz}\left(-\frac{h}{2}\right) \tag{2.15}$$

and the shear stresses on the top and bottom surfaces are taken as vanishing. Now multiply the first equilibrium equation (2.9) by z and integrate to get

$$\frac{\partial M_x}{\partial x}+\frac{\partial M_{xy}}{\partial y}+\int_{-h/2}^{h/2}\frac{\partial \sigma_{zx}^{(k)}}{\partial z}z\,dz=0 \tag{2.16}$$

The integrand in (2.16) can be written as

$$z\frac{\partial \sigma_{zx}^{(k)}}{\partial z}=\frac{\partial}{\partial z}\left(z\sigma_{zx}^{(k)}\right)-\sigma_{zx}^{(k)}$$

As before, assume

$$\sigma_{zx}^{(k)}\big|_{z=\pm h/2}=0$$

to write (2.16) as

$$\frac{\partial M_x}{\partial x}+\frac{\partial M_{xy}}{\partial y}-Q_x=0 \tag{2.17}$$

where (2.7) has been used. Similarly, it is found that

$$\frac{\partial M_y}{\partial y}+\frac{\partial M_{xy}}{\partial x}-Q_y=0 \tag{2.18}$$

Finally, taking the derivatives of (2.17) and (2.18) and substituting into (2.14) gives

$$\frac{\partial^2 M_x}{\partial x^2}+2\frac{\partial^2 M_{xy}}{\partial x\,\partial y}+\frac{\partial^2 M_y}{\partial y^2}+q=0 \tag{2.19}$$

These results are identical with those of homogeneous plate theory with the exception of the definitions of stress resultants (2.6)–(2.8).

Our next step is to express the governing equilibrium equations in terms of displacements. Substitute the relations of the form (2.1), written out in full in (1.14), into the resultants, (2.6) and (2.8), to obtain

$$\begin{bmatrix} N_x \\ N_y \\ N_{xy} \\ M_x \\ M_y \\ M_{xy} \end{bmatrix} = \left[\begin{array}{ccc|ccc} A_{11} & A_{12} & A_{16} & B_{11} & B_{12} & B_{16} \\ A_{21} & A_{22} & A_{26} & B_{21} & B_{22} & B_{26} \\ A_{61} & A_{62} & A_{66} & B_{61} & B_{62} & B_{66} \\ \hline B_{11} & B_{12} & B_{16} & D_{11} & D_{12} & D_{16} \\ B_{21} & B_{22} & B_{26} & D_{21} & D_{22} & D_{26} \\ B_{61} & B_{62} & B_{66} & D_{61} & D_{62} & D_{66} \end{array}\right] \begin{bmatrix} \varepsilon_{xx}^0 \\ \varepsilon_{yy}^0 \\ \varepsilon_{xy}^0 \\ \kappa_x \\ \kappa_y \\ \kappa_{xy} \end{bmatrix} \tag{2.20}$$

where

$$(A_{ij}, B_{ij}, D_{ij}) = \int_{-h/2}^{h/2} Q_{ij}^{(k)}(1, z, z^2)\, dz \tag{2.21}$$

At this point we should note that, in general, bending and stretching effects in (2.20) are coupled through the B_{ij} portion of the matrix. There is no such coupling in the classical theory. Later we see special cases where the B_{ij} terms vanish, providing an uncoupling effect. The coordinate rotation transformations for A_{ij}, B_{ij}, and D_{ij} follow those already found for Q_{ij}.

To get explicit equilibrium equations in terms of displacements, relations (2.20) are substituted into (2.12), (2.13), and (2.19). The resulting equations are

$$A_{11}u_{0,xx} + 2A_{16}u_{0,xy} + A_{66}u_{0,yy} + A_{16}v_{0,xx} + (A_{12} + A_{66})v_{0,xy} + A_{26}v_{0,yy}$$
$$- B_{11}w_{,xxx} - 3B_{16}w_{,xxy} - (B_{12} + 2B_{66})w_{,xyy} - B_{26}w_{,yyy} = 0 \tag{2.22}$$

$$A_{16}u_{0,xx} + (A_{12} + A_{66})u_{0,xy} + A_{26}u_{0,yy} + A_{66}v_{0,xx} + 2A_{26}v_{0,xy} + A_{22}v_{0,yy}$$
$$- B_{16}w_{,xxx} - (B_{12} + 2B_{66})w_{,xxy} - 3B_{26}w_{,xyy} - B_{22}w_{,yyy} = 0 \tag{2.23}$$

and

$$D_{11}w_{,xxxx}+4D_{16}w_{,xxxy}+2(D_{12}+2D_{66})w_{,xxyy}+4D_{26}w_{,xyyy}$$
$$+D_{22}w_{,yyyy}-B_{11}u_{0,xxx}-3B_{16}u_{0,xxy}-(B_{12}+2B_{66})u_{0,xyy}$$
$$-B_{26}u_{0,yyy}-B_{16}v_{0,xxx}-(B_{12}+2B_{66})v_{0,xxy}-3B_{26}v_{0,xyy}$$
$$-B_{22}v_{0,yyy}=+q \tag{2.24}$$

We now have three equations in the unknown displacement functions, $u_0(x,y)$, $v_0(x,y)$, and $w(x,y)$. As has already been mentioned, these equations have coupling between in-plane and out-of-plane effects, that is, membrane and bending effects. Boundary conditions appropriate to the theory involve the specification of one function from each of the following pairs

$$u_n \text{ or } N_n \qquad u_t \text{ or } N_{nt} \qquad w_{,n} \text{ or } M_n \qquad w \text{ or } M_{nt,t}+Q_n$$

where n and t designate normal and tangential coordinates along the edge.
Some special cases are considered.

Special Cases

Isotropy—Single Lamina. If there is only one lamina that is isotropic, the following special forms result

$$A_{11}=\frac{Eh}{1-\nu^2}=A$$
$$A_{12}=\nu A$$
$$A_{22}=A$$
$$A_{16}=A_{26}=0$$
$$A_{66}=(1-\nu)A$$
$$B_{ij}=0$$
$$D_{11}=\frac{Eh^3}{12(1-\nu^2)}=D$$
$$D_{12}=\nu D$$
$$D_{22}=D$$
$$D_{16}=D_{26}=0$$
$$D_{66}=(1-\nu)D$$

There is no coupling between membrane and bending effects.

Orthotropic—Single Lamina. For a single orthotropic lamina,

$$A_{ij} = Q_{ij}h$$

$$B_{ij} = 0$$

$$D_{ij} = \frac{Q_{ij}h^3}{12}$$

Again, there is no coupling between bending and membrane effects.

Specially Orthotropic—Single Lamina. The specially orthotropic single lamina is the same as the orthotropic case, but with the coordinate system taken such that

$$A_{16} = A_{26} = D_{16} = D_{26} = 0$$

With $D_{16} = 0$, $D_{26} = 0$, there will be no coupling between bending and twisting effects.

Symmetric Laminates. If the laminate is constructed such that it has complete symmetry of individual lamina thickness, properties, and orientations about the middle plane of the laminate, then from (2.21) it follows that

$$B_{ij} = 0$$

and there is no coupling between bending and membrane effects. In this type of laminate the coefficients A_{16}, A_{26}, D_{16}, and D_{26} may or may not be zero, depending on the type of construction. For example, a cross-ply laminate is composed of orthotropic laminae arranged at 90° relative orientations, in this case, $A_{16} = A_{26} = D_{16} = D_{26} = 0$, where the zx and yz planes are in the directions of the material symmetry planes. Alternatively, an angle ply laminate violates these conditions and it does not have the simplification of vanishing A_{16}, A_{26}, D_{16}, and D_{26} terms.

Asymmetric Laminates. One would certainly prefer to analyze and design with symmetric laminates, but there are cases where other design requirements dictate the use of an asymmetric laminate. There are no special simplifications that result when asymmetric laminates are used, and the full form of (2.20) is involved.

Although the classical plate theory is very satisfactory for most problems involving homogeneous plates, the validity of the classical theory for laminates is much less certain. To ascertain the validity or range of validity of the classical theory, we need to make comparisons with exact solutions. Fortunately, such exact solutions are available, as we see in the next section.

5.3 CYLINDRICAL BENDING

We consider a symmetric laminate with N layers of orthotropic laminae with planes of material symmetry parallel to the coordinate planes. A plane strain condition is assumed with no strain in the y direction. The problem is thus one-dimensional in x. One-dimensional problems of this type are referred to as cylindrical bending, and exact three-dimensional solutions for such problems have been obtained by Pagano [5.3], based on a generalization of the solution given by Timoshenko and Goodier [5.4].

Exact Solution

Under plane strain conditions

$$\varepsilon_2 = \varepsilon_{yy} = 0$$

$$\varepsilon_4 = \varepsilon_{yz} = 0$$

$$\varepsilon_6 = \varepsilon_{xy} = 0$$

leaving the stress-strain relations as

$$\begin{Bmatrix} \sigma_1 \\ \sigma_3 \\ \sigma_5 \end{Bmatrix} = \begin{bmatrix} C_{11} & C_{13} & 0 \\ C_{31} & C_{33} & 0 \\ 0 & 0 & C_{55} \end{bmatrix} \begin{Bmatrix} \varepsilon_1 \\ \varepsilon_3 \\ \varepsilon_5 \end{Bmatrix} \tag{3.1}$$

where $\varepsilon_1 = \varepsilon_{xx}$, $\varepsilon_3 = \varepsilon_{zz}$, and $\varepsilon_5 = \varepsilon_{zx}$. This relation can be inverted to give

$$\begin{Bmatrix} \varepsilon_1 \\ \varepsilon_3 \\ \varepsilon_5 \end{Bmatrix} = \begin{bmatrix} R_{11} & R_{13} & 0 \\ R_{31} & R_{33} & 0 \\ 0 & 0 & R_{55} \end{bmatrix} \begin{Bmatrix} \sigma_1 \\ \sigma_3 \\ \sigma_5 \end{Bmatrix} \tag{3.2}$$

where

$$R_{11}=\frac{C_{33}}{C_{11}C_{33}-C_{13}^2}$$

$$R_{13}=\frac{-C_{13}}{C_{11}C_{33}-C_{13}^2}$$

$$R_{33}=\frac{C_{11}}{C_{11}C_{33}-C_{13}^2}$$

$$R_{55}=\frac{1}{C_{55}} \tag{3.3}$$

The problem is that of the laminate loaded by a normal stress on the top surface, with all other surface tractions vanishing. Thus

$$\sigma_{zz}|_{z=h/2}=q_0\sin\frac{\pi x}{l}$$

$$\sigma_{zx}|_{z=h/2}=0$$

$$\sigma_{zz}|_{z=-h/2}=\sigma_{zx}|_{z=-h/2}=0 \tag{3.4}$$

Of course, continuity of stress and displacement at all interfaces between laminae is enforced.

Let there be a local coordinate at the mid-plane of each lamina. For the ith lamina let

$$\sigma_{xx}^{(i)}=f_i''(z)\sin\frac{\pi x}{l}$$

$$\sigma_{zz}^{(i)}=-\frac{\pi^2}{l^2}f_i(z)\sin\frac{\pi x}{l}$$

$$\sigma_{zx}^{(i)}=-\frac{\pi}{l}f_i'(z)\cos\frac{\pi x}{l} \tag{3.5}$$

where $f_i(z)$ is an unknown function to be determined and the primes denote derivatives with respect to z. The equilibrium equations are given by

$$\sigma_{xx,x}+\sigma_{zx,z}=0$$

and

$$\sigma_{zz,z} + \sigma_{zx,x} = 0 \tag{3.6}$$

Substituting (3.5) in (3.6) it is found that the latter are identically satisfied. The only compatibility equation that is not identically satisfied is given by

$$2\frac{\partial^2 \varepsilon_{zx}}{\partial z\,\partial x} = \frac{\partial^2 \varepsilon_{zz}}{\partial x^2} + \frac{\partial^2 \varepsilon_{xx}}{\partial z^2} \tag{3.7}$$

It is left as an exercise to show that this compatibility equation reduces to

$$R_{11}^{(i)} f_i''''(z) - \left[2R_{55}^{(i)} + 2R_{13}^{(i)}\right]\lambda^2 f_i''(z) + R_{33}^{(i)}\lambda^4 f_i(z) = 0 \tag{3.8}$$

where

$$\lambda = \frac{\pi}{l}$$

A solution of (3.8) for $f_i(z)$ is found to be given by

$$f_i(z) = \sum_{j=1}^{4} A_{ji} \exp(m_{ji} z_i), \qquad i = 1, 2 \ldots N \tag{3.9}$$

where

$$\left.\begin{matrix} m_{1i} \\ m_{2i} \end{matrix}\right\} = \pm\lambda\left(\frac{a_i + b_i}{c_i}\right)^{1/2}$$

$$\left.\begin{matrix} m_{3i} \\ m_{4i} \end{matrix}\right\} = \pm\lambda\left(\frac{a_i - b_i}{c_i}\right)^{1/2} \tag{3.10}$$

with

$$a_i = 2R_{55}^{(i)} + 2R_{13}^{(i)}$$

$$b_i = \left[a_i^2 - 4R_{11}^{(i)} R_{33}^{(i)}\right]^{1/2}$$

$$c_i = 2R_{11}^{(i)} \tag{3.11}$$

and where A_{ji} are constants to be determined.

The stresses corresponding to (3.9) are found to be given by

$$\sigma_{xx}^{(i)} = \sin\lambda x \sum_{j=1}^{4} A_{ji} m_{ji}^2 \exp(m_{ji} z_i)$$

$$\sigma_{zz}^{(i)} = -\lambda^2 \sin\lambda x \sum_{j=1}^{4} A_{ji} \exp(m_{ji} z_i)$$

$$\sigma_{zx}^{(i)} = -\lambda \cos\lambda x \sum_{j=1}^{4} A_{ji} m_{ji} \exp(m_{ji} z_i) \tag{3.12}$$

The displacements are given by

$$u^{(i)} = \frac{\cos\lambda x}{\lambda} \sum_{j=1}^{4} A_{ji}\left[R_{13}^{(i)}\lambda^2 - R_{11}^{(i)} m_{ji}^2 \right] \exp(m_{ji} z_i)$$

$$w^{(i)} = \sin\lambda x \sum_{j=1}^{4} A_{ji}\left[R_{13}^{(i)} m_{ji} - \frac{R_{33}^{(i)}\lambda^2}{m_{ji}} \right] \exp(m_{ji} z_i) \tag{3.13}$$

Expressions corresponding to (3.12) and (3.13) are given in [5.3] when the $y-z$ plane is a plane of transverse isotropy, rather than merely being orthotropic.

Continuity conditions at the interfaces as well as the surface conditions (3.4) lead to a system of $4N$ equations in the $4N$ unknown constants, A_{ji}. These linear algebraic equations are easily solvable on a computer. Before showing particular results for cylindrical bending, it is necessary to obtain the corresponding predictions from the classical theory of Section 5.2.

Classical Solution

From equations (2.22) and (2.24) the governing classical theory equations are

$$A_{11} u_{0,xx} - B_{11} w_{0,xxx} = 0$$

and

$$D_{11} w_{0,xxxx} - B_{11} u_{0,xxx} = q \tag{3.14}$$

where

$$(A_{11}, B_{11}, D_{11}) = \int_{-h/2}^{h/2} C_{11}(1, z, z^2)\, dz \tag{3.15}$$

For $q = q_0 \sin\lambda x$ the equations in (3.14) give

$$u_0 = \frac{B_{11}q_0}{(A_{11}D_{11} - B_{11}^2)\lambda^3}\cos\lambda x$$

$$w_0 = \frac{A_{11}q_0}{(A_{11}D_{11} - B_{11}^2)\lambda^4}\sin\lambda x \tag{3.16}$$

The only nonvanishing strain component is

$$\varepsilon_{xx} = u_{0,x} - zw_{0,xx}$$

which then has the solution

$$\varepsilon_{xx} = \frac{(A_{11}z - B_{11})q_0}{(A_{11}D_{11} - B_{11}^2)\lambda^2}\sin\lambda x \tag{3.17}$$

The stress $\sigma_{xx}^{(i)}$ is obtained by multiplying (3.17) by $C_{11}^{(i)}$. The classical theory assumptions do not suffice to determine the stresses σ_{zx} and σ_{zz}, but knowing $\sigma_{xx}^{(i)}$ they can be obtained by direct integration of the equilibrium equations.

Example

A numerical example is now taken to provide a comparison between the exact solution and the classical theory solution for cylindrical bending. Typical lamina properties are taken as

$$\frac{E_L}{E_T} = 25$$

$$\frac{\mu_{LT}}{E_T} = 0.5$$

$$\nu_{LT} = \nu_{TT} = 0.25$$

$$\frac{\mu_{TT}}{E_T} = 0.2$$

where the indices L and T refer to longitudinal (fiber) direction and transverse to fiber direction. The Q properties can be deduced from these data. Suitably normalized stresses and displacements are taken as

$$\bar{\sigma}_{xx} = \frac{\sigma_{xx}(l/2,z)}{q_0}$$

$$\bar{u}_0 = \frac{E_T u_0(0,z)}{hq_0}$$

The stress σ_{xx} and displacement u_0 are shown in Figs. 5.1 and 5.2 for the case of a symmetric three layer laminate with the L direction coinciding with the x direction in the outer layers, and the T direction parallel to x in the inner layer. In this example $l/h=4$; thus the wave length of the sinusoidal load is eight times the thickness of the plate. It is seen that the classical theory gives an inadequate representation of the solution at this wave length. At a ratio of $l/h=10$, the error in the maximum flexural stress is about 14%. As l/h increases, the error continuously decreases. We see from Fig 5.1 that the discontinuity in properties across interfaces causes a discontinuity in flexural stress, and from Fig. 5.2 we see that warpage of the cross section is quite strong. These effects are far more prevalent in laminates than in homogeneous plates, as we see next.

Figure 5.3 shows the results for a one layer plate of the specified properties with the fibers oriented in the x direction, again at $l/h=4$. It is seen that the homogeneous plate provides a far less critical test of the classical theory than does the laminate of this example. To put this another way, at a given wave length of loading, the laminate is less well modeled by the classical theory than is the homogeneous case. It should be remembered, however, that these results are for this particular example of a three layer laminate. As the number of layers is increased, in an alternating

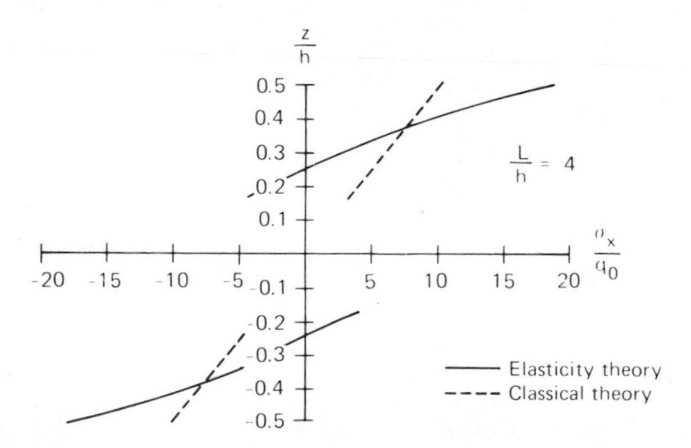

Fig. 5.1 Flexural stress distribution, three layer laminate. After Pagano [5.3].

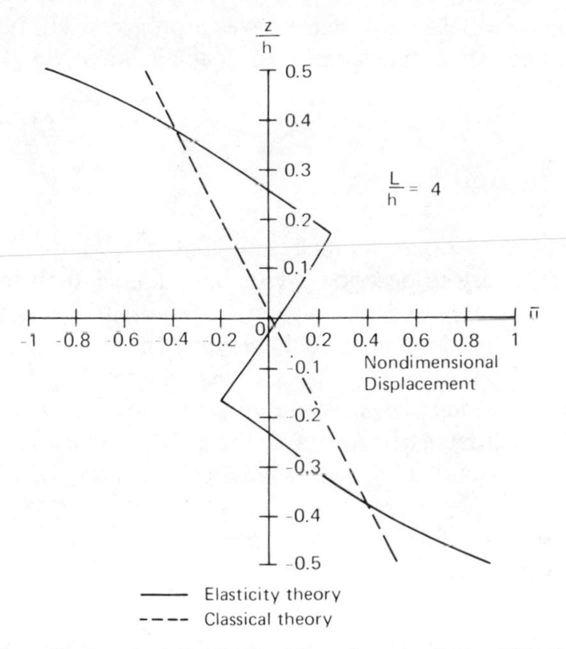

Fig. 5.2 In-plane displacement distribution, three layer laminate. After Pagano [5.3].

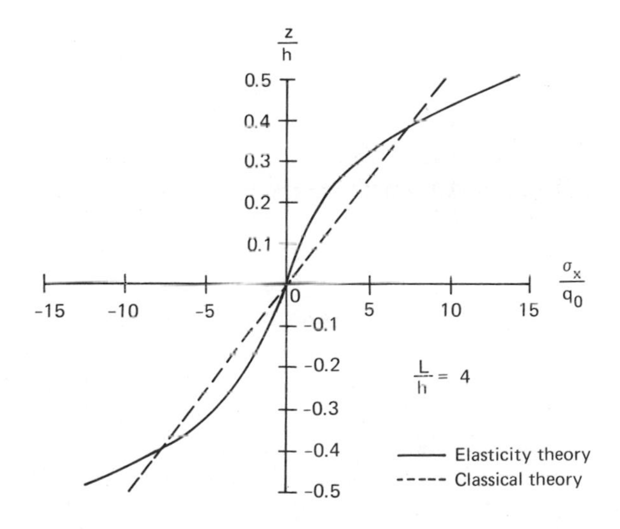

Fig. 5.3 Flexural stress distribution, one layer plate. After Pagano [5.3].

orientation pattern, classical theory gives a progressively better prediction of the response at a fixed ratio of loading wave length to laminate thickness.

5.4 HIGH-ORDER THEORY

In the preceding section we just saw that it is more difficult for the classical level theory to describe events in laminates than in homogeneous plates. This motivates us to inquire into the possibility of using a higher-order theory to model laminate behavior. There are many theories of higher order than that of the classical level that have been developed for use with homogeneous plates. Probably the highest-order theory developed is that of Lo, Christensen, and Wu [5.5]. This theory for homogeneous plates has been extended by them to model laminates [5.6]. We show some results from the latter theory here.

Before describing the high-order theory, let us place the well known shear deformation level theory into perspective so far as laminates are concerned. The kinematic assumptions in the shear deformation theory are of the form

$$
\begin{aligned}
u &= u^0(x,y) + z\psi_x(x,y) \\
v &= v^0(x,y) + z\psi_x(x,y) \\
w &= w(x,y) \qquad (4.1)
\end{aligned}
$$

We note that these shear deformation level kinematic equations have exactly the same form as in the classical level theory (2.2). Although this situation may seem paradoxical at first, the explanation is actually very simple. In the classical theory the ψ_x and ψ_y terms in (4.1) have the functional dependence upon $w_0(x,y)$ shown in (2.2). This assumption for the form of ψ_x and ψ_y is a constraint on the shear deformation theory that reduces it to the classical theory level. Despite the increased level of generality of the shear deformation theory, its flexural stress distributions show little improvement over those of the classical theory in the case of laminated plates. Accordingly, it is expected that higher-order effects than those involved in (4.1) are needed to model stress distributions for laminates, such as those shown in Fig. 5.1.

Taking the preceding observation one step further, we see in the result of Fig. 5.2 that there can be significant warping of the cross section in laminates. Consistent with this observation, we take a level of kinematic assumptions that includes warpage of the cross section. Thus displacement forms of the following type are assumed:

$$u = u^0(x,y) + z\psi_x(x,y) + z^2\zeta_x(x,y) + z^3\phi_x(x,y)$$

$$v = v^0(x,y) + z\psi_y(x,y) + z^2\zeta_y(x,y) + z^3\phi_y(x,y)$$

$$w = w^0(x,y) + z\psi_z(x,y) + z^2\zeta_z(x,y) \tag{4.2}$$

We do not here give the derivation of the theory based on the expansion (4.2). It is a long, tedious derivation that is fully outlined in [5.5, 5.6]. The derivation is based on the potential energy variational theorem, and it is completely straightforward. In contrast to the classical level theory, which is based upon stress-strain relations appropriate to states of plane stress, the theory to be stated here is based on the full three-dimensional form of the stress-strain relations. Naturally, at the level of (4.2) the full three-dimensional form of the stress-strain relations must be used when strain effects in the thickness direction are to be modeled.

The full theory suitable to laminated plates will be stated. The stress-strain relations are given by (1.3), appropriate to a material with one plane of symmetry, which we rewrite here in x,y,z notation:

$$\begin{bmatrix} \sigma_{xx} \\ \sigma_{yy} \\ \sigma_{zz} \\ \sigma_{yz} \\ \sigma_{zx} \\ \sigma_{xy} \end{bmatrix} = \begin{bmatrix} C_{11} & C_{12} & C_{13} & 0 & 0 & C_{16} \\ & C_{22} & C_{23} & 0 & 0 & C_{26} \\ & & C_{33} & 0 & 0 & C_{36} \\ & & & C_{44} & C_{45} & 0 \\ & & & & C_{55} & 0 \\ & & & & & C_{66} \end{bmatrix} \begin{bmatrix} \varepsilon_{xx} \\ \varepsilon_{yy} \\ \varepsilon_{zz} \\ \varepsilon_{yz} \\ \varepsilon_{zx} \\ \varepsilon_{xy} \end{bmatrix} \tag{4.3}$$

There are 11 generalized displacement functions in (4.2). The governing 11 equations of equilibrium are given by

$$N_{x,x} + N_{xy,y} + q_x = 0$$

$$N_{y,y} + N_{xy,x} + q_y = 0$$

$$Q_{x,x} + Q_{y,y} + q = 0$$

$$M_{x,x} + M_{xy,y} - Q_x + m_x = 0$$

$$M_{y,y} + M_{xy,x} - Q_y + m_y = 0$$

$$R_{x,x} + R_{y,y} - N_z + m = 0$$

$$P_{x,x}+P_{xy,y}-2R_x+n_x=0$$

$$P_{y,y}+P_{xy,x}-2R_y+n_y=0$$

$$S_{x,x}+S_{y,y}-2M_z+n=0$$

$$\overline{M}_{x,x}+\overline{M}_{xy,y}-3S_x+1_x=0$$

$$\overline{M}_{y,y}+\overline{M}_{xy,x}-3S_y+1_y=0 \tag{4.4}$$

where the resultants appearing in (4.4) are defined by

$$\begin{bmatrix} N_x & N_y & N_z & N_{xy} & Q_x & Q_y \\ M_x & M_y & M_z & M_{xy} & R_x & R_y \end{bmatrix} = \int_{-h/2}^{h/2} \begin{Bmatrix} 1 \\ z \end{Bmatrix} [\sigma_x \sigma_y \sigma_z \sigma_{xy} \sigma_{xz} \sigma_{yz}]\, dz$$

$$\begin{bmatrix} P_x & P_y & P_{xy} \\ \overline{M}_x & \overline{M}_y & \overline{M}_{xy} \end{bmatrix} = \int_{-h/2}^{h/2} \begin{Bmatrix} z^2 \\ z^3 \end{Bmatrix} (\sigma_x \sigma_y \sigma_{xy})\, dz \tag{4.5}$$

and

$$[S_x \quad S_y] = \int_{-h/2}^{h/2} z^2 (\sigma_{xz} \sigma_{yz})\, dz$$

where, for brevity, the notation $\sigma_x = \sigma_{xx}$, and so on, is employed. The effective load terms in (4.4) are given by

$$(q_x, n_x) = \left[\sigma_{xz}\left(\frac{h}{2}\right) - \sigma_{xz}\left(-\frac{h}{2}\right) \right] \left[1, \frac{h^2}{4} \right]$$

$$(q_y, n_y) = \left[\sigma_{yz}\left(\frac{h}{2}\right) - \sigma_{yz}\left(-\frac{h}{2}\right) \right] \left[1, \frac{h^2}{4} \right]$$

$$(m_x, 1_x) = \left[\sigma_{xz}\left(\frac{h}{2}\right) + \sigma_{xz}\left(-\frac{h}{2}\right) \right] \left[\frac{h}{2}, \frac{h^3}{8} \right]$$

$$(m_y, 1_y) = \left[\sigma_{yz}\left(\frac{h}{2}\right) + \sigma_{yz}\left(-\frac{h}{2}\right) \right] \left[\frac{h}{2}, \frac{h^3}{8} \right]$$

$$m = \frac{h}{2} \left[\sigma_z\left(\frac{h}{2}\right) + \sigma_z\left(-\frac{h}{2}\right) \right] \tag{4.6}$$

Along the edge of the plate, one member of each of the following 11 products must be prescribed:

$$\begin{aligned}
&N_n u_n^0, \quad N_{nt} u_t^0, \quad M_n \psi_n \\
&M_{nt} \psi_t, \quad P_n \zeta_n, \quad P_{nt} \zeta_t \\
&\overline{M}_n \phi_n, \quad \overline{M}_{nt} \phi_t, \quad Q_n w^0 \\
&R_n \psi_z, \quad \text{and} \quad S_n \zeta_z
\end{aligned} \tag{4.7}$$

where n and t are the directions normal and tangential to the edge of the plate.

The equations and boundary conditions given in (4.4)–(4.7) are independent of the properties of the materials of the plate and hence hold true for homogeneous isotropic as well as laminated plates.

In the laminated plate context the combination of relations (4.3)–(4.5) can be used to state the 11 governing equilibrium equations in terms of the 11 unknown displacement functions. These 11 equations, in matrix notation, are given by

$$[L_{ij}]\begin{bmatrix} u^0 \\ v^0 \\ w^0 \\ \psi_x \\ \psi_y \\ \psi_z \\ \zeta_x \\ \zeta_y \\ \zeta_z \\ \phi_x \\ \phi_y \end{bmatrix} = -\begin{bmatrix} q_x \\ q_y \\ -q \\ m_x \\ m_y \\ -m \\ n_x \\ n_y \\ -n \\ 1_x \\ 1_y \end{bmatrix} \qquad i,j=1\ldots 11 \tag{4.8}$$

where L_{ij} are symmetric differential operators having coefficients that involve the properties of the individual lamina. The L_{ij} forms are too lengthy to state here in their entirety, and reference should be made to [5.6] for the complete forms. Only typical forms are given here, but they are ones that will be of use to us in the next section. These typical terms of L_{ij}

are given by

$$
\begin{aligned}
L_{22} &= 2A_{66}\partial_{xx} + 4A_{26}\partial_{xy} + A_{22}\partial_{yy} \\
L_{26} &= 2A_{36}\partial_{x} + A_{23}\partial_{y} \\
L_{28} &= 2D_{66}\partial_{xx} + 4D_{26}\partial_{xy} + D_{22}\partial_{yy} \\
L_{66} &= -2D_{55}\partial_{xx} - 4D_{45}\partial_{xy} - 2D_{44}\partial_{yy} + A_{33} \\
L_{68} &= (2D_{36} - 4D_{45})\partial_{x} + (D_{23} - 4D_{44})\partial_{y} \\
L_{88} &= 2H_{66}\partial_{xx} + 4H_{26}\partial_{xy} + H_{22}\partial_{yy} - 4D_{44}
\end{aligned}
\tag{4.9}
$$

where

$$
(A_{ij}, B_{ij}, D_{ij}, F_{ij}, H_{ij}, K_{ij}, M_{ij}) = \int_{-h/2}^{h/2} (1, z, z^2, z^3, z^4, z^5, z^6) C_{ij}\, dz \tag{4.10}
$$

and ∂_{ij} are differential operators with regard to the coordinates specified, that is, $\partial_{xy} = \partial^2/\partial x\,\partial y$. The other L_{ij} terms have forms similar to those of (4.9). As seen from (4.10) there are seven different properties functions, $A_{ij} \ldots M_{ij}$ that enter the theory. Once the properties of the various laminae are specified, the generalized laminate properties are obtained from (4.10) by direct integration.

Several cylindrical bending examples have been solved in [5.6] to compare the prediction of the high-order theory with exact solutions. One such result is stated here, relative to the same three lamina cross-ply laminate example given in Section 5.3. The problem is that of the cylindrical bending of a laminate in a state of plane strain, subjected to a sinusoidal pressure loading on the top surface. The ratio of load half wave length to laminate thickness, $l/h = 4.0$, is the same as in the example of Section 5.3. The result for the flexural stress is shown in Fig. 5.4, where the exact result and the classical theory results from Fig. 5.1 are reproduced again.

It is seen from Fig. 5.4 that the high-order theory provides a far better comparison with the exact solution than does the classical theory. Furthermore, the high-order theory gives a very close value to that of the exact solution for the maximum flexural stress. Of course, these comparisons are being made under the fairly stringent condition that the load wave length is eight times the thickness of the laminate. As noted in the last section, as the load wave length becomes progressively longer relative to the thickness, the classical theory gives a more satisfactory prediction of behavior. Without doubt, the classical theory suffices in many practical design

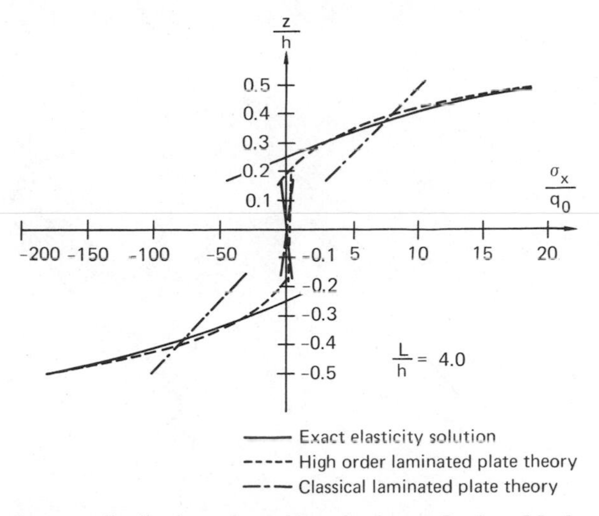

Fig. 5.4 Flexural stress distribution, three layer laminate. Section 5.3 data. After Lo, Christensen, and Wu [5.6].

problems. The present results show that the classical theory is inadequate in the region of short wave length loads and correspondingly near cutouts or other geometric irregularities. Furthermore, the results of this section and the preceding one show that the problem is more accentuated with laminates than it is with homogeneous plates. These situations call for a more detailed model than the classical theory can supply. A model such as is needed may be the high-order theory given here, or a detailed numerical solution from a three-dimensional elasticity point of view.

A completely different high-order theory has been given by Pagano [5.7]. In this method, the individual laminae are treated according to a shear deformation level theory. Continuity of displacements and stresses are prescribed at the interfaces between lamina. It may be noted that the high-order theory described here, which is based on continuous displacements, also satisfies stress continuity conditions at lamina interfaces, with the stresses determined by the method given in [5.8]. This point is discussed further in the next section.

5.5 INTERLAMINAR EDGE EFFECTS

Peculiar effects can arise in laminates that simply do not appear with homogeneous materials. We need go no further than the simple tension test to find unexpected conditions with laminates.

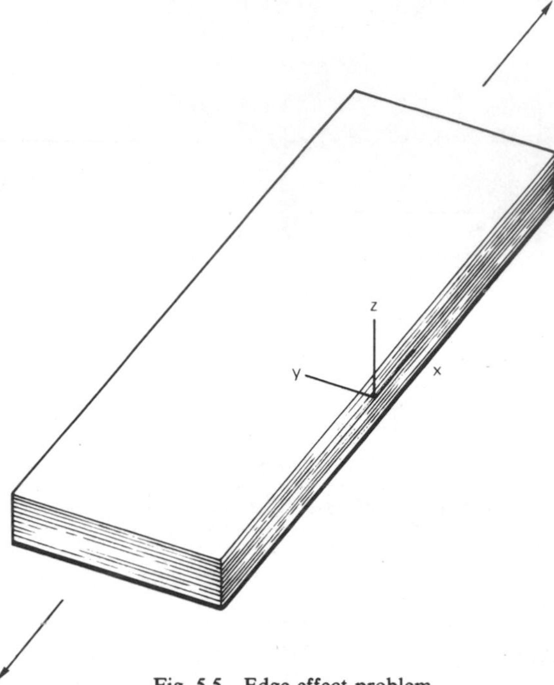

Fig. 5.5 Edge effect problem.

Consider the tension test shown in Fig. 5.5. If the various laminae were not constrained to remain in contact, the variation of mechanical properties between laminae would require that they contract by varying amounts in the y direction when subjected to an identical strain in the x direction. Furthermore, if the symmetry planes of the laminae are not coincident with the coordinate planes of Fig. 5.5, then there are differential displacements of the laminae in the x direction as well. However, in engineering laminates the various laminae are constrained to remain in contact, and obviously interlaminar stresses are required to maintain the state of contact. As we shall see, these interlaminar stresses occur only in the region of the free edge. If the laminate shown in Fig. 5.5 were infinitely wide, there would be no interlaminar stress effect. It is the introduction of a free edge into the problem that requires an edge zone of interlaminar stresses to keep the laminae in a state of kinematic compatibility. For a homogeneous material the state of uniaxial tension is an exact solution of the equilibrium equations. We begin to see the complications that can arise with laminates.

Rather than treating the finite width problem of Fig. 5.5, we assume that the specimen is sufficiently wide that the edges do not interact; thus we treat a single edge problem. The laminate is subjected to a strain ε in the x

direction, and we seek information concerning the state of stress in the laminate in the region of the stress free edge. The problem is two-dimensional, involving the y and z coordinates of Fig. 5.5. It would be extremely difficult to obtain an exact analytical solution of the problem, valid for general types of laminates. We approach the problem using the high-order laminate theory of the previous section.

The explicit problem to be studied here is that of a cross-ply laminate of symmetric construction. The behavior of the laminate, for this problem, involves the unknown displacements v and w in the y and z directions. The high-order theory displacement expansion is given as Eqs. (4.2). For the present problem of a symmetric laminate, only the symmetric deformation terms from (4.2) need be retained, as

$$\begin{aligned} v &= v^0 + z^2\zeta_y \\ w &= z\psi_z \end{aligned} \tag{5.1}$$

The governing equilibrium equations from (4.8) have the form

$$\begin{aligned} L_{22}v^0 + L_{26}\psi_z + L_{28}\zeta_y &= -q_y \\ L_{62}v^0 + L_{66}\psi_z + L_{68}\zeta_y &= m \\ L_{82}v^0 + L_{86}\psi_z + L_{88}\zeta_y &= -n_y \end{aligned} \tag{5.2}$$

where q_y, m, and n_y are the surface loading conditions, (4.6), and $L_{22} \ldots L_{88}$ are the differential operators given by (4.9). Combining (5.2) and (4.9) gives

$$\begin{aligned} A_{22}v^0_{,yy} + A_{23}\psi_{z,y} + D_{22}\zeta_{y,yy} &= 0 \\ A_{23}v^0_{,yy} - 2D_{44}\psi_{z,yy} + A_{33}\psi_z + (D_{23} - 4D_{44})\zeta_{y,y} &= 0 \\ D_{22}v^0_{,yy} + [D_{23} - 4D_{44}]\psi_{z,y} + H_{22}\zeta_{y,yy} - 8D_{44}\zeta_y &= 0 \end{aligned} \tag{5.3}$$

where the traction-free faces of the laminate give $q_y = m = n_y = 0$, and where from (4.10)

$$(A_{ij}, D_{ij}, H_{ij}) = \int_{-h/2}^{h/2} (1, z^2, z^4) C_{ij}(z)\, dz \tag{5.4}$$

We expect a solution that decays with distance from the free edge;

accordingly, we assume

$$v^0 = Ah\exp\left(\frac{-\lambda y}{h}\right)$$

$$\psi_z = B\exp\left(\frac{-\lambda y}{h}\right)$$

$$\zeta_y = \frac{C}{h}\exp\left(\frac{-\lambda y}{h}\right) \tag{5.5}$$

Substituting (5.5) into (5.3) gives

$$\begin{aligned}
&+A_{22}\lambda A && -A_{23}B && +D_{22}\frac{\lambda}{h^2}C=0 \\
&-A_{23}\lambda A && \left[-2D_{44}\frac{\lambda^2}{h^2}+A_{33}\right]B && -[D_{23}-4D_{44}]\frac{\lambda}{h^2}C=0 \\
&D_{22}\lambda^2 A && -[D_{23}-4D_{44}]\lambda B && +\left[H_{22}\frac{\lambda^2}{h^2}-8D_{44}\right]C=0
\end{aligned} \tag{5.6}$$

For a solution to this set of equations to exist, the determinant of the coefficients must vanish. The form of this characteristic equation can be shown to be

$$a\lambda^4 + 2b\lambda^2 + c = 0$$

where a, b, c, are expressible in terms of the coefficients in (5.6). Thus, there are two roots for λ^2 and in general two acceptable roots for λ; any root that does not give a decaying solution must be discarded. Having found the roots λ_k, the ratios

$$\frac{B_k}{A_k} \quad \text{and} \quad \frac{C_k}{A_k}$$

follow from (5.6). The solutions for (5.1) are written as

$$v = \sum_k \left(hA_k + z^2\frac{C_k}{h}\right)\exp\left(-\lambda_k\frac{y}{h}\right)$$

and

$$w = z\sum_k B_k\exp\left(-\lambda_k\frac{y}{h}\right) \tag{5.7}$$

where $k = 1,2$.

The terms (5.7) represent the edge effect that must be superimposed on the field caused by the applied strain in the x direction. Only the displacement v need be accordingly modified, as

$$v = \sum_k \left(hA_k + z^2 \frac{C_k}{h} \right) \exp\left(-\lambda_k \frac{y}{h} \right) + \alpha y + \beta \tag{5.8}$$

where α is a parameter to be determined, denoting the lateral strain effect away from the edge, and β is just a rigid body translation term, which we need not carry along further.

From (4.7) the force-free edge conditions are specified by

$$N_y = \int_{-h/2}^{h/2} \sigma_{yy}|_{y=0}\, dz = 0$$

$$P_y = \int_{-h/2}^{h/2} z^2 \sigma_{yy}|_{y=0}\, dz = 0$$

$$R_y = \int_{-h/2}^{h/2} z \sigma_{yz}|_{y=0}\, dz = 0 \tag{5.9}$$

The stresses shown in (5.9) must now be obtained. First, the relevant strains are

$$\varepsilon_{yy} = \frac{\partial v}{\partial y}$$

$$\varepsilon_{zz} = \frac{\partial w}{\partial z}$$

$$\varepsilon_{yz} = \frac{1}{2} \left(\frac{\partial v}{\partial z} + \frac{\partial w}{\partial y} \right) \tag{5.10}$$

Using the solutions (5.7) and (5.8) in (5.10) gives the strains to be used in the following stress-strain forms, obtained from (4.3)

$$\sigma_{yy} = C_{12}\varepsilon + C_{22}\varepsilon_{yy} + C_{23}\varepsilon_{zz}$$

$$\sigma_{yz} = C_{44}\varepsilon_{yz} \tag{5.11}$$

where ε is the imposed strain in the x direction. Combining (5.10) and

(5.11) with the solutions (5.8) and the last of (5.7) then gives

$$\sigma_{yy} = C_{12}\varepsilon + C_{22}\alpha + \sum_k \left\{ C_{23}B_k - \frac{\lambda_k}{h} C_{22}\left(hA_k + z^2 \frac{C_k}{h} \right) \right\} \exp\left(-\lambda_k \frac{y}{h} \right)$$

$$\sigma_{yz} = \frac{zC_{44}}{2h} \sum_k (2C_k - \lambda_k B_k) \exp\left(-\lambda_k \frac{y}{h} \right) \tag{5.12}$$

Substituting (5.12) into (5.9) gives the final forms

$$\int_{-h/2}^{h/2} \left\{ \alpha C_{22} - C_{22} \sum_k \left[C_{23}B_k - \lambda_k C_{22}\left(A_k + z^2 \frac{C_k}{h^2} \right) \right] \right\} dz = -\varepsilon \int_{-h/2}^{h/2} C_{12}\, dz$$

$$\int_{-h/2}^{h/2} z^2 \left\{ \alpha C_{22} - C_{22} \sum_k \left[C_{23}B_k - \lambda_k C_{22}\left(A_k + z^2 \frac{C_k}{h^2} \right) \right] \right\} dz = -\varepsilon \int_{-h/2}^{h/2} z^2 C_{12}\, dz$$

$$\sum_k [2C_k - \lambda_k B_k] = 0 \tag{5.13}$$

Recalling that B_k/A_k and C_k/A_k are known from (5.6), then (5.13) is seen to comprise three equations in the three unknowns, A_1, A_2, and α.

Once the properties of the various laminae in the composite are specified, the coefficients shown in (5.4) are found by integration. The characteristic equation of relations (5.6) is solved next, and finally the boundary condition equations can be solved, and then the stresses are known throughout the laminate. Actually, the procedure for determining the stresses can be done one of two ways. The direct way has already been mentioned, as given by formulas (5.12). When this procedure is carried out, it is found that the stresses σ_{yz}, and σ_{zz} as well, are not continuous across the interfaces between the laminae. This is a consistent result of the theory, since it assumes continuous displacement fields and satisfies equilibrium only in terms of resultants. One could take the mean value of stresses, σ_{yz} and σ_{zz} at interfaces, but there is a much better way to overcome this result of the theory. The improved procedure was pointed out by Lo, Christensen, and Wu [5.8] in connection with the high-order plate theory. The best way to determine the out-of-plane stresses σ_{yz} and σ_{zz}, is to determine them directly by integrating the equations of equilibrium with the in-plane stress, σ_{yy}, given by the high-order theory solution just found. The constants of integration obtained in each lamina are found by satisfying stress continuity at the interfaces between laminae. The success of the method is virtually assured; note that it is the only way these out-of-plane stress components can be found in the classical theory.

The study of an example is left as a possible exercise. It should be noted that edge problems of this type can be even more complicated than the cross-ply example considered here. If the planes of material symmetry do not coincide with the coordinate axes, Fig. 5.5, then the edge effect also has a component of displacement in the x direction, and a stress component σ_{zy} is of importance. For unsymmetric laminates bending effects must also be involved. There have been many efforts to study these edge effect problems. We mention the work of Pagano [5.9] and Pipes and Pagano [5.10] in this regard. Much of the work has centered on determining the maximum value of the interlaminar stresses right at the edge. With each more refined analysis the maximum value of the stress is shown to rise. Finite element analyses have been employed to study the stress concentration, and the stresses are found to be increased with continuously decreasing element size. The quest, apparently, is to show that a stress singularity exists at the edge. It hardly seems worth the trouble to indirectly demonstrate the existence of a stress singularity in the context of the present free edge problem of laminates. It is well known that stress singularities are prevalent at the corners of boundaries joining dissimilar materials, from a linear elasticity point of view. The singularities at the corners of material junctions allow the problem to be viewed in a manner similar to that of fracture mechanics. Of course, it must be remembered that with fiber reinforced laminates, there are no perfect planes of separation between materials. Thus any stress conditions viewed on a dimensional scale smaller than that of the fiber diameter are fictitious.

With the above stated understanding, the present solution must be viewed as a demonstration of the existence of the edge effect, but the stress magnitudes obtained by it are not useful for design. Nor are the stress predictions of any other analysis useful in practice when, in fact, they have surpressed a singular or near singular type of behavior by making approximations. Useful qualitative results, however, can be obtained from such analyses. For example, lamina stacking arrangements can be found that provide a compressive interlaminar normal stress, as opposed to other arrangements that could produce tensile stresses. The one thing that can be said with certainty is that edge effect stresses exist in laminates, they have some effect on static strength of laminates, and they probably have a very great effect on the long term strength of laminates under fatigue conditions.

PROBLEMS

1. Verify the invariant relations

$$Q'_{11}+Q'_{22}+2Q'_{12}=Q_{11}+Q_{22}+2Q_{12}$$

$$Q'_{66}-Q'_{12}=Q_{66}-Q_{12}$$

of Section 5.1.

2. Given the normalized set of orthotropic properties

$$Q_{ij}=\begin{bmatrix} 10 & 1 & 0 \\ 1 & 1 & 0 \\ 0 & 0 & 1 \end{bmatrix}$$

Determine Q'_{ij} for a 45° rotation of the in-plane axes.

3. An unsymmetrical laminate of given properties A_{xx}, B_{xx} and D_{xx} (Eq. 2.21) must be used to resist a uniform lateral pressure p and an in-plane compressive load N_x, force per unit length. The plate is taken to be of infinite extent in the y direction and of length L in the x direction. The edges are simply supported. Using the classical theory, obtain the maximum deflection of the laminate.
4. Discuss situations in which cross-ply and angle-ply laminates have advantages over each other.
5. Derive the compatibility equation (3.8) from (3.7).
6. With regard to edge effects, take the free edge to be along the direction of the x axis. The laminate has a uniform, normal strain state in the x direction. A symmetric laminate is to be made of three cross-ply layers with an equal fraction of fibers running in both directions. Should the layer(s) with the fibers in the x direction be on the inside or outside of the laminate? Explain your reasoning.
7. As a project, formulate a shear deformation level theory for laminates. Examine the literature on the subject.
8. With regard to the bending of symmetric laminates, what is the lowest order theory that accounts for warping of the cross section? The term *order* refers to the order of the terms retained in a power series expansion of displacements in the coordinate normal to the middle

plane. What order theory would you need to use to account for transverse normal strain effects in contact problems involving smooth rigid indentors and laminates?

9. Explain the difficulty in devising a theory that assumes stress distributions rather than displacement distributions across the thickness of a laminate.

REFERENCES

5.1. S. W. Tsai and N. J. Pagano, "Invariant properties of composite materials," in *Composite Materials Workshop*, Technomic Publ. Co., Westport, Conn., 1968.

5.2. R. F. S. Hearmon, *An Introduction to Applied Anisotropic Elasticity*, Oxford, University Press, New York, 1961.

5.3. N. J. Pagano, "Exact solutions for composite laminates in cylindrical bending," *J. Comp. Mater.*, vol. 3, 398 (1969).

5.4. S. P. Timoshenko and J. N. Goodier, *Theory of Elasticity*, 3rd., McGraw-Hill, New York, 1970.

5.5. K. H. Lo, R. M. Christensen, and E. M. Wu, "A high order theory of plate deformation—Part 1: Homogeneous plates," *J. Appl. Mech.*, vol. 44, 663 (1977).

5.6. K. H. Lo, R. M. Christensen, and E. M. Wu, "A high order theory of plate deformation—Part 2: Laminated plates," *J. Appl. Mech.*, vol. 44, 669 (1977).

5.7. N. J. Pagano, "Stress fields in composite laminates," *Int. J. Solids Structures*, vol. 14, 385 (1978).

5.8. K. H. Lo, R. M. Christensen, and E. M. Wu, "Stress solution determination for high order plate theory," *Int. J. Solids Structures*, vol. 14, 655 (1978).

5.9. N. J. Pagano, "On the calculation of interlaminar normal stress in composite laminate," *J. Comp. Mater.*, vol. 8, 65 (1974).

5.10. R. B. Pipes and N. J. Pagano, "Interlaminar stresses in composite laminates—an approximate elasticity solution," *J. Appl. Mech.*, vol. 41, 668 (1974).

CHAPTER VI
ANALYSIS, STRENGTH, AND DESIGN

In this chapter we consider a collection of related but separate topics involving analysis, strength, and design. The common line between these topics is that of seeking information that is directly helpful in designing structures made from composite materials.

If a heterogeneous body is macroscopically isotropic, then, insofar as design is concerned, little distinction need be made between its behavior and that of a homogeneous isotropic body. It is in the area of anisotropic behavior that special attention must be given to the problems of designing with composite materials. There are many facets to creative design. No longer is it sufficient simply to make a structure massive enough to withstand all conceivable loads. Innovative design must employ a full knowledge of the properties of the material and the loads or constraints the structure must withstand.

The design process then places a premium on knowing the detailed conditions of stress and deformation within the material, and matching these to allowable conditions. The need is obvious, reliable methods of analysis must be used to understand conditions within the material, and realistic failure or yield criteria must be at hand for the final design evaluation. The developments of the preceding chapter, on laminates, comprise an example of analysis appropriate to a particular but very important type of geometry. However, we did not complete the design cycle by designing to some failure or service criterion.

In the following sections we consider both of the ingredients of design (analysis and failure) and we conclude with two realistic design examples.

Most of the developments in this chapter are motivated by fiber composite examples; however, the methods admit adaptation to other types of heterogeneous media.

6.1 A BOUNDARY LAYER THEORY OF ANISOTROPIC ELASTICITY

For purposes of analysis, heterogeneous materials are idealized as equivalent homogeneous materials. The analysis of composite materials then reduces to that appropriate to a homogeneous but anisotropic type of media. For many uses this procedure is further restricted to the class of behavior appropriate to anisotropic elasticity, although later we consider inelastic effects. Anisotropic elasticity theory is a self-contained topic in itself, and we do not attempt to summarize the results here in a few pages. For treatments of this subject, see Lekhnitskii [6.1] and Hearmon [6.2]. However, some special results from anisotropic elasticity theory are of great importance to us and we cover these.

Specifically, we consider the behavior of a single lamina of highly anisotropic elastic media. This characterization is suggested by the practical case of high modulus aligned fibers in a compliant matrix phase. To be sure, fiber composites are seldom used in this form; rather they are typically used in a laminated geometry with various angular orientations of the various lamina. Nevertheless, even in laminate geometry the behavior of the single lamina is fundamental, and we examine the single lamina case in some detail.

Our interest here then is on the behavior of highly anisotropic elastic media under two-dimensional conditions of plane stress or plane strain. If the media were isotropic, the problem would be governed by a biharmonic stress function. The general anisotropic case would be expected to be more complicated. However, consider for a moment the limiting case of a material with infinite rigidity in one direction, as suggested by infinitely stiff fiber lines in the plane of the problem. With this type of constraint, it is seen that the only possible mode of deformation is that of shear, say ε_{12}, where x_1 is in the direction of the infinitely stiff fiber lines. Recalling that problems of shear deformation in elasticity are governed by harmonic functions, we see that a great simplification has occurred in going from isotropic behavior with biharmonic functions, to the last described behavior involving only harmonic functions. The complication here, however, is that we do not have infinitely stiff fibers; rather we have very stiff fibers, which make the material highly anisotropic. It appears that there should be some simplification of the formulation in the highly anisotropic case over the formulation for the general anisotropic case, or even that for the isotropic case. We find this conjecture to be true; in fact, we show that the two-dimensional behavior in the case of highly anisotropic media is, in fact, governed by a harmonic function. The key element that contributes to the success of this simplified formulation is the introduction of a boundary layer concept. This concept was first introduced in the composites context by Everstine and Pipkin [6.3]. We here follow the treatment of Spencer [6.4].

Two-Dimensional Theory

Our treatment is restricted to two-dimensional conditions of plane stress or plane strain. In either case the stress-strain relations can be written as

$$\begin{bmatrix} \sigma_{xx} \\ \sigma_{yy} \\ \sigma_{xy} \end{bmatrix} = \begin{bmatrix} L & M & 0 \\ M & N & 0 \\ 0 & 0 & 2\mu_L \end{bmatrix} \begin{bmatrix} \varepsilon_{xx} \\ \varepsilon_{yy} \\ \varepsilon_{xy} \end{bmatrix} \tag{1.1}$$

where we must relate L, M, N, and μ_L to our more familiar full form of the stress-strain relations. Specifically, for an orthotropic material we have

$$\sigma_i = C_{ij}\varepsilon_j \tag{1.2}$$

using contracted notation, where C_{ij} is given by (V-1.4). In the case of plane strain we have

$$\begin{aligned} L &= C_{11} \\ M &= C_{12} \\ N &= C_{22} \\ 2\mu_L &= C_{66} \end{aligned} \tag{1.3}$$

and in the case of plane stress we have

$$\begin{aligned} L &= C_{11} - \frac{C_{13}^2}{C_{33}} \\ M &= C_{12} - \frac{C_{13}C_{23}}{C_{33}} \\ N &= C_{22} - \frac{C_{23}^2}{C_{33}} \\ 2\mu_L &= C_{66} \end{aligned} \tag{1.4}$$

where the 1,2 plane is the plane of deformation. The C_{ij} terms are related to the engineering properties of transversally isotropic materials by relations (III-1.7).

The equations of equilibrium

$$\frac{\partial \sigma_{xx}}{\partial x}+\frac{\partial \sigma_{xy}}{\partial y}=0$$

and

$$\frac{\partial \sigma_{xy}}{\partial x}+\frac{\partial \sigma_{yy}}{\partial y}=0 \tag{1.5}$$

when combined with (1.1) become

$$L\frac{\partial^2 u}{\partial x^2}+(M+\mu_L)\frac{\partial^2 v}{\partial x\,\partial y}+\mu_L\frac{\partial^2 u}{\partial y^2}=0$$

and

$$\mu_L\frac{\partial^2 v}{\partial x^2}+(M+\mu_L)\frac{\partial^2 u}{\partial x\,\partial y}+N\frac{\partial^2 v}{\partial y^2}=0 \tag{1.6}$$

Let

$$\frac{\mu_L}{L}=\varepsilon^2$$

$$\frac{\mu_L}{M}=d^2$$

$$\frac{\mu_L}{N}=c^2 \tag{1.7}$$

Then equations (1.6) have the form

$$\frac{1}{\varepsilon^2}\frac{\partial^2 u}{\partial x^2}+\left(1+\frac{1}{d^2}\right)\frac{\partial^2 v}{\partial x\,\partial y}+\frac{\partial^2 u}{\partial y^2}=0$$

and

$$\frac{\partial^2 v}{\partial x^2}+\left(1+\frac{1}{d^2}\right)\frac{\partial^2 u}{\partial x\,\partial y}+\frac{1}{c^2}\frac{\partial^2 v}{\partial y^2}=0 \tag{1.8}$$

Thus far, no restriction has been placed on the properties, other than that of orthotropy. However, at this point we wish to recognize the general character of a lamina of aligned fibers that has much stiffer properties for deformation in one certain direction, than for other deformations. Accordingly, we wish to let the moduli entering the present formulation reflect that circumstance. Take the modulus L in (1.1) as that suggested by the placement of stiff fibers in the x direction. With L being very large compared with the other moduli measures, we see that ε in (1.7) may be taken to be very small.

Consider the physics of the situation for a moment. If the fibers were infinitely stiff, a single fiber, or line idealization of a fiber, could carry a concentrated load without disturbing the neighboring media. This would be called a singular line. Now in the case where the fibers are not infinitely stiff, but still are very stiff compared with the other modes of deformation, then one would expect a situation close to that of the singular line, which approaches the singular line behavior in the limit. We now see the possibilities of stress transmission along narrow zones that decay very quickly with lateral dimensions. This circumstance suggests the type of treatment followed in boundary layer theory, where the lateral coordinate is "stretched" to accommodate the rapid variation in the lateral direction.

Proceeding along these lines, we stretch the y coordinate by letting

$$y = \varepsilon\eta \tag{1.9}$$

With this coordinate change the stress displacement relations (1.1) become

$$\begin{aligned}
\frac{\sigma_{xx}}{\mu_L} &= \frac{1}{\varepsilon^2}\frac{\partial u}{\partial x} + \frac{1}{\varepsilon d^2}\frac{\partial v}{\partial \eta} \\
\frac{\sigma_{yy}}{\mu_L} &= \frac{1}{d^2}\frac{\partial u}{\partial x} + \frac{1}{\varepsilon c^2}\frac{\partial v}{\partial \eta} \\
\frac{\sigma_{xy}}{\mu_L} &= \frac{1}{\varepsilon}\frac{\partial u}{\partial \eta} + \frac{\partial v}{\partial x}
\end{aligned} \tag{1.10}$$

With the change (1.9) the equilibrium equations have the form

$$\begin{aligned}
&\frac{1}{\varepsilon^2}\frac{\partial^2 u}{\partial x^2} + \frac{1}{\varepsilon}\left(1 + \frac{1}{d^2}\right)\frac{\partial^2 v}{\partial x\,\partial \eta} + \frac{1}{\varepsilon^2}\frac{\partial^2 u}{\partial \eta^2} = 0 \\
&\frac{\partial^2 v}{\partial x^2} + \frac{1}{\varepsilon}\left(1 + \frac{1}{d^2}\right)\frac{\partial^2 u}{\partial x\,\partial \eta} + \frac{1}{\varepsilon^2 c^2}\frac{\partial^2 v}{\partial \eta^2} = 0
\end{aligned} \tag{1.11}$$

Thus to the terms of the lowest order in ε (1.11) become

$$\frac{\partial^2 u}{\partial x^2}+\frac{\partial^2 u}{\partial \eta^2}=0$$

and (1.12)

$$\frac{\partial^2 v}{\partial \eta^2}=0$$

and we write these as

$$\nabla^2 u=0$$

and (1.13)

$$v=g(x)+\eta h(x)$$

where $g(x)$ and $h(x)$ are to be determined and ∇^2 is the harmonic operator in (1.12). It is assumed that the terms d and c in (1.11) are of a magnitude that does not conflict with the process of neglecting terms to obtain (1.12). Such conditions are assured if $d=O(1)$ and $c=O(1)$ or $O(\varepsilon)$. Also it remains to be verified from the boundary layer solution that the neglected terms in going from (1.11) to (1.12) are acceptably small.

Reference should be made to Spencer [6.4] for discussion of the appropriate boundary conditions of the theory. For our purposes we consider the boundary conditions along with the examples of interest here.

Concentrated Force Example

Consider next a simple example to illustrate the method. Take a half plane loaded by a concentrated force, as shown in Fig. 6.1.

$$\text{at } x=0, \qquad \sigma_{xx}=-P\delta(y)$$

$$= \qquad \sigma_{xy}=0 \tag{1.14}$$

The very stiff property direction is taken as that of the x coordinate. Position $y=0$ is the center of the boundary layer. If the property L were perfectly rigid, the line at $y=0$ would be a singular line supporting infinite stress.

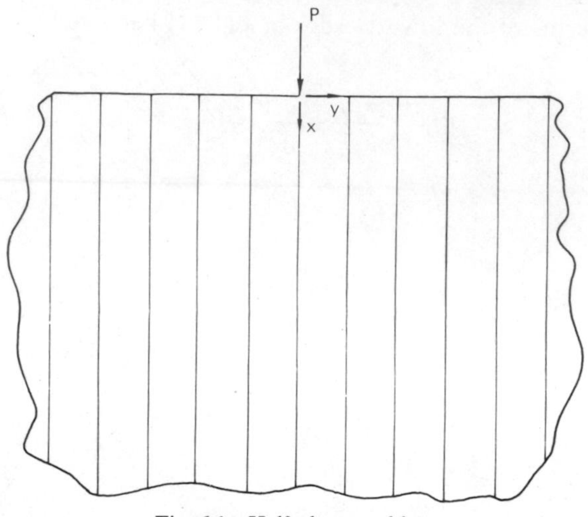

Fig. 6.1 Half plane problem.

The stress expressions (1.10) to lowest order in ε are given by

$$\frac{\sigma_{xx}}{\mu_L} = \frac{1}{\varepsilon^2}\frac{\partial u}{\partial x}$$

$$\frac{\sigma_{yy}}{\mu_L} = \frac{1}{\varepsilon c^2}\cdot\frac{\partial v}{\partial \eta}$$

$$\frac{\sigma_{xy}}{\mu_L} = \frac{1}{\varepsilon}\frac{\partial u}{\partial \eta} \tag{1.15}$$

The boundary condition (1.14) in terms of the stretched coordinate η is given by

$$\text{at } x=0, \qquad \sigma_{xx} = -\frac{1}{\varepsilon}P\delta(\eta)$$

thus to lowest order in ε the boundary conditions of σ_{xx} and σ_{xy} are given by

$$\text{at } x=0, \qquad \frac{\partial u}{\partial x} = -\varepsilon\frac{P}{\mu_L}\delta(\eta)$$

$$\frac{\partial u}{\partial \eta} = 0$$

For $\eta \to \pm\infty$ we must have $\sigma_{xx} = \sigma_{yy} = \sigma_{xy} = 0$; thus

$$\text{as } \eta \to \pm\infty \qquad \begin{cases} \dfrac{\partial u}{\partial x} \to 0 \\ \dfrac{\partial u}{\partial \eta} \to 0 \\ \dfrac{\partial v}{\partial \eta} \to 0 \end{cases}$$

Also, for vanishing stresses as $x \to \infty$ we have

$$\text{as } x \to \infty \qquad \begin{cases} \dfrac{\partial u}{\partial x} \to 0 \\ \dfrac{\partial u}{\partial \eta} \to 0 \\ \dfrac{\partial v}{\partial \eta} \to 0 \end{cases}$$

To satisfy the preceding requirements on v, along with $v = 0$ at $y = 0$, and for v to have the form of (1.13), it follows that

$$v = 0$$

is the solution.

Next for u we take the derivative of (1.13) to obtain

$$\nabla^2 \left(\frac{\partial u}{\partial x} \right) = 0$$

The solution of this equation that satisfies the preceding conditions at $x = 0$, $\eta \to \pm\infty$, and $x \to \infty$ is given by

$$\frac{\partial u}{\partial x} = -\frac{\varepsilon P}{\pi \mu_L} \frac{x}{(x^2 + \eta^2)}$$

The integral of this form is then

$$u = \frac{-\varepsilon P}{2\pi \mu_L} \ln (x^2 + \eta^2) + A \tag{1.16}$$

where A is a constant.

The stresses may now be found from (1.15) as

$$\sigma_{xx} = -\frac{P}{\varepsilon\pi}\frac{x}{(x^2+\eta^2)}$$

$$\sigma_{yy} = 0$$

$$\sigma_{xy} = -\frac{P}{\pi}\frac{\eta}{(x^2+\eta^2)} \tag{1.17}$$

We note that σ_{xy} does not vanish at $x=0$; however, this is a consistent result of the theory since σ_{xy} is of higher order in $(1/\varepsilon)$ than σ_{xx}.

It remains to be verified that the resultant force at any section normal to x balances the applied force P. Thus we write

$$\begin{aligned}
-\int_{-\infty}^{\infty}\sigma_{xx}\,dy &= -\varepsilon\int_{-\infty}^{\infty}\sigma_{xx}\,d\eta \\
&= \frac{P}{\pi}\int_{-\infty}^{\infty}\frac{x}{(x^2+\eta^2)}\,dx \\
&= \frac{P}{\pi}\left[\tan^{-1}\frac{\eta}{x}\right]_{-\infty}^{\infty} \\
&= P
\end{aligned}$$

and the proper result is established.

The exact solution of this problem is available and Spencer has given a comparison of the stress σ_{xx} at $y=0$ between the two solutions. The two expressions differ only in a coefficient that is about 14% different for a typical carbon fiber composite material.

If the problem of Fig. 6.1 is modified so that the concentrated force is in the y direction, then no boundary layers exist, and there is no simplification in the problem. Other problems of this type are also solved by Spencer [6.4].

Crack Example

A second example of the use of boundary layer theory is now given. This problem involves the deformation of a highly anisotropic medium containing a crack, with stiff fibers running parallel to the crack. The specific problem is shown in Fig. 6.2. The state of simple shearing stress is applied at infinite distances from the crack. We decompose the problem into the

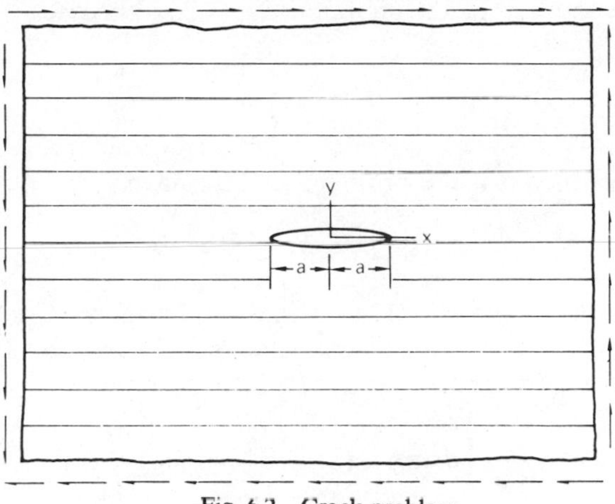

Fig. 6.2 Crack problem.

sum of two parts (a) a state of uniform shearing stress throughout the entire medium, and (b) a problem with oppositely directed shearing stresses applied to the boundary of the crack, which vanish at large distances from the crack. We now proceed with problem (b).

The boundary conditions of the problem are stated by

$$\text{at } y=0, \quad -a<x<a \qquad \begin{cases} \sigma_{xy}=-\tau \\ \sigma_{yy}=0 \end{cases} \tag{1.18}$$

and the stresses and displacements go to zero as $x^2+y^2 \to \infty$. To satisfy the condition at infinity and with v to have the form of the solution in (1.13), it is necessary that

$$v=0 \tag{1.19}$$

From relations (1.10) the stresses are then given by

$$\begin{aligned} \frac{\sigma_{xx}}{\mu_L} &= \frac{1}{\varepsilon^2}\frac{\partial u}{\partial x} \\ \frac{\sigma_{yy}}{\mu_L} &= \frac{1}{d^2}\frac{\partial u}{\partial x} \\ \frac{\sigma_{xy}}{\mu_L} &= \frac{1}{\varepsilon}\frac{\partial u}{\partial \eta} \end{aligned} \tag{1.20}$$

The boundary condition (1.18) on σ_{xy}, using (1.20), becomes

$$\text{at } y=0, \quad |x|<a, \qquad \left\{ \frac{\partial u}{\partial \eta} = -\varepsilon \frac{\tau}{\mu_L} \right. \tag{1.21}$$

From the antisymmetry of the problem we also have

$$\text{at } y=0, \quad |x|>a, \qquad \{ u=0 \tag{1.22}$$

Our problem then, from (1.12), is to find the harmonic function $u(x,\eta)$ that satisfies (1.21) and (1.22) and vanishes as $x^2+y^2 \to \infty$. The solution from potential theory is given by

$$u = -\varepsilon \frac{\tau}{\mu_L} \left[\rho \sin\phi - (\rho_1\rho_2)^{1/2} \sin\tfrac{1}{2}(\phi_1 - \phi_2) \right] \tag{1.23}$$

where for convenience complex notation is employed such that

$$\begin{aligned} x + i\eta &= \rho e^{i\phi} \\ x - a + i\eta &= \rho_1 e^{i\phi_1} \\ x + a + i\eta &= \rho_2 e^{i\phi_2} \end{aligned} \tag{1.24}$$

Let us consider the determination of σ_{xx} from (1.20). We seek the explicit solution for $y=0$, $|x|<a$. In this case

$$\text{at } y=0, \quad |x|<a, \qquad \begin{cases} \phi = 0 & \\ \phi_1 = \pi & \rho_1 = a - x \\ \phi_2 = 0 & \rho_2 = x + a \end{cases}$$

and (1.23) reduces to

$$u = -\varepsilon \frac{\tau}{\mu_L} (a^2 - x^2)^{1/2}, \qquad \text{at } y=0, \quad |x|<a \tag{1.25}$$

Taking the derivative of (1.25) for use in (1.20) finally gives

$$\sigma_{xx} = -\frac{\tau}{\varepsilon} \frac{x}{(a^2 - x^2)^{1/2}}, \qquad \text{at } y=0, \quad |x|<a \tag{1.26}$$

We see that the solution leads to a square root singularity for stress at the crack tip. The other components of stress also involve the singularity.

Thus even in the situation of highly anisotropic materials, the classical square root singularity emerges. The method and solution obtained here is far less complicated than that which would be involved in dealing directly with the full anisotropic elasticity formulation. Other crack problems are given in the work by Spencer [6.4].

The boundary layer method has been used by Everstine and Pipkin [6.5] to solve the problem of a highly anisotropic cantilever beam under end loading. The problem is somewhat more complicated than that with which we have been concerned. In the beam problem one must be concerned with both the boundary layer solution and the region interior to the boundary layers (where the fibers are treated as perfectly inextensible). In the examples studied here the boundary layer solution itself contains all the desired information.

The analysis and examples of this section illustrate methods that can be used to deduce the stress and deformation conditions within anisotropic elastic media. Unless special assumptions are made, the general problem is simply that of solving the full formulation of anisotropic elasticity. Sometimes, however, even more general assumptions are invoked. For example, some organic fibers exhibit markedly different tensile and compressive modulus type behavior, even in the small strain range. Special theories have been devised to accommodate such behavior; see, for example, Bert [6.6].

We henceforth assume that the stress-deformation solutions are known, whether as exact solutions, approximate analytical solutions, or numerical solutions. We turn our attention now to the corresponding problem of criteria for strength and failure of materials, assuming that the stress conditions are known.

6.2 STRENGTH AND FAILURE CRITERIA

The use of any material in a design application requires assurance that it will function safely and efficiently. Thus the design process must involve the comparison of the state of stress and deformation, for a particular design, with a criterion of strength for the material. The criteria for strength or failure are often expressed in terms of stresses, but such criteria do not necessarily refer only to a state of complete rupture of the material. Failure criteria may, in fact, refer to the initial events, as evidenced in yielding, that are the precursors of ultimate failure. Thus with failure criteria being so widely interpreted, it is not possible to be highly definitive in their characterization. In fact, a failure criterion for a given material is in essence an empirical statement of the process of degradation and failure of the material. Despite these complications and reservations, we can state failure criteria that have met with success and that have reasonable generality.

As noted earlier, failure criteria are not derived; they are stated or postulated, and then tested against experience in the field. In the realm of isotropic material behavior there is a large amount of experience in dealing with failure criteria. The two most common criteria are those of the Mises and the Tresca type. For macroscopically isotropic composite materials, these same criteria are the logical ones to be tested and utilized. Unfortunately, however, the most common problems encountered with composite materials involve their use in anisotropic forms. Accordingly, failure criteria must be obtained appropriate to anisotropic material behavior. Furthermore, most of the common problems in this class are those appropriate to fiber reinforced materials; therefore our approach here is strongly guided by applications to fiber reinforced materials.

Generalized Mises Criterion

We first state what is probably the simplest reasonable generalization of an isotropic failure criterion to anisotropic conditions. Specifically, Hill [6.7], generalized the Mises criterion to the anisotropic form

$$F(\sigma_2-\sigma_3)^2+G(\sigma_3-\sigma_1)^2+H(\sigma_1-\sigma_2)^2+2L\sigma_4^2+2M\sigma_5^2+2N\sigma_6^2=1 \quad (2.1)$$

where the contracted notation (I-1.8) is used. The constants $F\cdots N$ must be evaluated by comparison with experimental data for failure in various states of stress. For isotropy, $F=G=H=L=M=N$, and with (2.1) written only in terms of principle stresses it takes the familiar form of the Mises criterion. This generalization of the Mises criterion assumes an independence from hydrostatic stress and it also does not distinguish between tensile and compressive states. Both of these effects are known to be of importance for composite materials and the need for more general criteria is obvious.

Tensor Polynomial Criterion

A general criterion has been given by Tsai and Wu [6.8]. In the most general form this criterion is expressed as

$$(F_i\sigma_i)^\alpha+(F_{ij}\sigma_i\sigma_j)^\beta+(F_{ijk}\sigma_i\sigma_j\sigma_k)^\gamma+\cdots=1 \quad (2.2)$$

where again contracted tensor notation is used with coefficients

$F_i, F_{ij}, F_{ijk} \ldots$ and exponents $\alpha, \beta, \gamma \ldots$ to be determined to give the best representation of experimental data. The form of (2.2) is, of course, a generalization to tensor conditions of the usual scalar polynomial form. The form F_i is a second order tensor, as

$$F_i \sigma_i = \tilde{F}_{kl} \sigma_{kl}$$

Thus F_i has six independent components. Similarly, F_{ij} is a fourth order tensor with 21 independent components. All higher-order tensors follow the same general character. There does not appear to be any increased generality to be gained by the exponents in (2.2); henceforth we take $\alpha = \beta = \cdots = 1$.

Further reduction of the $F_i, F_{ij}, \ldots$ tensors in (2.2) follows from material symmetry conditions that we now consider.

Orthotropy. Consider the case of orthotropy and take the coordinate planes parallel to the symmetry planes. The term $F_i \sigma_i$ is written out as

$$F_i \sigma_i = F_1 \sigma_1 + F_2 \sigma_2 + F_3 \sigma_3 + F_4 \sigma_4 + F_5 \sigma_5 + F_6 \sigma_6 \tag{2.3}$$

The shear stress terms involving σ_4, σ_5, and σ_6 give a physically unacceptable effect due to the sign of the shear stress, unless the related coefficients vanish, as

$$F_4 = F_5 = F_6 - 0 \tag{2.4}$$

This simplification would not be allowed if the material symmetry planes were not parallel to the coordinate planes. Now, in the second term of (2.2), for the coupling terms between normal stress and shear stress to be independent of the sign of the shear stress, it again follows that

$$F_{14} = F_{15} = F_{16} = 0$$

$$F_{24} = F_{25} = F_{26} = 0$$

$$F_{34} = F_{35} = F_{36} = 0 \tag{2.5}$$

Furthermore, the shear strengths are assumed to be uncoupled; thus to be independent of the sign of the shear stress

$$F_{45} = F_{46} = F_{56} = 0 \tag{2.6}$$

Using (2.5) and (2.6) in (2.2) it becomes

$$(F_1\sigma_1+F_2\sigma_2+F_3\sigma_3)+\big(F_{11}\sigma_1^2+F_{22}\sigma_2^2+F_{33}\sigma_3^2+2F_{12}\sigma_1\sigma_2+2F_{23}\sigma_2\sigma_3$$
$$+2F_{13}\sigma_1\sigma_3+F_{44}\sigma_4^2+F_{55}\sigma_5^2+F_{66}\sigma_6^2\big)+\cdots=1 \tag{2.7}$$

where, as already mentioned, the exponents are all taken as having the value 1. It is instructive to see the reduction of the tensor polynomial form (2.7) to the special form (2.1) postulated by Hill as a generalization of the Mises criterion. To effect this reduction, first neglect the Bauschinger effect by taking

$$F_1=F_2=F_3=0 \tag{2.8}$$

in (2.7). Next restrict the terms shown in (2.7) to be independent of hydrostatic pressure by taking

$$\frac{\partial}{\partial\sigma_p}(F_{ij}\sigma_i\sigma_j)=0 \tag{2.9}$$

where

$$\sigma_p=\sigma_1+\sigma_2+\sigma_3$$

It can be shown that the relation (2.9) leads to

$$F_{11}=-(F_{21}+F_{31})$$
$$F_{22}=-(F_{12}+F_{32})$$
$$F_{33}=-(F_{13}+F_{23}) \tag{2.10}$$

Substituting (2.8) and (2.10) into (2.7), the latter truncated at the level shown, then the remaining terms have the form of relation (2.1), the criterion given by Hill. We now see that the criterion (2.1) involves several restrictions, that is, neglect of Bauschinger effect, independence of hydrostatic stress, and an orthotropic material characterization with the coordinate planes taken parallel to the symmetry planes.

We see from (2.7) that in the case of orthotropy, F_i has three independent components and F_{ij} has nine independent components.

Transverse Isotropy. The case of a transversely isotropic material is a special case of orthotropy. Take the x_2x_3 plane to be the plane of symmetry. Then from symmetry

$$F_2 = F_3$$

and

$$F_{12} = F_{13}$$

$$F_{22} = F_{33}$$

$$F_{55} = F_{66}$$

Also, the shear condition gives

$$F_{44} = 2(F_{22} - F_{23})$$

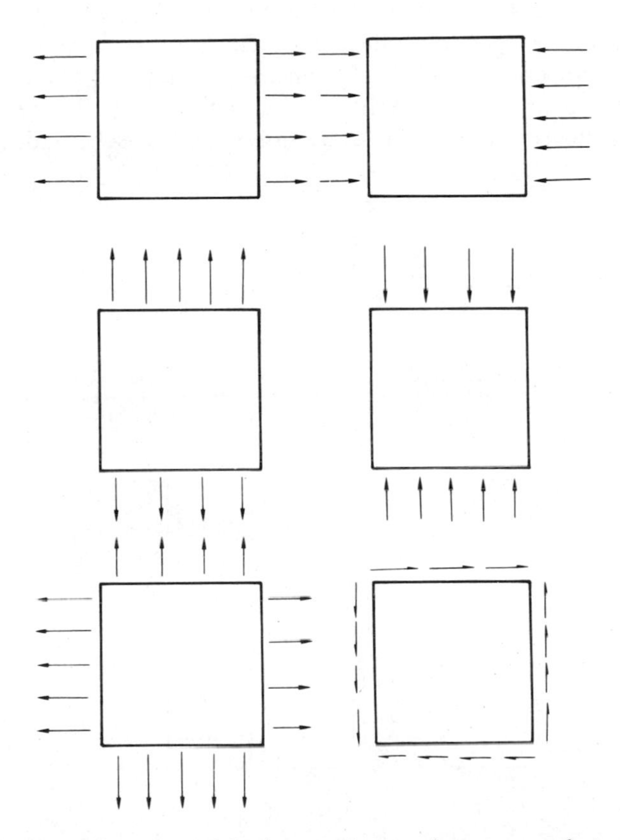

Fig. 6.3 Failure characterization experiments, plane stress, orthotropic.

These further restrictions on the orthotropic results reduce (2.7) to

$$[F_1\sigma_1+F_2(\sigma_2+\sigma_3)]+[F_{11}\sigma_1^2+F_{22}(\sigma_2^2+\sigma_3^2)+2F_{12}(\sigma_1\sigma_2+\sigma_1\sigma_3)$$
$$+2F_{23}\sigma_2\sigma_3+F_{44}\sigma_4^2+F_{55}(\sigma_5^2+\sigma_6^2)]+\cdots=1 \tag{2.11}$$

Plane Stress, Orthotropic. For plane stress we have by definition

$$\sigma_3=\sigma_4=\sigma_5=0 \tag{2.12}$$

leaving (2.7) as

$$[F_1\sigma_1+F_2\sigma_2]+[F_{11}\sigma_1^2+F_{22}\sigma_2^2+2F_{12}\sigma_1\sigma_2+F_{66}\sigma_6^2]+\cdots=1 \tag{2.13}$$

Thus in this case F_i has two components and F_{ij} has four. Transverse isotropy makes no further reduction in this case so long as the axis of symmetry is not normal to the 1,2 plane. At the level of truncation shown in (2.13) there are six constants to be determined. The experimental program to determine these six constants is shown schematically in Fig. 6.3.

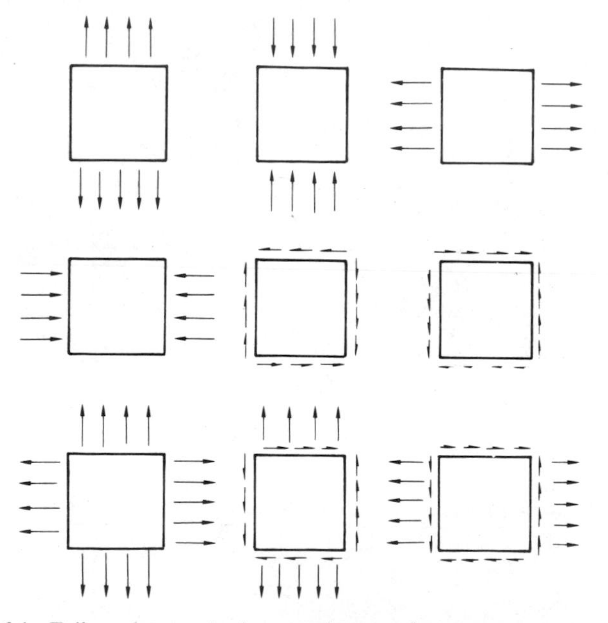

Fig. 6.4 Failure characterization experiments, plane stress, no symmetry.

Plane Stress, No Symmetry. Consider now the case where either the material is not orthotropic, or if it is orthotropic, it does not have the coordinate planes taken parallel to the planes of symmetry. The only simplification here for (2.2) is that afforded by (2.12), which gives

$$\left[F_1\sigma_1+F_2\sigma_2+F_6\sigma_6\right]+\left[F_{11}\sigma_1^2+F_{22}\sigma_2^2+F_{66}\sigma_6^2\right.$$
$$\left.+2F_{12}\sigma_1\sigma_2+2F_{16}\sigma_1\sigma_6+2F_{26}\sigma_2\sigma_6\right]+\cdots=1 \tag{2.14}$$

The nine experiments needed to determine the constants in (2.14) are shown schematically in Fig. 6.4. Whereas the relation (2.13) is that motivated by the symmetry properties of a single lamina of aligned fibers, the relation (2.14) has no symmetry assumptions involved in it. As such, it is applied to any anisotropic configuration, so long as plane stress conditions prevail. In this sense, then, the result (2.14) applies to laminates as well as a lamina. There are, however, complications that arise in considering laminates. It is not clear that a polynomial approach such as that of (2.14) models the complicated interlaminar conditions that may contribute to ultimate failure.

The coefficients in the polynomial criterion have been evaluated from test data. Results up to the quadratic terms explicitly shown here have been discussed by Wu [6.9]. Tennyson, MacDonald, and Nanyaro [6.10] have shown the advantages of not truncating the tensor polynomial expansion until after the cubic terms. It has been suggested that the coefficients in the tensor polynomial representation be allowed to have different values in different quadrants or regions. In this manner the existence of corners or vertices on the failure surface would be allowed.

Other Criteria and Considerations

There are other simple failure criteria in use that do not possess the generality of the tensor polynomial method, but that may offer some simple advantages. These are the maximum stress criterion and the maximum strain criterion. Both of these are typically used only for plane stress conditions. The maximum stress criterion simply has the form

$$F_x^c \leqslant \sigma_{xx} \leqslant F_x^T$$
$$F_y^c \leqslant \sigma_{yy} \leqslant F_y^T$$
$$F_{xy}' \leqslant \sigma_{xy} \leqslant F_{xy}''$$

Similarly, for the maximum strain criterion we have

$$E_x^c \leqslant \varepsilon_{xx} \leqslant E_x^T$$

$$E_y^c \leqslant \varepsilon_{yy} \leqslant E_y^T$$

$$E'_{xy} \leqslant \varepsilon_{xy} \leqslant E''_{xy}$$

The maximum strain criterion can be expressed in terms of stresses using the appropriate forms of the stress-strain relations. These simple criteria are quite widely used because of their simplicity, but they only have significance for one-dimensional states, and their use in multidimensional stress or strain states must be made with caution.

Thus far, we have characterized strength as a macroscopic property of the composite material, without regard to the individual constituents. In Chapters II, III, and IV we found many useful means of predicting the macroscopic stiffness properties in terms of the moduli of the individual phases. One would expect that a similar attack on the composite strength characterization could be mounted. The desired result would be the prediction of the composite strength property in terms of the strength properties of the individual constituents. Unfortunately, this approach to strength characterization is nowhere nearly as highly developed as the corresponding stiffness characterization. Nevertheless, the strength problem has been studied in special situations, and it is sometimes claimed that for a unidirectional fiber composite, the uniaxial strength in the direction of the fibers is adequately represented by the rule of mixtures. Although a result such as this may apply to a particular system, there is no evidence that it applies to a general class of materials.

A problem related to that just posed concerns the effect of the statistical variability of material properties and their effect on macroscopic strength. This particular problem has received wide attention in the context of the uniaxial strength of aligned fiber systems. The problem may be posed as follows. If the fibers are perfectly uniform, with no geometric or material variability, the composite uniaxial strength should be directly obtainable from the ideal strength of the fiber material. The reality of the situation is that composite strengths are found to be far lower than the ideal strength of the fiber material as measured on very small fiber lengths. The cause of the discrepancy is the variability of fiber properties along its length, as caused by the original manufacturing operation, or by subsequent handling, with the introduction of flaws and individual fiber breakage. This problem is best characterized in a statistical sense and there are many statistical treatments of the problem. Early work on this problem was given by Daniels [6.11] and later by Coleman [6.12], with a still later approach by Zweben and Rosen [6.13]. Perhaps the most general formulation of this type is that of Harlow and Phoenix [6.14]. It is beyond the scope intended here to cover these statistical theories of failure, which is a self-contained topic in itself. It is merely noted that it is a rapidly developing field, and one that may have considerable practical utility for design.

There are many other aspects to strength and failure criteria that we only mention here. Consider a laminate with the various lamina arranged at different angular orientations. The resin phase in a lamina, in a state of transverse tension, typically begins cracking at a much lower strain level than that at which fiber failure occurs in a lamina with fibers aligned with the direction of the load. This circumstance shows that the resin phase begins failing before the fiber phase in many practical laminate constructions. Typically, failure begins with the resin cracking in a single lamina, and this is referred to as first ply failure. It does not mean that the entire laminate has failed, or even that a significant number of fibers have failed. The laminate can still carry load, and ultimate failure may occur at a much higher load level than that at which first ply failure occurs. It is a matter of design philosophy, to choose between a failure criterion based on ultimate failure versus one based on first ply failure, or some other mode of incipient degradation of the composite structure.

Other design criteria may take precedence over that of simple static strength. For example, fatigue is one of the primary design factors in many practical situations. There is no substitute for knowledge of the actual fatigue data for realistic specimens designed to model practical situations. The sensitivity to fatigue varies widely between different types of composite materials. For example, carbon fiber composites are far less fatigue sensitive than are glass fiber composites. Crack growth in composites can comprise an important design variable. Crack sensitivity in composites in general, and laminates in particular, is a very complex topic, far more complicated than that in homogeneous isotropic materials. Crack growth can occur as that in an individual lamina (as in the resin cracking mentioned earlier), or as a crack entirely through the laminate thickness, or even as a planar region of debonding, which leads to delamination. In general, crack growth in composites combines the most difficult aspects of anisotropy along with heterogeneity.

Still further aspects of design toward a failure criterion involve various environmental effects. Since one or more phases of many composite materials are based on organic material utilization, temperature sensitivity often becomes of prime importance. Not only do elevated temperature states induce thermal stresses but the basic mechanical properties of the organic material phases undergo changes with temperature, which alter their capacity to bear load. Reduced temperature states embrittle the materials and make crack sensitivity an important variable. Moisture effects in composites often have a similar effect to that of elevated temperature states. Moisture diffusion can cause swelling of organic materials, which induces internal stresses, and basic properties are also changed because of increased moisture content. Often moisture presence has a particularly degrading effect on the integrity of the chemical bond at the interfaces between phases. Yet another related effect is that of shrinkage stresses incurred during the curing step in manufacturing operations. Careful design can be used to minimize all these deleterious effects.

Finally, mention should be made of dynamic effects. For example, gas turbine blades made from composite materials have the response to high rate loading as their primary design condition. Failure under high rate conditions can be very different from failure in quasi-static loading situations. Again, the time and rate dependent properties of organic materials play a role in the response characteristic of the composite material. Other effects may be of importance also. The ability of a material to absorb energy during failure is referred to as *toughness*. The dynamic toughness of some types of fiber reinforced materials relates to the energy dissipated because of the mechanism of fiber pullout from the matrix, with associated bond breakage and friction effects.

The above discussed physical effects all can relate to the ability of the material to perform its service function, and all can be determining design criteria. We have only attempted to give an analytical characterization to the criteria of quasi-static strength. Many of the other criteria admit useful and practical analytical characterizations that are of use in design. Typical design problems involve not just a single design criterion, but a whole spectrum of requirements. Thus a final design is usually a compromise between various requirements. We do not attempt to give a full design example, which typically involves an iteration process to reach the final design. However, we give two practical design examples that illustrate the interaction needed between analysis and failure criteria to reach a final design. These examples show the challenge and opportunity offered in the area of design with composite materials.

6.3 DESIGN EXAMPLE I: FIBER REINFORCED PRESSURE VESSELS

In this section we are concerned with some of the design aspects for fiber reinforced pressure vessels. We confine attention to the case of pressure vessels having thicknesses that are very small compared with the radii of

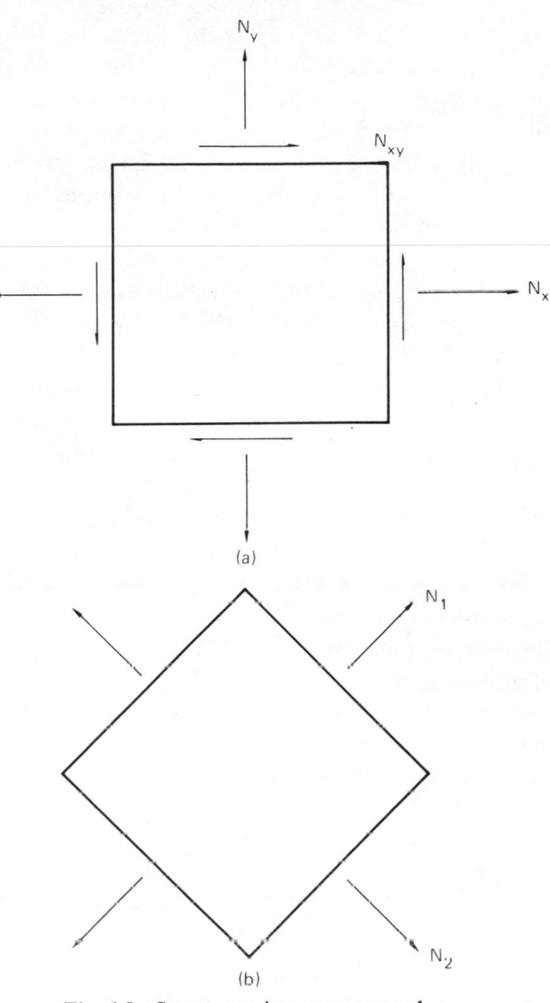

Fig. 6.5 Stress state in pressure vessel.

curvature. Under this restriction the local effect of the radius of curvature can be neglected, and considerations can be based on the behavior of a flat laminate. Thus, before we restrict our attention to specific types of pressure vessels, we must note that we assume the load state is known locally. The problem then is one of deciding how to design the composite material to resist the known state of internal loads.

We are given a state of in-plane resultant forces per unit length, as in Fig. 6.5*a*. First restrict attention to a single lamina. The simplest level design problem is to find how the fibers should be oriented to resist the loads of Fig. 6.5*a*. The first step is to find the principal resultants corresponding to the resultants of Fig. 6.5*a*. The resultants, designated as N_1 and N_2, are shown in Fig. 6.5*b*. Although it is intuitively appealing to expect the fibers to be oriented in the direction of greatest principal stress, that is not necessarily true. Brandmaier [6.15] has shown that in the case of the failure criterion (2.1) derived by Hill, the optimal fiber direction may be at some angle to the principal direction. Whether this is the case depends on the ratio of the shear strength to the transverse tensile strength. In either case it is not particularly helpful to pursue this example further because a biaxial stress state clearly calls for the use of a multidirectional orientation of fiber lamina, as is considered next.

Relative to the state of resultants shown in Fig. 6.5*b*, the fibers can be taken with an infinite variety of orientation arrangements. As in the preceding problem, it is by no means obvious what the optimal orientation program is relative to the principal directions. The term *optimal design* is taken to mean the minimal amount of material to resist the given loading state. Naturally, the solution of this optimality problem depends on the failure criterion used. It is, nevertheless, conventional practice to orient the plies in the principal directions. The relative number of individual plies in the two directions is determined by the ratio of principle loads, N_1/N_2. The total number of plies would be taken such that the failure criterion is satisfied in terms of stresses in individual lamina, or in terms of average stress through the thickness of the laminate. We next consider a design problem that shows rather clearly an example of optimal fiber orientation other than in the principal directions of stress.

Cylindrical Vessel

Consider a cylindrical pressure vessel with closed ends, as shown in Fig. 6.6*a*. For thin walled assumptions, the longitudinal stress and hoop stress in the cylindrical portion of the vessel, away from the ends, are given by

$$\sigma_{zz} = \frac{pa}{2h}$$

$$\sigma_{\theta\theta} = \frac{pa}{h} \tag{3.1}$$

where h is the thickness. Consider the helical winding pattern shown in Fig. 6.6*b*. Assume the stress in the fiber direction in each lamina is very large so that the other components of stress can be neglected in comparison. As shown in Fig. 6.6*c* with σ the stress in the fiber direction, the resisting horizontal component of stress is $\sigma \cos^2 \phi$. Equating this to the

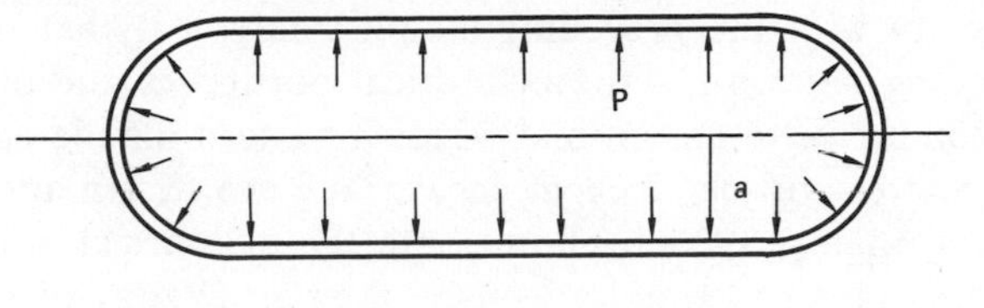

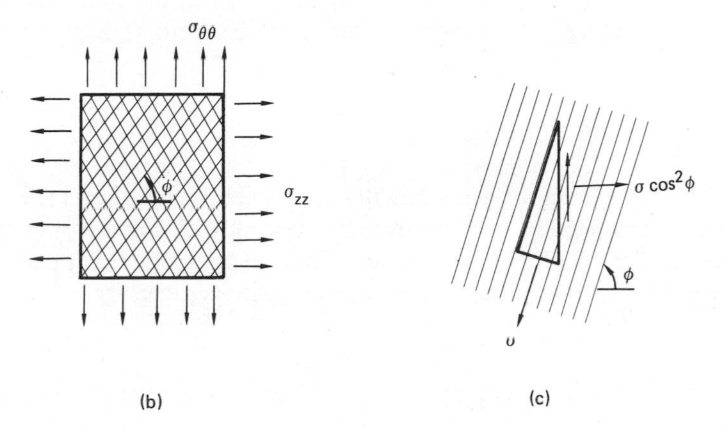

Fig. 6.6 Pressure vessel wrapping design.

longitudinal stress gives

$$\sigma \cos^2 \phi = \frac{pa}{2h} \tag{3.2}$$

Similarly, in the hoop direction it is found that

$$\sigma \sin^2 \phi = \frac{pa}{h} \tag{3.3}$$

Combining (3.2) and (3.3) gives

$$\tan^2 \phi = 2 \tag{3.4}$$

and from the point of view of this simple analysis, the optimum winding angle is about $\phi \simeq 55°$, which is certainly not in the principal direction. The shear force shown in Fig. 6.6*c* is balanced by the oppositely directed shear force of the adjacent lamina, such that the net shear force over the entire thickness vanishes. One would simply apply enough layers at the helical winding angle so that the failure criterion is satisfied at a given value of pressure p.

The difficulty with the preceding simple analysis is that it suggests an optimal winding pattern for the cylindrical portion of the pressure vessel that is not conducive to an optimal winding pattern for the end domes. As a practical matter, vessels such as in Fig. 6.6 are first made with a very small winding angle ϕ, which allows both the cylindrical portion and the end domes to be formed simultaneously; then an overlay of hoop wraps is applied to the cylindrical portion of the vessel to provide the extra stiffness needed to resist the hoop stress (3.1). Thus, again, the practical procedure involves building up a laminate with the fibers running in the principal stress directions, or as near to it as is possible.*

Spherical Vessel

Thus far we have not been explicitly concerned with failure criteria. Let us conclude this section with an example that does involve a specific failure criterion. For simplicity, we consider the case of a spherical pressure vessel. With point symmetry the isotropic stress state is given by

$$\sigma = \frac{pa}{2h} \tag{3.5}$$

where again thin walled conditions are assumed with h being the thickness, a the radius of the vessel, and p the pressure. Clearly, the optimal fiber layup procedure involves a quasi-isotropic type of construction. It is convenient for us to know the conditions within an individual shell of aligned fibers. Specifically, the membrane strains in the vessel are also those in the individual aligned fiber shell. The appropriate plane stress relation between isotropic stress and isotropic strain is given by

$$\sigma = \frac{E}{1-\nu}\varepsilon \tag{3.6}$$

In the case where the fiber phase is very stiff compared with the matrix phase, from Section 4.4 the effective properties E and ν are given by

$$E \simeq \frac{cE_f}{3}$$

$$\nu \simeq \tfrac{1}{3} \tag{3.7}$$

where c is the volume fraction of the fiber phase. The combination of

*I am appreciative of M. A. Hamstad and E. M. Wu's helpful discussion of these matters.

(3.5)–(3.7) gives

$$\varepsilon = \frac{pa}{cE_f h} \tag{3.8}$$

With the strain known in the individual shell of aligned fibers, the failure criterion can now be applied.

At this point the specific failure criterion must be specified. If the criterion is gross failure of the fiber phase, then the allowable strain at failure for fibers ε_f can be used in (3.8) to solve for the appropriate thickness of the vessel h. Alternatively, if cracking of the resin phase is to be prevented, then a different allowable strain ε_m must be prescribed, and the thickness h determined. Typically, the strain at failure for a lamina in transverse tension is much smaller than the strain at failure for the lamina in longitudinal tension. Thus if resin cracking is to be prevented, ε_m, defined earlier, must be used in (3.8) to size h. Interestingly, this resin cracking criterion is often the most critical condition because cracking of the matrix phase provides a network for gaseous and liquid diffusion, which may have highly deleterious effects.

Hybrid Systems

As the last item to be mentioned here, we consider the possibility of mixing different types of fibers within a composite to achieve certain design objectives. These types of systems are called *hybrid systems*, and the rationale for their use is by no means clear cut. Obviously, one fiber phase has a lower strain at failure than the other fiber component. With regard to the spherical pressure vessel problem, this simply means that the fiber phase with the lower strain at failure fails first and failure of the other fiber phase follows. From the point of view of just this single criterion, obviously there is no rationale for the hybrid system, and whichever individual fiber performs better should be used by itself. However, when the design requirements involve more than one specification, the possibilities for hybrid composites begin to emerge. Particularly when material cost is one of the design criteria, small quantities of an expensive fiber can be justified in return for providing mechanical performance advantages.

6.4 DESIGN EXAMPLE II: FIBER REINFORCED FLYWHEELS

A flywheel can be used as a device for storing kinetic energy. For many practical purposes the objective is to store the maximum amount of kinetic energy per unit weight of the flywheel. At first thought this would seem to favor the most dense material, since kinetic energy is proportional to mass density. However, that would be an incomplete consideration. The maximum speed of the flywheel is determined by some failure criterion, typically expressed in terms of stresses due to centrifugal body force, which in turn is proportional to density. If the failure criterion was first order in stress, the energy stored at maximum speed would be independent of the density of the material. It then follows that the energy per unit mass would be inversely proportional to the density. We thus arrive at the somewhat surprising conclusion that the design criterion of maximizing the kinetic energy per unit mass may favor the use of the lightest possible material, so long as the material can tolerate reasonably high stresses at failure. To put this another way, for two materials that withstand the same stress at failure, the one that is the least dense provides the greatest energy stored per unit mass. There are many fiber reinforced materials that not only tolerate high stresses at failure, but are very light compared with metals. Among these light, high strength fiber composites we include those composites with carbon fibers, aramid fibers, and high quality glass fibers. Obviously, fiber composites offer attractive advantages for flywheel design. We now follow the flywheel design methodology given by Christensen and Wu [6.16].

For isotropic materials the optimal design is often argued to be that of the constant stress flywheel, in which the shape is tapered to give the same stresses, in the plane of the flywheel, at all radial positions. Thus, in principle, the material will fail simultaneously at all points in the medium, and this is argued to provide the greatest kinetic energy storage at incipient failure. We wish to proceed along the same lines in the case of fiber reinforced flywheels. We seek a design that provides material failure simultaneously at all points in the medium. Within this framework then, we seek to determine the degree of anisotropy and the shape of the flywheel that provide the greatest energy stored per unit mass.

Analysis

The problem is one of axial symmetry and accordingly we restrict attention to that of a cylindrically orthotropic material. Furthermore, we restrict attention to that of thin flywheels such that plane stress conditions can be assumed. We thus write the stress-strain relations as

$$\sigma_{rr} = Q_r \varepsilon_{rr} + Q_{r\theta} \varepsilon_{\theta\theta}$$

$$\sigma_{\theta\theta} = Q_{\theta r} \varepsilon_{rr} + Q_\theta \varepsilon_{\theta\theta} \tag{4.1}$$

where r and θ are polar coordinates having

$$Q_{\theta r}=Q_{r\theta}$$

and the Poisson's ratios are defined by

$$\nu_{r\theta}=\frac{Q_{r\theta}}{Q_r}$$

$$\nu_{\theta r}=\frac{Q_{\theta r}}{Q_\theta} \tag{4.2}$$

and r and θ are polar coordinates.

At this point we must specify the failure criterion. As discussed in Section 6.2, there are many different criteria we could employ. We here use what is probably the simplest possible, realistic failure criterion. Specifically, one assumes that throughout the flywheel

$$\varepsilon_{rr}(r)=\varepsilon_{\theta\theta}(r)=\varepsilon(\text{constant}) \tag{4.3}$$

The physical interpretation of this criterion is very simple. We are designing toward failure of the fiber phase, rather than the matrix phase. That is, by taking the value ε in (4.3) to be the strain level at which fibers fail in uniaxial tension, we ensure that in the flywheel the fibers in both the radial and circumferential directions will fail simultaneously. Following the rationale of the preceding section, the resin cracking would have began at a lower speed. Thus we are designing toward ultimate failure of the flywheel, rather than some condition of first damage.

Under plane stress conditions, the compatibility equation has the form

$$r\frac{d\varepsilon_{\theta\theta}}{dr}+\varepsilon_{\theta\theta}-\varepsilon_{rr}=0 \tag{4.4}$$

Obviously, this governing condition is satisfied by (4.3). It remains to satisfy the equilibrium equation, under steady state conditions, which has the form

$$\frac{d}{dr}(hr\sigma_{rr})-h\sigma_{\theta\theta}+\rho\omega^2hr^2=0 \tag{4.5}$$

where the thickness of the disk is an unknown function of radius

$$h=h(r)$$

ρ is the density, and ω the angular speed.

Combining (4.1)–(4.3) and (4.5) gives

$$\left(r\frac{dh}{dr}+h\right)Q_r(1+\nu_{r\theta})\varepsilon-hQ_\theta(1+\nu_{\theta r})\varepsilon+\rho\omega^2hr^2=0 \tag{4.6}$$

This equation can be put into the form

$$\frac{dh}{h}=-(1-\alpha)\frac{dr}{r}-\frac{\rho\omega^2r\,dr}{Q_r(1+\nu_{r\theta})\varepsilon} \tag{4.7}$$

where

$$\alpha=\frac{(1+\nu_{\theta r})Q_\theta}{(1+\nu_{r\theta})Q_r} \tag{4.8}$$

Equation (4.7) may be integrated to obtain

$$h=kr^{-(1-\alpha)}e^{-\lambda r^2} \tag{4.9}$$

where

$$\lambda=\frac{\rho\omega^2}{2Q_r(1+\nu_{r\theta})\varepsilon} \tag{4.10}$$

and k is the constant of integration.

The objective here is to maximize the kinetic energy stored per unit mass of the flywheel. The kinetic energy is given by

$$T=\tfrac{1}{2}\rho\omega^2\int\int h(r)r^2(r\,d\theta\,dr) \tag{4.11}$$

where the integral is over the plane of the disk. Similarly, the total mass is given by

$$M=\rho\int\int h(r)(r\,d\theta\,dr) \tag{4.12}$$

Forming the ratio of (4.11) and (4.12) gives

$$\frac{T}{M}=\frac{\omega^2}{2}\frac{\int hr^3\,dr}{\int hr\,dr} \tag{4.13}$$

where the polar integration has been performed. Substituting the thickness function, (4.9), into (4.13) gives

$$\frac{2}{\omega^2}\frac{T}{M}=\frac{\int_0^\infty r^{(2+\alpha)}e^{-\lambda r^2}\,dr}{\int_0^\infty r^\alpha e^{-\lambda r^2}\,dr} \tag{4.14}$$

The problem now is to evaluate these integrals.

A change of variable facilitates integral evaluation. Let

$$r^2=u \tag{4.15}$$

Inserting (4.15) into (4.14) yields

$$\frac{2}{\omega^2}\frac{T}{M}=\frac{\int_0^\infty u^{(1+\alpha)/2}e^{-\lambda u}\,du}{\int_0^\infty u^{(-1+\alpha)/2}e^{-\lambda u}\,du} \tag{4.16}$$

These integrals are the standard form for gamma functions, and (4.16) can be written as

$$\frac{2}{\omega^2}\frac{T}{M}=\frac{1}{\lambda}\frac{\Gamma\left[\frac{3}{2}+(\alpha/2)\right]}{\Gamma\left[\frac{1}{2}+(\alpha/2)\right]} \tag{4.17}$$

Now the following identity for gamma functions is useful:

$$\Gamma(z+1)=z\Gamma(z) \tag{4.18}$$

and (4.17) can be put into the form

$$\frac{2}{\omega^2}\frac{T}{M}=\frac{1}{2\lambda}(1+\alpha) \tag{4.19}$$

Next, inserting the definitions of α from Eq. (4.8) and λ from Eq. (4.10) into (4.19) gives

$$\frac{2\rho}{\varepsilon}\frac{T}{M}=(1+\nu_{r\theta})Q_r+(1+\nu_{\theta r})Q_\theta \tag{4.20}$$

Finally, (4.2) is inserted into (4.20) to produce

$$\frac{2\rho}{\varepsilon}\frac{T}{M}=Q_r+Q_\theta+2Q_{r\theta} \tag{4.21}$$

It is the right-hand side of (4.21) that is to be maximized by varying the degree of the anisotropy. The thickness variation has already been obtained through the restriction that the material fail simultaneously throughout the entire disk. An examination of the terms in (4.21) reveals that there is, in fact, no further optimization to be performed. The combination of terms $Q_r + Q_\theta + 2Q_{r\theta}$ is invariant with respect to coordinate rotation even though the individual terms vary depending on the orientation of an individual lamina relative to the given coordinate system. This property of invariance was derived in Section 5.1, Eq. (V-1.19); thus whether the entire reinforcement is radial, or circumferential, or any combination, the same maximum energy density is still attainable. Of course, to achieve this maximum energy density, the disk must have the thickness variation dictated by (4.9) and very different shapes are involved, depending on the type of anisotropy involved.

Equation (4.21) can be put into a form using more familiar properties:

$$Q_r = \frac{E_r}{1 - \nu_{r\theta}\nu_{\theta r}}$$

$$Q_\theta = \frac{E_\theta}{1 - \nu_{r\theta}\nu_{\theta r}}$$

$$Q_{r\theta} = \frac{\nu_{\theta r} E_r}{1 - \nu_{r\theta}\nu_{\theta r}} \tag{4.22}$$

with E_r and E_θ representing the corresponding modulus measures. Inserting (4.22) into (4.21) yields

$$\frac{2\rho}{\varepsilon}\frac{T}{M} = \left[\frac{1 + (\nu_{\theta r}/\nu_{r\theta}) + 2\nu_{\theta r}}{1 - \nu_{r\theta}\nu_{\theta r}}\right] E_r \tag{4.23}$$

The corresponding isotropic material results can be extracted from (4.23) by letting

$$\nu_{r\theta} = \nu_{\theta r} = \nu$$

$$E_r = E$$

with E and ν being the corresponding isotropic properties. Thus for the isotropic (I) case, the result is

$$\left(\frac{T}{M}\right)_{\mathrm{I}} = \frac{E\varepsilon}{(1-\nu)\rho} \tag{4.24}$$

Evaluation of Results

A first order evaluation of the results just obtained is given by considering limiting cases. The most obvious limiting case is that of a fiber phase material in which the fibers are so stiff in comparison with the matrix phase that the latter contribution can be neglected.

With a matrix phase of negligible shear modulus, $\mu_m = 0$, and also using $\nu_r = \nu_m$ there results, for the case of radial reinforcement,

$$
\begin{aligned}
E_\theta &= 0 \\
E_r &= cE_f \\
\nu_{r\theta} &= \nu_f \\
\nu_{\theta r} &= 0
\end{aligned}
\tag{4.25}
$$

where c is the volume fraction of the fiber phase and the subscripts f and m refer to fiber and matrix phases, respectively. Combining (4.25) with (4.23) produces

$$\frac{\rho}{\varepsilon}\frac{T}{M} = \tfrac{1}{2}(cE_f) \tag{4.26}$$

With vanishing matrix phase modulus, the two-dimensional random fiber orientation scheme gives

$$
\begin{aligned}
E &= \tfrac{1}{3}cE_f \\
\nu &= \tfrac{1}{3}
\end{aligned}
\tag{4.27}
$$

These results follow from Section 4.4. Inserting (4.27) into (4.24) yields for the isotropic results the same form, (4.26), which, of course, results from the invariant property of the right-hand side of (4.21). It is important to note that (4.26) is exactly the same as that obtained from the analysis of a rotating ring.

Design Considerations

The final results of the optimization analysis are both (4.23), which gives the kinetic energy stored per unit mass as a function of the material properties, and the corresponding shape solution, (4.9), which also depends on the material properties. A remarkable result emerges in the derivation.

As long as the appropriate optimum shape is used for a given ratio of radial to circumferential reinforcement, the corresponding energy storage capacity is the same for the entire spectrum of optimum shapes. The next problem, then, is how to select the "best" shape with the corresponding type of anisotropy. The answer to this question lies in the realm of practical design. Consider the shape solution of (4.9). For $\alpha<1$, radial reinforcement dominates, and it is seen that the thickness solution is singular at $r=0$. On the other hand, for $\alpha>1$, circumferential reinforcement dominates and from (4.9) it is seen that at $r=0, h=0$. It is only in the case of $\alpha=1$, which corresponds to equal radial and circumferential stiffness, that there is not an unappealing and, in fact, an impossible design condition at $r=0$. In the case of predominant circumferential reinforcement, it is interesting to note from (4.9) that the optimal shape involves a section that thickens with increasing radius near the origin but has a maximum thickness at some determinable radius and thereafter tapers to vanishing thickness. Obviously, the case of equal radial and circumferential reinforcement is the most practical configuration.

A flywheel could be constructed with equal radial and circumferential reinforcement. This certainly is more general than the isotropic case. However, in the limiting case results of the preceding section, it was shown that identical results emerge for the energy storage whether the system is isotropic or just of equal stiffness in the radial and circumferential directions. (The latter case is orthotropic but not, in general, isotropic.)

One question to be raised at this point is whether the fiber reinforced flywheel, with fibers arranged to give an effectively isotropic property characterization, has any advantage over a ring type fiber reinforced flywheel. The limiting case for fiber effects, shown in the preceding section, revealed these two configurations to give the same energy storage. First, the solid wheel has a design advantage over the ring: it is more efficient in space use, having less swept volume. Second, it is likely to be more efficient and stable under conditions of acceleration and deceleration because the attachment of the ring type wheel to spokes and hubs causes problems.

The most basic question to be answered here is which configuration uses matrix properties in the most efficient manner. (The comparisons of the preceding section neglected matrix properties.) To answer this question we next examine a flywheel made by laminating individual uniaxial laminae at various angles to obtain effective isotropic properties. An asymptotic estimate of the resulting isotropic modulus was derived in Section 4.4 as Eq. (IV-4.40).

$$E=\frac{c}{3}E_f+\frac{1-c}{3}E_m+\frac{19}{27}\frac{(1+c)E_f+(1-c)E_m}{(1-c)E_f+(1+c)E_m}E_m \tag{4.28}$$

Insertion of (4.28) into (4.24) gives a prediction of the energy storage capacity in terms of both fiber and matrix phase properties. Equation (4.28) has been evaluated for aramid fiber, epoxy matrix systems in two separate cases. For a fiber volume fraction of $c=\frac{2}{3}$, the results are as follows:

$$\frac{E}{E_f}=0.24 \qquad \text{for} \qquad \frac{E_m}{E_f}=\frac{1}{190}$$

$$\frac{E}{E_f}=0.31 \qquad \text{for} \qquad \frac{E_m}{E_f}=\frac{5}{190}$$

The former result corresponds to a very flexible epoxy resin whereas the latter result is for a stiff epoxy resin. For the fiber case alone,

$$\frac{E}{E_f}=0.22 \qquad \text{for} \qquad \frac{E_m}{E_f}=0$$

The matrix phase is seen to provide a considerable stiffening effect in the case of the stiff epoxy. These results should be compared with ring behavior, which is governed by the rule of mixtures. According to the rule of mixtures, the matrix phase increases the stiffness by 1.3% over that of fibers alone in the case of $E_m/E_f=5/190$. Compare this result with the 38% improvement shown by the data for the case of the isotropic solid disk design. A caveat must be inserted at this point. For a laminated system producing isotropic properties and using a stiff matrix phase, the matrix phase would fail before the fiber phase; thus the fiber phase could not be used to full capacity. However, this situation can be reversed by going to a more flexible matrix phase, such as the one in the first example, in which $E_m/E_f=1/190$. In this case the matrix phase still provides for more reinforcement than does the matrix phase in the ring configuration.

There are many other design approaches to flywheels, the most common of which is the circumferentially wound ring, as already mentioned. The present design example, however, serves well to illustrate a rational approach to design using composite materials. There are many challenging problems in the field of design with composite materials.

PROBLEMS

1. Can the boundary layer theory of Section 6.1 be used to solve the crack problems shown in Figs. 6.7*a* and *b*?

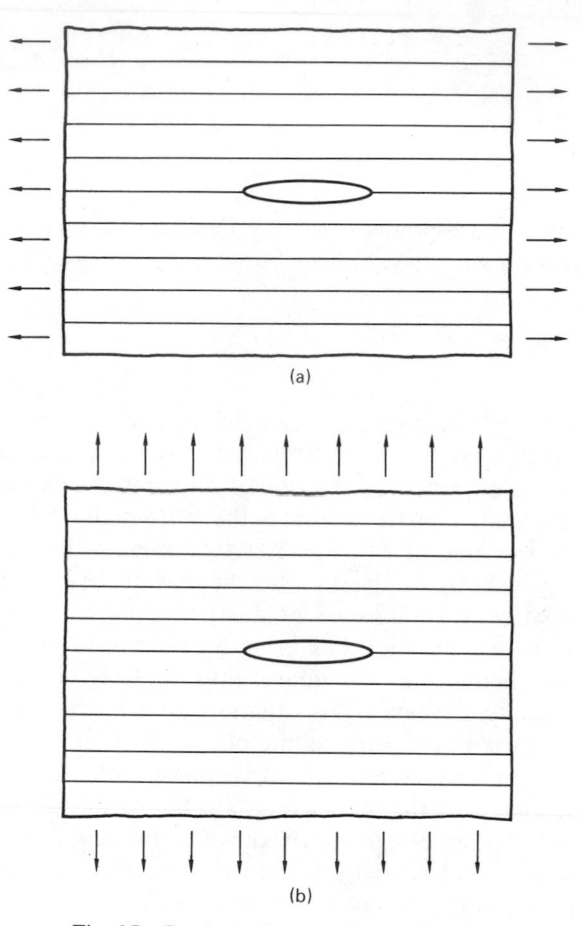

Fig. 6.7 Crack problems under simple tension.

2. In the boundary layer theory of Section 6.1, the fiber phase is assumed to be very stiff compared with the matrix phase properties. Derive the appropriate form of the linear theory governing differential equations when the fiber lines are inextensible (rigid fibers).
3. With reference to the failure criteria work of Section 6.2, derive expressions (2.10) from (2.9).

4. Would two-dimensional failure criteria under plane strain conditions be expected to be different from those of plane stress conditions?
5. Assume the condition of single-size cylindrical fibers in a hexagonal packing arrangement. Assume the fibers are perfectly rigid compared with the matrix phase. Use the basic geometry of the problem to derive the strain concentration factor in terms of the volume fraction of the fiber phase, when the composite is in a state of transverse tension.
6. Discuss the microscale effects that can contribute to the failure of composite materials. Include aspects of interfacial bond failure, load transfer around broken fibers, and so on.
7. Discuss the methods to be followed to account for stress concentration effects at cutouts in composite material pressure vessels. Are there aspects to the problem that are special to composite materials, over and above the considerations for homogeneous, isotropic materials?
8. In the flywheel design procedure of Section 6.4, we assumed properties that are independent of radial position. Is there anything to be gained by letting the properties depend on radial position?

REFERENCES

6.1 S. G. Lekhnitskii, *Theory of Elasticity of an Anisotropic Elastic Body*, Holden-Day, San Francisco, 1963.

6.2 R. F. S. Hearmon, *An Introduction to Applied Anisotropic Elasticity*, Oxford University Press, New York, 1961.

6.3 G. C. Everstine and A. C. Pipkin, "Stress channelling in transversely isotropic elastic composites," *ZAMP*, vol. 22, 825 (1971).

6.4 A. J. M. Spencer, "Boundary layers in highly anisotropic plane elasticity," *Int. J. Solids Structures*, vol. 10, 1103 (1974).

6.5 G. C. Everstine and A. C. Pipkin, "Boundary layers in fiber-reinforced materials," *J. Appl. Mech.*, vol. 40, 518 (1973).

6.6 C. W. Bert, "Models for fibrous composites with different properties in tension and compression," *J. Eng. Mater. Technol.*, ASME, vol. 99, 344 (1977).

6.7 R. Hill, "A theory of the yielding and plastic flow of anisotropic metals," *Pro. R. Soc.*, vol. A193, 281 (1948).

6.8 S. W. Tsai and E. M. Wu, "A general theory of strength for anisotropic materials," *J. Comp. Mater.*, vol. 5, 58 (1971).

6.9 E. M. Wu, "Phenomenological anisotropic failure criterion," in *composite materials*, vol. 2, G. P. Sendeckyj, Ed. Academic, New York, 1974.

6.10 R. C. Tennyson, D. MacDonald, and A. P. Nanyaro, "Evaluation of the tensor polynomial failure criterion for composite materials," *J. Comp. Mater.*, vol. 12, 63 (1978).

6.11 H. E. Daniels, "The statistical theory of the strength of bundles of threads. I" *Proc. R. Soc.*, vol. A183, 405 (1945).

6.12 B. D. Coleman, "Statistics and time dependence of mechanical breakdown in fibers," *J. Appl. Phys.*, vol. 19, 968 (1958).

6.13 C. Zweben and B. W. Rosen, "A statistical theory of material strength with application to composite materials," *J. Mech. Phys. Solids*, vol. 18, 189 (1970).

6.14 D. G. Harlow and S. L. Phoenix, "The chain-of-bundles probability model for the strength of fibrous materials I: analysis and conjectures," *J. Comp. Mater.* vol. 12, 195 (1978).

6.15 H. E. Brandmaier, "Optimum filament orientation criteria," *J. Comp. Mater.* vol. 4, 422 (1970).

6.16 R. M. Christensen and E. M. Wu, "Optimal design of anisotropic (fiber-reinforced) flywheels," *J. Comp. Mater.*, vol. 11, 395 (1977).

CHAPTER VII
WAVE PROPAGATION

Composite materials are often used in situations involving the sudden application of loads. The dynamic response of the structure ensues after load application, and a state of stress may be generated that leads to failure. It is necessary to understand the response characteristics of the material body to account properly for all important effects.

The dynamic response of deformable heterogeneous materials can be broadly classified into one of two groupings. If the wave length of the characteristic response of the material is very long compared with the scale of the inhomogeneity, then the material response is governed by the effective properties of the equivalent homogeneous medium. In this case the methods of structural response and wave propagation are identical to those of homogeneous materials. We are only briefly concerned with this situation in the next section.

In the second grouping of response classification we have the situation in which the wave length of the response is not ideally long with respect to the characteristic dimension of the inhomogeneity. In this situation very complicated dynamic effects can occur. The interfaces between material phases cause wave reflection and refraction. It would be very difficult to model wave propagation retaining full information on all wave effects at all interfaces in typical composite materials. For this reason, highly idealized geometric models are employed to study basic effects. We are concerned here with several analyses of this type.

The geometric model of most common application is that of periodically layered elastic media. We examine several aspects of wave response for this type of media, as being representative of the behavior of more complicated geometric models. Nevertheless, one analysis is given of the full response of a three-dimensional medium with discrete inclusions.

Perfectly periodic types of heterogeneity are assumed in the layered media problems and in the three-dimensional discrete inclusion problem. In Section 7.6 we relax the restriction of periodicity, and consider a certain type of randomly heterogeneous media. In this case we find a basic attenuation effect due to incoherent wave scattering, an effect absent in periodic type media. Finally, in the last section we study a dynamic instability problem as an application of mixture theory.

7.1 WAVE CHARACTER IN EQUIVALENT HOMOGENEOUS MEDIA

We now consider wave propagation in heterogeneous elastic media under conditions of very long wave lengths. With this wave length restriction, we can assume that the medium responds as an equivalent homogeneous medium. The infinite wave length behavior corresponds to effectively static behavior. The governing properties are the effective moduli C_{ijkl} in

$$\sigma_{ij} = C_{ijkl}\varepsilon_{kl} \tag{1.1}$$

At this point we do not restrict the type of anisotropy that characterizes the heterogeneous material. The only restrictions are the usual ones on C_{ijkl},

$$C_{ijkl} = C_{jikl} = C_{ijlk} = C_{klij}$$

The equations of motion are given by

$$\sigma_{ij,j} = \rho\frac{\partial^2 u_i}{\partial t^2} \tag{1.2}$$

where ρ is the volumetrically averaged mean density of the material. The propagation of time harmonic waves will be analyzed. The displacement form for a plane harmonic wave is given by

$$u_i = A_i \exp\left[i\omega\left(\frac{n_i x_i}{v} - t\right)\right] \tag{1.3}$$

where ω is the real, circular frequency, $\mathbf{n}$ is the unit vector giving the direction of propagation, A is the wave amplitude, and v is the velocity of propagation.

Using the strain displacement relation $\varepsilon_{ij} = \frac{1}{2}(u_{i,j} + u_{j,i})$ in (1.1) and that result in (1.2) along with u_i from (1.3) gives

$$\left[\frac{C_{ijkl}n_j n_l}{v^2} - \rho\delta_{ik}\right]A_k = 0 \tag{1.4}$$

For a nontrivial solution of Eqs. (1.4) the determinant of the coefficients must vanish giving

$$\left|\frac{C_{ijkl}n_j n_l}{v^2} - \rho\delta_{ik}\right| = 0 \tag{1.5}$$

For a given direction of propagation $\mathbf{n}$, relation (1.5) leads to a cubic equation in the velocity of propagation v. In general, there are three distinct roots for $v^{(k)}$ and the three corresponding displacement vectors, $A_i^{(k)}$, which are orthogonal. For an isotropic material only two of the roots $v^{(k)}$ are distinct and these are the phase velocities for longitudinal and transverse waves, as

$$v^{(1)} = \left(\frac{\lambda + 2\mu}{\rho}\right)^{1/2}$$

$$v^{(2)} = v^{(3)} = \left(\frac{\mu}{\rho}\right)^{1/2}$$

In the anisotropic case the displacements cannot be identified as being longitudinal or transverse except when the direction of propagation corresponds to a symmetry direction for the material.

The fact that the velocities of the three waves depend on the direction of propagation is a great complication of the behavior of anisotropic materials. It also follows that similar effects occur in two-dimensional problems such as the propagation of flexural waves in anisotropic plates. The directional dependency of the flexural wave velocity means that there is no symmetry of response, even when the applied loads themselves are of an axisymmetric nature, as in the impact of an object on a plate.

The essential feature of the present solution is that the wave velocities $v^{(k)}$ are independent of frequency ω. Thus for plane waves a pulse shape, composed of a spectrum of frequencies, can propagate without distortion of the shape of the pulse. When we relax the very long wave length restriction, in heterogeneous media, we find that the velocity of propagation depends on the frequency of the wave. This is the effect of dispersion, and we are intimately concerned with this fundamental effect in the following sections.

In relaxing the long wave length restriction, we begin by going to the opposite extreme by considering short waves in the next section.

7.2 TRANSMISSION AND REFLECTION IN LAYERED MEDIA

In the preceding section we deduced the harmonic wave characteristics in anisotropic, but effectively homogeneous media. The wave speed under very long wave length conditions is governed by the effective moduli of the medium and the average density. Let us begin here by explicitly determining this wave speed in a particular case.

Wave Speeds

Take the case of periodically layered elastic media as shown in Fig. 7.1. The Lamé constants of the two materials are as shown in the figure. We consider plane waves propagating normal to the direction of the layering. Because of the symmetry of the medium, there will only be two distinct wave speeds, that of longitudinal waves and that of transverse (shear) waves. The medium is naturally arranged as a series model; thus the effective static properties for longitudinal and shear waves are given by

$$\lambda+2\mu=\frac{h}{h_1/(\lambda_1+2\mu_1)+h_2/(\lambda_2+2\mu_2)}$$

and

$$\mu=\frac{h}{h_1/\mu_1+h_2/\mu_2} \tag{2.1}$$

These, of course, are just two of the five independent properties that govern the transversely isotropic media of Fig. 7.1. The full set of the five effective properties were given in Section 4.5.

The wave speeds corresponding to (2.1) are

$$v_{\text{LONG}}^2=\frac{h^2}{[h_1/(\lambda_1+2\mu_1)+h_2/(\lambda_2+2\mu_2)](h_1\rho_1+h_2\rho_2)}$$

and

$$v_{\text{TRANS}}^2=\frac{h^2}{(h_1/\mu_1+h_2/\mu_2)(h_1\rho_1+h_2\rho_2)} \tag{2.2}$$

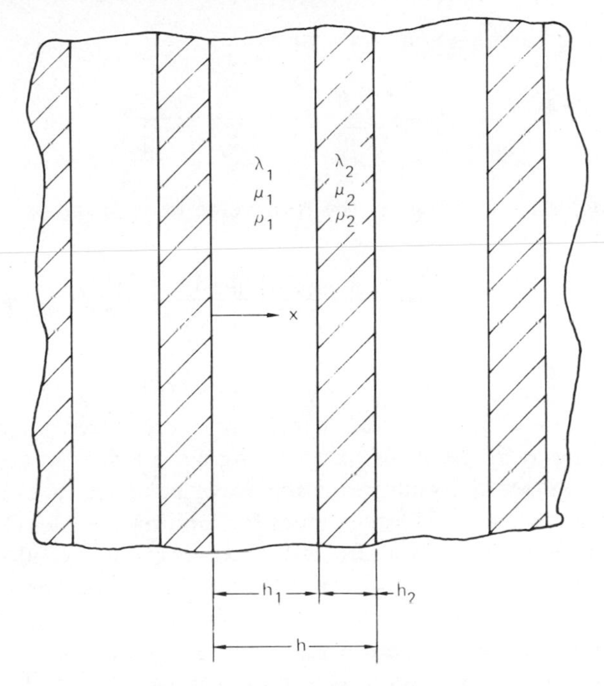

Fig. 7.1 Layered elastic media.

where the mean density is $(h_1\rho_1 + h_2\rho_2)/h$. These wave speeds govern the behavior of (infinitely) long waves in the problem of interest here.

In homogeneous media we know that phase velocities are just given by $\sqrt{(\lambda+2\mu)/\rho}$ and $\sqrt{\mu/\rho}$ for longitudinal and transverse waves, respectively. Let us use this information to calculate the corresponding wave speed in the medium of Fig. 7.1. Of course, we know that part of any wave striking the interface must be reflected. At this point then we are concerned just with the wave speed of the part of the wave that is transmitted through the interface. We return to the reflection problem a little later. The wave speed, which we calculate by knowledge of the wave speeds in the two materials, thus gives the maximum speed at which a transient disturbance can propagate through the heterogeneous medium.

First, we consider shear waves. Write the wave speed as

$$v_{\text{TRANS}}^{\text{M}} = \frac{h_1 + h_2}{\Delta t} \tag{2.3}$$

where Δt is the transit time of the wave across thickness h. Knowing the wave speed in each material gives Δt as

$$\Delta t = \frac{h_1}{\sqrt{\mu_1/\rho_1}} + \frac{h_2}{\sqrt{\mu_2/\rho_2}} \tag{2.4}$$

Substituting (2.4) in (2.3) gives the transverse wave speed as

$$v_{\text{TRANS}}^{\text{M}} = \frac{\sqrt{\mu_1 \mu_2/\rho_1\rho_2}\,(h_1+h_2)}{h_1\sqrt{\mu_2/\rho_2} + h_2\sqrt{\mu_1/\rho_1}} \tag{2.5}$$

Similar results follow for longitudinal waves. It is easy to prove that the wave speed in (2.5) is greater than the wave speed v_{TRANS} in (2.2) corresponding to the behavior of very long wave length waves.

We begin to see the outlines of the wave behavior. Wave speed must depend on wave length, with the values for limiting wave lengths of infinite and zero length given by (2.2) and (2.5), respectively, along with the corresponding forms for v_{LONG}. We do not know the behavior of wave velocity v as a function of wave length, except in these limiting cases. In the next section we are concerned with deducing wave velocity as a function of wave length. At this point we can recognize that a transient disturbance, composed of all wave lengths, must have a very complicated behavior. Despite this great complication, we shall succeed in unraveling the actual transient behavior in some practical cases.

Wave Reflections

Next we turn to the problem of wave reflections at interfaces. Consider the interface between material 1 and 2 with the wave going from 1 to 2. The continuity of stress and displacement at the interface requires

$$\begin{aligned} \sigma_I + \sigma_R &= \sigma_T \\ u_I + u_R &= u_T \end{aligned} \tag{2.6}$$

where subscripts I refers to the incoming wave, R to the reflected wave, and T to the transmitted wave. The balance of momentum in both layers requires

$$\frac{\partial \sigma_n}{\partial x} = \rho_n \frac{\partial^2 u_n}{\partial t^2}, \qquad n = 1, 2 \tag{2.7}$$

where x is the direction of propagation of the longitudinal or transverse wave.

Take forms of solutions for particle velocity and stress as

$$\dot{u}_n = Af(x \pm v_n t)$$

$$\sigma_n = Bf(x \pm v_n t) \tag{2.8}$$

where A and B are amplitudes and v_n are the phase velocities. Substitution of (2.8) into (2.7) gives

$$Bf'(x \pm v_n t) = \rho_n v_n A f'(x \pm v_n t) \tag{2.9}$$

where the prime denotes a derivative. Relation (2.9) gives

$$\sigma_n = \rho_n v_n \dot{u}_n \tag{2.10}$$

Writing the second of (2.6) in terms of velocities and using velocities from (2.10) then yields

$$\frac{\sigma_R}{\rho_1 v_1} - \frac{\sigma_T}{\rho_2 v_2} = -\frac{\sigma_I}{\rho_1 v_1} \tag{2.11}$$

Combining this equation with the first of (2.6), and solving for σ_R and σ_T gives, in the case of shear waves,

$$\frac{\sigma_R}{\sigma_I} = \frac{\sqrt{\rho_1 \mu_1} - \sqrt{\rho_2 \mu_2}}{\sqrt{\rho_1 \mu_1} + \sqrt{\rho_2 \mu_2}}$$

$$\frac{\sigma_T}{\sigma_I} = \frac{2\sqrt{\rho_1 \mu_1}}{\sqrt{\rho_1 \mu_1} + \sqrt{\rho_2 \mu_2}} \tag{2.12}$$

Similar results follow for longitudinal waves.

The terms of the form $\sqrt{\rho_n \mu_n}$ are known as impedances. When the impedances of the two layers are matched, there is no wave reflection, and the entire wave is transmitted through the interface. It is the impedance mismatch that partially blocks the passage of waves and causes multiple reflections.

The scale of observation here is focused on the mechanical effects occurring in each layer. In solving a general transient problem, we could conceivably keep account of each and every wave reflection and transmission at each and every interface. The job is staggeringly large, but it has been done with the aid of digital computers, up to a certain number of interfaces. However, as the number of interfaces is further increased, numerical accuracy becomes a problem. Fortunately, there is a more elegant approach to the problem, as we see in the next two sections.

7.3 DISPERSION RELATIONS

We now take a radically different point of view from that of the preceding section. We consider the passage of time harmonic waves through layered media. In so doing, we obtain the dependence of wave velocity on wave length, mentioned in the preceding section. These types of problems have long been of importance in geophysics. Rytov [7.1] gave an early account of the basic effects. The presentation here follows that of Sun, Achenbach, and Herrmann [7.2] and Brekhovskikh [7.3].

The equations of motion are written as

$$G_n \frac{\partial^2 u}{\partial x^2} = \rho_n \frac{\partial^2 u}{\partial t^2}, \qquad n = 1,2 \tag{3.1}$$

where in the case of longitudinal waves, $G_n = \lambda_n + 2\mu_n$ and for transverse waves, $G_n = \mu_n$, and displacement u is in the direction of wave motion for longitudinal waves and normal to it for transverse waves. We first consider the case of waves propagating in a direction normal to that of the layering.

Wave Direction Normal to Layers

The media characterization of Fig. 7.1 still applies and we are now considering the transmission of plane harmonic waves in the x direction, normal to the interfacial planes. We seek a traveling wave solution in the form

$$u_n(x,t) = U_n(x) \exp\left[i\omega\left(t + \frac{x}{v} \right) \right] \tag{3.2}$$

which we can write as

$$u_n(x,t) = U_n(x) \exp[\, i(\omega t + kx)\,] \tag{3.3}$$

where ω is the circular frequency, taken to be real, v is the wave velocity, k is the wave number, and $U_n(x)$ is taken to be strictly periodic with

$$U_n(x+h) = U_n(x)$$

The frequency, wave speed, and wave number are interrelated by

$$k = \frac{\omega}{v} \tag{3.4}$$

The form of the solution (3.2) or (3.3), with $U_n(x)$ being periodic, is that appropriate to the type of wave behavior known as Floquet theory. We now find the function $U_n(x)$ to satisfy the equation of motion.

We know the solution of (3.1) is given by terms of the type

$$\exp\left[i\omega\left(t \pm \frac{x}{v_n}\right)\right]$$

where v_n are the phase velocities in the two materials. Thus we take $U_n(x)$ in the form

$$\exp\left[i\omega\left(-\frac{x}{v} \pm \frac{x}{v_n}\right)\right]$$

and we have the general solution of $U_n(x)$ as

$$U_n(x) = A_n \exp(i\alpha_1 x) + B_n \exp(i\alpha_2 x) \tag{3.5}$$

where

$$\alpha_1 = -\frac{\omega}{v} + \frac{\omega}{v_n}$$

$$\alpha_2 = -\frac{\omega}{v} - \frac{\omega}{v_n} \tag{3.6}$$

and A_n and B_n are complex constants.

Next we must require that (3.5) be periodic in x, as

$$U_1(x)|_{x=h_1} = U_2|_{x=-h_2}$$

which gives

$$A_1 \exp\left[i\omega h_1\left(-\frac{1}{v} + \frac{1}{v_1}\right)\right] + B_1 \exp\left[i\omega h_1\left(-\frac{1}{v} - \frac{1}{v_1}\right)\right]$$

$$= A_2 \exp\left[-i\omega h_2\left(-\frac{1}{v} + \frac{1}{v_2}\right)\right] + B_2 \exp\left[i\omega h_2\left(\frac{1}{v} + \frac{1}{v_2}\right)\right] \tag{3.7}$$

Similarly, the periodicity of stress leads to the condition

$$\frac{G_1}{v_1}A_1\exp\left[i\omega\left(-\frac{1}{v}+\frac{1}{v_1}\right)h_1\right]+\frac{G_2}{v_2}B_1\exp\left[i\omega\left(-\frac{1}{v}-\frac{1}{v_1}\right)h_1\right]$$
$$=\frac{G_2}{v_2}A_2\exp\left[-i\omega\left(-\frac{1}{v}+\frac{1}{v_2}\right)h_2\right]+\frac{G_2}{v_2}B_2\exp\left[i\omega\left(\frac{1}{v}+\frac{1}{v_2}\right)h_2\right] \tag{3.8}$$

The continuity of displacement and stress at $x=0$ gives

$$A_1+B_1=A_2+B_2 \tag{3.9}$$

and

$$\frac{G_1}{v_1}A_1+\frac{G_1}{v_1}B_1=\frac{G_2}{v_2}A_2+\frac{G_2}{v_2}B_2 \tag{3.10}$$

For a nontrivial solution of (3.7)–(3.10) to exist, it is necessary that the determinant of the coefficients vanish. This procedure gives the characteristic equation

$$\cos\frac{\omega h}{v}=\cos\lambda_1\cos\lambda_2-\chi\sin\lambda_1\sin\lambda_2 \tag{3.11}$$

where

$$\lambda_n=\frac{\omega h_n}{v_n}$$
$$\chi=\frac{\rho_1 G_1+\rho_2 G_2}{2\sqrt{\rho_1\rho_2 G_1 G_2}} \tag{3.12}$$

Using the wave number from (3.4), (3.11) can be written in the form

$$\cos kh=\cos\frac{vkh_1}{v_1}\cos\frac{vkh_2}{v_2}-\chi\sin\frac{vkh_1}{v_1}\sin\frac{vkh_2}{v_2} \tag{3.13}$$

The solution of (3.11) gives wave velocity as a function of frequency ω, and (3.13) determines wave velocity as a function of wave number k. The wave length is just $2\pi/k$; thus we would thereby know the wave velocity as a function of wave length.

At this point we must verify that the wave speed from (3.11) at vanishing frequency, $\omega=0$, corresponds to the results (2.2) deduced from the effective static properties of the media. Also, we wish to deduce the derivative

$$\left.\frac{d^2v}{d\omega^2}\right|_{\omega=0}$$

to have a measure of the variation of wave speed v in the low frequency range. To obtain these results, expand the terms in (3.11) as power series in ω. Let v have the expansion

$$v=v_0+\frac{\alpha}{2}\omega^2+\cdots \tag{3.14}$$

To terms of $O(\omega^2)$ (3.11) gives

$$\frac{\omega^2h^2}{v_0^2}=\frac{\omega^2h_1^2}{v_1^2}+\frac{\omega^2h_2^2}{v_2^2}+2\chi\frac{\omega^2h_1h_1}{v_1v_2} \tag{3.15}$$

From this relation it may be shown that v_0 is identical with the expressions in (2.2) in the respective cases of longitudinal and transverse waves. We write the result symbolically as

$$v_0^2=\frac{G_{\text{STATIC}}}{\rho_{\text{MEAN}}}$$

Next, retaining terms in (3.11) up to $O(\omega^4)$ and using the result (3.15) gives α in (3.14) as

$$\alpha=\frac{v_0^3h_1^2h_2^2}{3h^2v_1^2v_2^2}(1-\chi^2) \tag{3.16}$$

From (3.12) it can be shown that $\chi\geqslant 1$, thus (3.14) and (3.16) show that on a plot of wave speed versus frequency, the wave speed decreases with increasing frequency in the low frequency range.

Relations such as (3.11) and (3.13) relating wave speed to frequency or wave number are known as *dispersion relations*. The physical character of wave speed as a function of wave length is said to be of a dispersive nature. That is, pulse shapes cannot be propagated with unchanged shape, as in homogeneous elasticity, because the waves of various wave lengths are dispersed as the pulse progresses. An example of the dispersion curves are given after we first obtain the corresponding results for waves propagating in the direction of the layering.

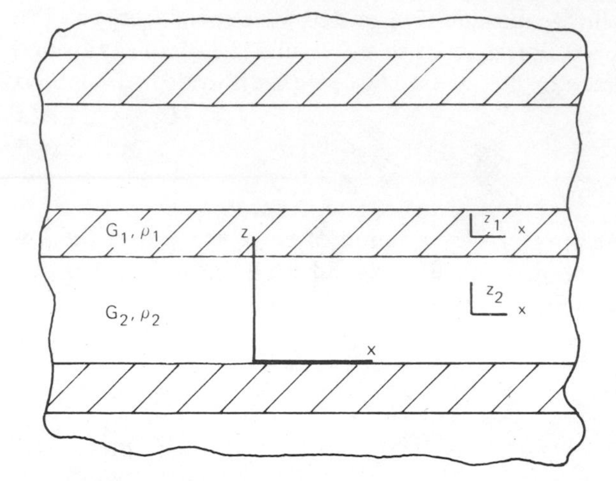

Fig. 7.2 Layered media coordinate convention for waves traveling in direction of layering.

Wave Direction Along Layers

Now we take the wave direction to be rotated by angle $\pi/2$ from that shown in Fig. 7.1, such that the wave moves in the direction of the layering. The coordinate system is as shown in Fig. 7.2. Because the coordinate describing the layer properties is different from the coordinate of propagation, this case is a little more complicated than the preceding case. Nevertheless, we obtain comparable results.

Apply the Helmholtz decomposition to write the displacement vector as

$$\mathbf{u} = \nabla\phi + \nabla \times \boldsymbol{\psi} \tag{3.17}$$

where ∇ is the gradient operator. The equations of motion for homogeneous media are satisfied if the scalar and vector potentials satisfy

$$\nabla^2\phi = \frac{\rho}{\lambda + 2\mu}\ddot{\phi}$$

$$\nabla^2\boldsymbol{\psi} = \frac{\rho}{\mu}\ddot{\boldsymbol{\psi}} \tag{3.18}$$

We must consider three separate cases: (1) horizontally polarized shear waves, (2) vertically polarized shear waves, and (3) longitudinal waves. By a horizontally polarized shear wave, we mean one with shear stress components in the x,y plane, Figure 2.2, whereas a vertically polarized shear wave has a shear stress component in the x,z plane.

First we give attention to horizontally polarized shear waves. Let u_y be the only nonzero displacement component. Then

$$u_y = u_y(x,z,t)$$

and we let $\phi=0$ in (3.18) with ψ_z, the only nonzero component of ψ. Taking the derivative of the resulting equation for ψ_z gives

$$\nabla^2 u_y = \frac{\rho}{\mu}\ddot{u}_y \tag{3.19}$$

Although we could have deduced this equation directly, we used the vector potential to illustrate its use. It will be necessary for us to use potential characterizations shortly.

Take coordinates at the center of each layer as shown in Fig. 7.2. The harmonic wave solution of (3.19) for each layer is given by

$$u_y^{(n)} = [A_n \sin\eta_n z_n + B_n \cos\eta_n z_n]\exp[ik(x+vt)], \qquad n=1,2$$

where

$$\eta_n^2 = k_n^2 - k^2 \tag{3.20}$$

with $k_n = \omega/v_n$. For symmetric motion about $z_n=0$ we have $A_n=0$, whereas for antisymmetric motion we have $B_n=0$.

For symmetric motion the periodicity requirement on displacement (and stress)

$$u_y^{(1)}|_{z_1=h_1/2} = u_y^{(2)}|_{z_2=-h_2/2}$$

is automatically satisfied. The continuity requirements on displacement and stress at the interface $z_1=-h_1/2$, $z_2=h_2/2$ give the characteristic equation as

$$\mu_1\left(\frac{k_1^2}{k^2}-1\right)^{1/2}\tan\left(\frac{\eta_1 h_1}{2}\right)+\mu_2\left(\frac{k_2^2}{k^2}-1\right)^{1/2}\tan\left(\frac{\eta_2 h_2}{2}\right)=0 \tag{3.21}$$

Similarly, for the antisymmetrical terms, it is found that

$$\mu_2\left(\frac{k_2^2}{k^2}-1\right)^{1/2}\tan\left(\frac{\eta_1 h_1}{2}\right)+\mu_1\left(\frac{k_1^2}{k^2}-1\right)^{1/2}\tan\left(\frac{\eta_2 h_2}{2}\right)=0 \tag{3.22}$$

These two relations (3.21) and (3.22) give the relationship between frequency ω and wave number k, for the types of waves specified. Using $\omega = kv$, these results can be converted to wave speed versus wave number or frequency.

Next we consider longitudinal and vertically polarized shear waves. Take the displacement components as

$$u_x = u_x(x,z,t)$$

$$u_y = 0$$

$$u_z = u_z(x,z,t) \tag{3.23}$$

The scalar potential $\phi(x,z)$ in (3.17) has components u_x and u_z, as does the vector component $\psi_y(x,z)$ in ψ. Using the first expression of (3.18) it follows that

$$\nabla^2 u_x = \frac{\rho}{\lambda + 2\mu} \ddot{u}_x$$

and

$$\nabla^2 u_z = \frac{\rho}{\lambda + 2\mu} \ddot{u}_z \tag{3.24}$$

for dilatational waves specified by

$$\frac{\partial u_x}{\partial z} + \frac{\partial u_z}{\partial x} = 0 \tag{3.25}$$

From the second expression of (3.18) it follows that

$$\nabla^2 u_x = \frac{\rho}{\mu} \ddot{u}_x$$

$$\nabla^2 u_z = \frac{\rho}{\mu} \ddot{u}_z \tag{3.26}$$

for solenoidal waves specified by

$$\frac{\partial u_x}{\partial x} + \frac{\partial u_z}{\partial z} = 0 \tag{3.27}$$

Consider first the dilatational wave case, and take

$$u_{1x} = \left[A_1 \cos \eta_1 \left(z - \frac{h_1}{2} \right) + B_1 \sin \eta_1 \left(z - \frac{h_1}{2} \right) \right] \exp[ik(x + vt)] \tag{3.28}$$

where

$$\eta_n\left[\frac{\omega^2}{\dfrac{(\lambda_n+2\mu_n)}{\rho}}\right]-k^2 \tag{3.29}$$

For layer 2, take

$$u_{2x}=\left[A_2\cos\eta_2\left(z+\frac{h_2}{2}\right)+B_2\sin\eta_2\left(z+\frac{h_2}{2}\right)\right]\exp[ik(x+vt)] \tag{3.30}$$

Determine u_{1z} and u_{2z} from the two equations of the form (3.25) for each layer.

Next specify the distortional waves. Let

$$u_{1x}=\left[C_1\cos\beta_1\left(z-\frac{h_1}{2}\right)+D_1\sin\beta_1\left(z-\frac{h_1}{2}\right)\right]\exp[ik(x+vt)]$$

and

$$u_{2x}=\left[C_2\cos\beta_2\left(z+\frac{h_1}{2}\right)+D_2\sin\beta_2\left(z+\frac{h_2}{2}\right)\right]\exp[ik(x+vt)] \tag{3.31}$$

where

$$\beta_n^2=\left(\frac{\omega^2}{\mu_n/\rho}\right)-k^2 \tag{3.32}$$

Determine u_{1z} and u_{2z} from the equations of the form (3.27) in the two layers.

Combine the results for the two types of waves. The unknown amplitudes of the total wave motion are the eight constants, A_1, A_2, B_1, B_2, C_1, C_2, D_1, and D_2. The interface conditions are

$$\text{at } z=0, \qquad u_{1x}=u_{2x}$$

$$u_{1z}=u_{2z}$$

$$\sigma_{1z}=\sigma_{2z}$$

$$\sigma_{1zx}=\sigma_{2zx} \tag{3.33}$$

The periodicity requirements are

$$u_{1x}|_{z=h_1} = u_{2x}|_{z=-h_2}$$

$$u_{1z}|_{z=h_1} = u_{2z}|_{z=-h_2}$$

$$\sigma_{1z}|_{z=h_1} = \sigma_{2z}|_{z=-h_2}$$

$$\sigma_{1zx}|_{z=h_1} = \sigma_{2zx}|_{z=-h_2} \tag{3.34}$$

Conditions (3.33) and (3.34) result in eight homogeneous equations in the eight unknown constants. The determinant of the coefficients is set equal to zero to assure a solution. The system divides naturally into two independent forms associated with longitudinal and transverse waves.

For the longitudinal wave motion, take $B_1 = B_2 = C_1 = C_2 = 0$. Consider a typical term that remains, as

$$u_{1x} \to D_1 \sin\left(z - \frac{h_1}{2}\right)$$

The corresponding shear strain term is seen to be odd with respect to $z = h_1/2$; thus when averaged over the entire layer the average shear strain is zero. Similarly, all other sources of shear strain average to zero when integrated over the thickness. The resulting dispersion equation for longitudinal waves is found to be

$$\begin{aligned} 4(\mu_1 - \mu_2)K_1K_2 + \omega^2\rho_1\left[\frac{\omega^2\rho_1}{k^2} - 4(\mu_1 - \mu_2)\right]K_2 \tan\frac{\beta_1 h_1}{2} \\ + \omega^2\rho_2\left[\frac{\omega^2\rho_2}{k^2} + 4(\mu_1 - \mu_2)\right]K_1 \tan\frac{\beta_2 h_2}{2} \\ - \frac{\omega^4\rho_1\rho_2}{k^2}\left[L_1 \tan\frac{\beta_2 h_2}{2} + L_2 \tan\frac{\beta_1 h_1}{2}\right] = 0 \end{aligned} \tag{3.35}$$

where

$$\begin{aligned} K_1 &= k^2 \tan\frac{\beta_1 h_1}{2} + \eta_1 \beta_1 \tan\frac{\eta_1 h_1}{2} \\ K_2 &= k^2 \tan\frac{\beta_2 h_2}{2} + \eta_2 \beta_2 \tan\frac{\eta_2 h_2}{2} \\ L_1 &= k^2 \tan\frac{\beta_1 h_1}{2} - \eta_2 \beta_2 \tan\frac{\eta_1 h_2}{2} \\ L_2 &= k^2 \tan\frac{\beta_2 h_2}{2} - \eta_1 \beta_2 \tan\frac{\eta_2 h_1}{2} \end{aligned} \tag{3.36}$$

For vertically polarized shear waves take $A_1 = A_2 = D_1 = D_2 = 0$, which results in a mechanical behavior of no average volume change over the entire thickness. The dispersion equation is given by

$$\frac{\mu_2 \beta_2}{\mu_1 \beta_1}\left(\tan^2 \frac{\beta_1 h_1}{2} + \tan^2 \frac{\beta_2 h_2}{2}\right) + \left[1 + \left(\frac{\mu_2 \beta_2}{\mu_1 \beta_1}\right)\right] \tan \frac{\beta_1 h_1}{2} \tan \frac{\beta_2 h_2}{2} = 0 \tag{3.37}$$

Typical Dispersion Curves

Typical dispersion curves are given for the data

$$\frac{h_1}{h_1 + h_2} = 0.8$$

$$\frac{\rho_1}{\rho_2} = 3$$

$$\nu_1 = 0.3$$

$$\nu_2 = 0.35$$

which example is from Sun, Achenbach, and Herrmann [7.2]. Figure 7.3 shows the lowest branch of the dispersion curve for the longitudinal mode of wave motion in the direction of the layering. Figure 7.4 shows the lowest branch in the case of longitudinal wave motion propagating in the direction normal to the layering. The lowest branches are known as the acoustic modes. There are an infinite number of branches.

An interesting effect is seen when the dispersion relations are plotted in the form of a frequency spectrum, as frequency versus wave number. As shown by Sve [7.4], in the case of waves moving normal to the layering a structure of the type shown in Fig. 7.5 occurs. There are stop bands such that for real wave numbers k, certain frequencies are not admitted. The heterogeneous material thus acts as a wave filter. Another common effect connected with harmonic wave propagation in heterogeneous media is that of the cutoff frequency. The cutoff frequency is that frequency at which the wave number k undergoes a transition from real values to imaginary or complex values. The complex wave number case implies the existence of a nonpropagating, standing wave.

Yet another aspect of wave behavior should be clarified at this point. We note that the higher branches of the dispersion curves have associated velocities that are larger than the phase velocities in either material. This

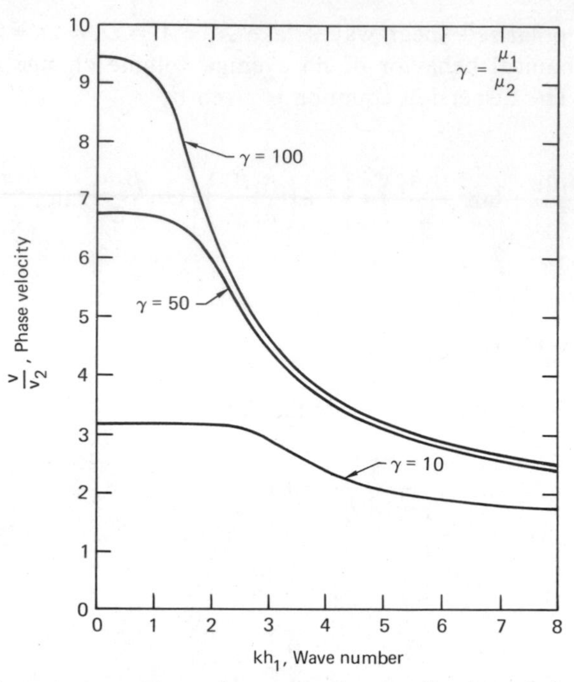

Fig. 7.3 Lowest symmetric mode propagating in direction of layering. After Sun, Achenbach, and Herrmann [7.2].

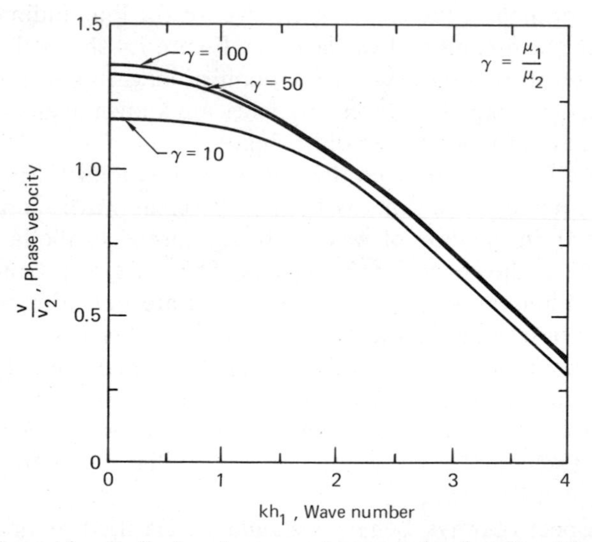

Fig. 7.4 Lowest longitudinal mode propagating normal to layering. After Sun, Achenbach, and Herrmann [7.2].

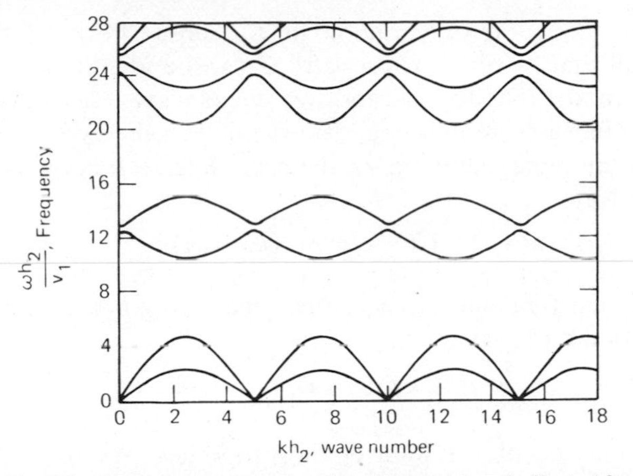

Fig. 7.5 Propagation normal to layering, stop band effect. After Sve [7.4].

feature at first seems paradoxical, until we realize that energy is not transported at the velocities inherent in the dispersion relations. Energy is transported at the group velocity, which is defined as $v_g = d\omega/dk$, which never exceeds the phase velocities in the constituent materials. See Brillouin [7.5] for a discussion of the group velocity.

7.4 TRANSIENT WAVE PROPAGATION: LAYERED MEDIA

We are now ready to face realistic problems of transient wave propagation in heterogeneous media. In this section we consider the propagation of a pulse in layered elastic media, in the direction normal to that of the layering.

There are many different approaches to problems of these types. Several of these different approaches are mentioned at the end of this section. The approach followed here is that due to Kohn [7.6]. Any analytical method must make some approximations. The approach used by Kohn is based on Fourier synthesis along with a low frequency expansion technique. The method thus requires long wave length behavior, but dispersion effects are modeled in a manner to be seen.

General Solution

For one-dimensional wave propagation in heterogeneous media, we write the equation of motion as

$$\frac{\partial}{\partial x}\left[G(x)\frac{\partial u(x,t)}{\partial x}\right]-\rho(x)\frac{\partial^2 u(x,t)}{\partial t^2}=0 \tag{4.1}$$

where x is the direction of variation of the properties and the basic geometry is still that of Fig. 7.1. Property G can be identified with either $\lambda+2\mu$ or μ to model longitudinal and transverse waves, respectively, and displacement u has a corresponding identification in the two cases.

We begin in the same manner as in the preceding section by assuming a solution in the form

$$u(x,t)=\hat{U}(x,k)\exp[i(kx-\omega t)] \tag{4.2}$$

where ω is the real frequency, k is the real wave number, and $\hat{U}(x,k)$ is taken to be periodic in x as

$$\hat{U}(x+h,k)=\hat{U}(x,k)$$

where h is the period of variation of the properties. We recall from the preceding section that frequency ω, wave number k, and wave speed must be interrelated as in (3.4).

Substituting (4.2) into (4.1) gives

$$\left[\left(G\frac{d^2}{dx^2}+\frac{dG}{dx}\frac{d}{dx}\right)+ik\left(2G\frac{d}{dx}+2\frac{dG}{dx}\right)+(ik)^2G\right]\hat{U}(x,k)$$
$$+\rho\omega^2\hat{U}(x,k)=0 \tag{4.3}$$

For small values of the wave number k (long wave lengths) expand $\hat{U}(x,k)$ as a power series in (ik) to obtain

$$\hat{U}(x,k)=1+(ik)u_1(x)+(ik)^2u_2(x)+\cdots \tag{4.4}$$

where at this point we lose no generality in taking the first term as 1. Writing frequency as a function of wave number, we also take a power series expansion

$$\omega^2(k)=v_0^2k^2-\beta k^4+\cdots \tag{4.5}$$

where v_0 and β are to be determined to model long wave behavior from the dispersion relations. Combining (4.4) and (4.5) into (4.3) and setting coefficients of powers of k^2 equal to zero gives the first three forms as

$$\left(G\frac{d^2}{dx^2}+\frac{dG}{dx}\frac{d}{dx}\right)1=0$$

$$\left(G\frac{d^2}{dx^2}+\frac{dG}{dx}\frac{d}{dx}\right)u_1(x)+\frac{dG}{dx}=0$$

$$\left(G\frac{d^2}{dx^2}+\frac{dG}{dx}\frac{d}{dx}\right)u_2(x)+\left(2G\frac{d}{dx}+\frac{dG}{dx}\right)u_1(x)+G(x)+\rho(x)v_0^2=0 \tag{4.6}$$

For disturbances propagating in opposite directions, we write

$$u(x,t)=u^{(+)}(x,t)+u^{(-)}(x,t) \tag{4.7}$$

We use Fourier synthesis to write the general form of $u^{(+)}$ and $u^{(-)}$ as

$$u^{(\pm)}(x,t)=\frac{1}{2\pi}\int_{-\infty}^{\infty}A^{(\pm)}(k)\,\hat{U}(x,k)\exp\{i[kx\pm\omega(k)t]\}\,dk \tag{4.8}$$

Thus we are synthesizing the pulse from waves of all wave lengths, where $A^{(+)}$ and $A^{(-)}$ are functions of wave number and $\omega(k)$ is the square root of that in (4.5). Now use the expansion for $\hat{U}(x,k)$ from (4.4) in (4.8)

$$\begin{aligned}u^{(\pm)}(x,t)&=\frac{1}{2\pi}\int_{-\infty}^{\infty}A^{(\pm)}(k)[1+(ik)u_1(x)+\cdots]\exp\{i[kx\pm\omega(k)t]\}\,dk\\&=\left[1+u_1(x)\frac{\partial}{\partial x}+\cdots\right]U^{(\pm)}(x,t)\end{aligned} \tag{4.9}$$

where

$$U^{(\pm)}(x,t)=\frac{1}{2\pi}\int_{-\infty}^{\infty}A^{(\pm)}(k)\exp\{i[kx\pm\omega(k)t]\}\,dk \tag{4.10}$$

We now verify that $U^{(\pm)}(x,t)$ satisfies

$$\frac{\partial^2 U^{(\pm)}(x,t)}{\partial t^2}=\left[v_0^2\frac{\partial^2}{\partial x^2}+\beta\frac{\partial^4}{\partial x^4}+\cdots\right]U^{(\pm)}(x,t) \tag{4.11}$$

Substitute $U^{(\pm)}(x,t)$ from (4.10) in (4.11) to get

$$\begin{aligned}&\int_{-\infty}^{\infty}A^{(\pm)}(k)[-\omega^2(k)]\exp\{i[kx\pm\omega(k)t]\}\,dk\\&\qquad=\frac{1}{2\pi}\int_{-\infty}^{\infty}A^{(\pm)}[-v_0^2k^2+\beta k^4+\cdots]\exp\{i[kx\pm\omega(k)t]\}\,dk\end{aligned}$$

Now substitute for $\omega^2(k)$ from (4.5) into the left-hand side integrand and observe the term-by-term identity of the equation.

Thus we have the solution

$$u(x,t)=\left[1+u_1(x)\frac{\partial}{\partial x}+u_2(x)\frac{\partial^2}{\partial x^2}+\cdots\right]U(x,t) \tag{4.12}$$

where $U(x,t)$ satisfies (4.11) and from (4.6) $u_1(x)$ satisfies

$$\left[G(x)\frac{d^2}{dx^2}+\frac{dG(x)}{dx}\frac{d}{dx}\right]u_1(x)+\frac{dG(x)}{dx}=0 \tag{4.13}$$

and similar forms follow for $u_2(x)$ and higher-order terms. Write (4.13) as

$$\frac{d}{dx}\left[G(x)\frac{du_{1(x)}}{dx}+G(x)\right]=0$$

and integrate to obtain

$$G(x)\frac{du_{1(x)}}{dx}+G(x)=C \tag{4.14}$$

where C is a constant. Integrate again to get

$$u_1(x)=\int_{\text{CONSTANT}}^{x}\left(\frac{C}{G(\eta)}-1\right)d\eta+D \tag{4.15}$$

where D is a constant we can discard with no loss in generality. The lower limit in (4.15) can be chosen arbitrarily and we take it as zero.

For $u_1(x)$ to be periodic, as $u_1(x+h)=u_1(x)$, (4.15) must have the form

$$u_1(x)=\int_0^x\left(\frac{1}{\overline{G^{-1}}\,G(\eta)}-1\right)d\eta$$

where

$$\overline{G^{-1}}=\frac{1}{h}\int_0^h\frac{1}{G(\eta)}\,d\eta \tag{4.16}$$

It is a simple matter to prove the periodicity of (4.15). Equation (4.15) can be written out as

$$u_1(x)=\frac{h\displaystyle\int_0^x\frac{1}{G(\eta)}\,d\eta}{\displaystyle\int_0^h\frac{1}{G(\eta)}\,d\eta}-x$$

Clearly, this form is periodic; furthermore,

$$u_1(nh)=0,\qquad n=0,1,2\ldots$$

thus, to the order considered,

$$u(x,t) = U(x,t) \qquad \text{at } x = nh, \quad n = 0, 1, 2 \ldots$$

Now find the local states of stress and strain. For strain we have

$$\frac{\partial u}{\partial x} = \frac{\partial}{\partial x}\left\{\left[1 + u_1(x)\frac{\partial}{\partial x} + u_2(x)\frac{\partial^2}{\partial x^2} + \cdots\right]U(x,t)\right\}$$

$$= \left(1 + \frac{du_1}{dx}\right)\frac{\partial U}{\partial x} + \left(u_1 + \frac{du_2}{dx}\right)\frac{\partial^2 U}{\partial x^2} + \cdots$$

Using $u_1(x)$ from (4.15) gives

$$\frac{\partial u}{\partial x} = \left[\frac{1}{\overline{G^{-1}}\,G(x)}\right]\frac{\partial U}{\partial x} + \cdots \tag{4.17}$$

Thus strain is seen to be a rapidly varying function of x because of the presence of the term $G(x)$. Note that $U(x,t)$ provides an envelope function for $\partial u/\partial x$. The local stress is given by

$$\sigma = G(x)\frac{\partial u}{\partial x}$$

and using (4.17) this becomes

$$\sigma = \frac{1}{\overline{G^{-1}}}\frac{\partial U}{\partial x} + \cdots \tag{4.18}$$

Thus, in contrast to strain, stress (at this level of approximation) varies slowly with x. This behavior is consistent with the continuity of stress, whereas strain across interfaces need not be continuous.

Let us see if we can deduce a direct governing equation for stress rather than determining it from displacements. From (4.18) we see that we can write the general form of σ as

$$\sigma(x,t) = \alpha\frac{\partial U(x,t)}{\partial x} + \gamma(x)\frac{\partial^2 U(x,t)}{\partial x^2} + \cdots \tag{4.19}$$

where although α is independent of x, coefficient $\gamma(x)$ is not independent of x. It readily follows from (4.11) that if we truncate (4.19) between the first and second terms, then we have

$$\frac{\partial^2\sigma(x,t)}{\partial t^2}=\left[v_0^2\frac{\partial^2}{\partial x^2}+\beta\frac{\partial^4}{\partial x^4}+\cdots\right]\sigma(x,t) \tag{4.20}$$

However, if we truncate (4.19) at a higher level, then (4.20) is not valid, because coefficient $\gamma(x)$ depends on position. Henceforth, we retain only the lowest order term in (4.19); thus (4.20) is appropriate for use.

The general form of solution for stress then in comparison with (4.10) is given by

$$\sigma(x,t)=\frac{1}{2\pi}\int_{-\infty}^{\infty}\{A^{(+)}(k)\exp\{i[kx+\omega(k)t]\}$$
$$+A^{(-)}(k)\exp\{i[kx-\omega(k)t]\}\}\,dk \tag{4.21}$$

where the terms $A^{(+)}$ and $A^{(-)}$ are different from those in the displacement expression. Take nondimensional variables as

$$\frac{x}{h}=\xi$$

and

$$\frac{tv_0}{h}=\tau$$

then $\sigma(\xi,\tau)$ is given by

$$\sigma(\xi,\tau)=\frac{1}{2\pi}\int_{-\infty}^{\infty}A^{(+)}(k)\exp\left\{i\left[kh\xi+\omega(k)\frac{h}{v_0}\tau\right]\right\}$$
$$+A^{(-)}(k)\exp\left\{i\left[kh\xi-\omega(k)\frac{h}{v_0}\tau\right]\right\}dk \tag{4.22}$$

Initial Value Problems

We wish to solve the initial value problem of

$$\text{at } \tau=0,\qquad \sigma=\sigma_0\delta(\xi)$$
$$\frac{d\sigma}{d\tau}=0 \tag{4.23}$$

where $\delta(\xi)$ is the Dirac delta function. From (4.22) we have

$$\left.\frac{d\sigma}{d\tau}\right|_{\tau=0}=\frac{1}{2\pi}\int_{-\infty}^{\infty}\omega(k)\left[A^{(+)}(k)\exp(ikh\xi)-A^{(-)}\exp(ikh\xi)\right]dk=0$$

which gives

$$A^{(+)}=A^{(-)}$$

Using this result then to satisfy the first condition of (4.23), we have

$$\int_{-\infty}^{\infty}A^{(+)}(k)\exp(ikh\xi)\,dk=\pi\sigma_0\delta(\xi)$$

Taking the Fourier transformation of this equation, we readily find

$$A^{(+)}=\frac{\sigma_0 h}{2}$$

Thus the general solution is given by

$$\sigma(\xi,\tau)=\frac{\sigma_0 h}{4\pi}\int_{-\infty}^{\infty}\left\{\exp\left[i\left(kh\xi-\omega(k)\frac{h}{v_0}\tau\right)\right]\right.$$

$$\left.+\exp\left[i\left(kh\xi-(k)\frac{h}{v_0}\tau\right)\right]\right\}dk \tag{4.24}$$

We need $\omega(k)$ for use in (4.24). Applying the binomial theorem to (4.5) gives

$$\omega=v_0 k\left(1-\frac{1}{2}\frac{\beta}{v_0^2}k^2+\cdots\right) \tag{4.25}$$

Substituting (4.25), truncated after the second term, into the solution (4.24) gives

$$\sigma(\xi,\tau)=\frac{h\sigma_0}{4\pi}\int_{-\infty}^{\infty}\left[\exp[ikh(\xi+\tau)]\exp\left(-\frac{i\beta hk^3}{2v_0^2}\tau\right)\right.$$

$$\left.+\exp[ikh(\xi-\tau)]\exp\left(\frac{i\beta hk^3}{2v_0^2}\tau\right)\right]dk \tag{4.26}$$

This result can be converted to

$$\sigma(\xi,\tau)=\frac{h\sigma_0}{2\pi}\int_0^\infty\left\{\cos\left[kh(\xi+\tau)-\frac{1}{2}\frac{\beta hk^3}{v_0^2}\tau\right]\right.$$

$$\left.+\cos\left[kh(\xi-\tau)+\frac{1}{2}\frac{\beta hk^2}{v_0^2}\tau\right]\right\}dk$$

These terms have the form of Airy functions, and we can write

$$\sigma(\xi,\tau)=\frac{\sigma_0}{2(3\beta\tau/2v_0^2h^2)^{1/3}}\left\{\mathrm{Ai}\left[\frac{-(\xi+\tau)}{(3\beta\tau/2v_0^2h^2)^{1/3}}\right]+\mathrm{Ai}\left[\frac{(\xi-\tau)}{(3\beta\tau/2v_0^2h^2)^{1/3}}\right]\right\} \tag{4.27}$$

where

$$\mathrm{Ai}\left[\frac{y}{3(a)^{1/3}}\right]=\frac{(3a)^{1/3}}{\pi}\int_0^\infty\cos(ak^3+yk)\,dk$$

The solution (4.27) is of the form of two pulses moving in opposite directions. The properties of Airy functions show that the pulses decay as they propagate. After the two pulses have propagated large distances, their effects are distinct, and (4.27) can be approximated by

$$\sigma(\xi,\tau)\simeq\frac{\sigma_0}{2(3\beta\tau/2v_0^2h^2)^{1/3}}\mathrm{Ai}\left(\frac{\xi-\tau}{(3\beta\tau/2v_0^2h^2)^{1/3}}\right),\qquad \varepsilon\geqslant 0 \tag{4.28}$$

for values of ξ near the main part of the pulse.

The general character of the solution (4.27) is shown schematically in Fig. 7.6. As the pulses propagate, the wave dispersion causes the pulses to broaden and flatten. This is not an attenuation effect since no energy has been taken out of the total wave structure. Rather the dispersive wave velocity effect simply redistributes the energy in a manner that provides a psuedo attenuation effect. The long time response of a half space to a unit step function boundary stress can be obtained from (4.28) by integration, to give

$$\sigma(\xi,\tau)=\tfrac{1}{3}-\int_0^\lambda\mathrm{Ai}(\eta)\,d\eta \tag{4.29}$$

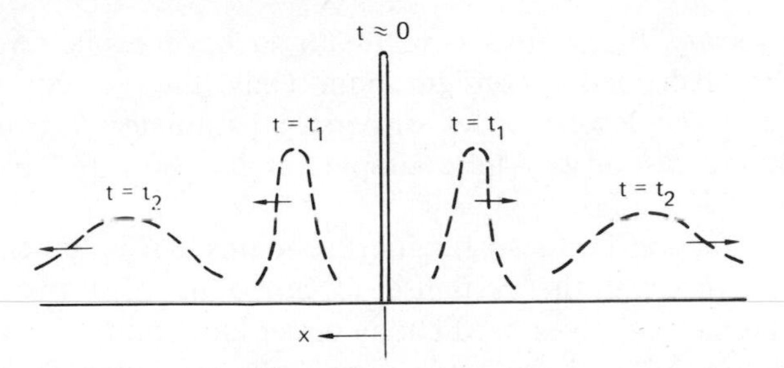

Fig. 7.6 Pulse propagation in dispersive media. At $t=0$ have superposition of two delta functions, each of strength $\frac{1}{2}\delta(x)$.

where

$$\lambda = \frac{\xi - \tau}{(3\beta\xi/2v_0^2h^2)^{1/3}} \tag{4.30}$$

and where the constant of integration is evaluated such that $\sigma \rightarrow 1$ as $\tau \rightarrow \infty$. In obtaining (4.29) by integrating the response in (4.28), the variable τ in the cube root terms has been replaced by ξ, which is permissible for long times near the main part of the pulse.

The result (4.29) was first obtained by Peck and Gurtman [7.7] in an analysis they called the *head of the pulse* method, which is also based on Fourier synthesis for long wave lengths. The form of the solution (4.29) is shown in Fig. 7.7. The response is seen to include an overshoot effect suggestive of a resonance behavior in the material. We see from (4.29) and (4.30) that the form of the solution shown in Fig. 7.7 applies at all positions. From (4.30) we see that the pulse fluctuations broaden as the wave progresses, but the amount of the overshoot remains the same. One

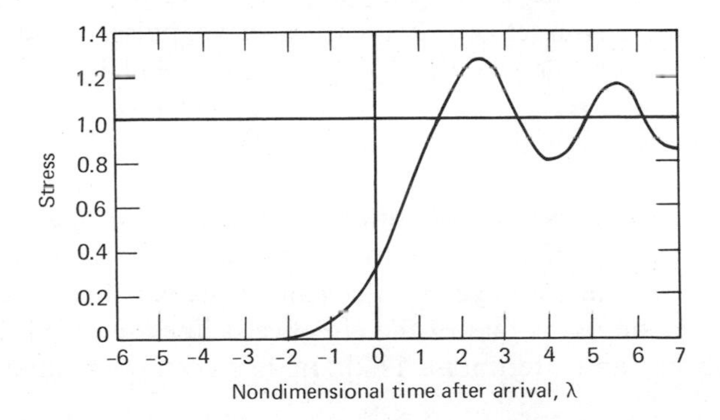

Fig. 7.7 Transient pulse response. After Peck and Gurtman [7.7].

need only specify β, v_0, and h in (4.30) to have explicit results for a particular layered media configuration. Only the velocity v_0 at zero frequency and the lowest order dispersive parameter β from (4.5) are involved in the response. These simple results show the power of the Fourier synthesis method.

The delta function and step function solutions can be taken as Green's functions to generate the response to any type of transient loading. Although the method gives predictions under all conditions, it could only be expected to be realistic under the long wave length assumptions inherent in the derivation. The present results show that the short wave length components become dispersed as the wave propagates, and the long wave length waves provide the major portion of the response. An experimental comparison with the theoretical prediction of the type of (4.29) was given by Whittier and Peck [7.8] with very satisfactory results.

The dispersion parameter β in the expansion (4.5) can be obtained either from theoretical predictions or from experimental measurements of the wave velocity as a function of wave length (frequency). It can be shown that a theoretical prediction for β in periodically layered elastic media is given by

$$\beta = \frac{v_0^6 h_1^2 h_2^2 (\rho_1 G_1 - \rho_2 G_2)^2}{12 h^2 G_1^2 G_2^2} \tag{4.31}$$

where the two elastic layers have the two thicknesses and properties shown. The zero frequency wave speed v_0 was deduced in the Section 7.2 as the formulas (2.2) for longitudinal and transverse waves.

We now see that the fundamental characteristic in the present method is the knowledge of the dispersive character of harmonic waves. Although we have been examining only a single type of heterogeneous medium, in fact, all types of heterogeneous media exhibit a dispersive wave character. It is precisely the dispersion effect that causes wave pulses to diminish in amplitude and to broaden as they progress. The delta function solution (4.28) shows these effects very clearly. In the next section we apply the Fourier synthesis method to a more general type of heterogeneous medium.

There have been a great many different approaches to the modeling of the wave propagation effects in heterogeneous media. The approach shown here is one of the simplest; yet it is elegant in its simplicity. A different method in common use is that of the effective stiffness method devised by Sun, Achenbach, and Herrmann [7.2]. In this method assumptions are made for the local displacement functions within each phase or cell, still retaining the dependence of the displacement functions on the coordinates of the macroscopic problem. Hamilton's principle is used to obtain the governing differential equations of the system. The coefficients of the terms in the equations incorporate the materials properties and the geometric features of the heterogeneous media. The method has been highly developed, as, for example, by Drumheller and Bedford [7.9] and Bedford and Drumheller [7.10]. A general account of the method has been given by Achenbach [7.11]. In the latter reference the similarity of the approach to general microstructure theories, such as multipolar theories, is noted.

In the theory of interacting continua, Hegemeier and colleagues have developed a method based on expansions for both stresses and displacements, and the use of asymptotic expansion techniques. The method, as originally given by Hegemeier and Nayfeh [7.12], has been applied in many papers, not only to layered media but also to fiber reinforced media. The method produces very good results, both in comparison with computer solutions and with experimental results. Yet another approach, also using asymptotic expansions, is that of Ben-Amoz [7.13]. The explicit use of mixture theories to study wave propagation has been made by Bedford and Stern [7.14], and Tiersten and Jahanmir [7.15]. The application of mixture theories is mentioned further in Sections 7.6 and 7.7. Finally, within the limited context of the study and determination of dispersion relations, we mention the work of Nemat-Nasser and Minagawa [7.16] for layered elastic media, and Mukherjee and Lee [7.17] for layered viscoelastic media. Useful review articles in the area of wave propagation in composite materials are those of Achenbach [7.18] and Moon [7.19].

7.5 TRANSIENT WAVE PROPAGATION: THREE-DIMENSIONALLY PERIODIC MEDIA

Having developed a simple method of Fourier synthesis in the previous section, we now seek to apply the method to a more general situation than that of layered media. Specifically, we consider the propagation of transient disturbances in three-dimensionally periodic media. The complications of nonperiodicity are considered in the next section. The method here, as previously, follows the work of Kohn [7.20].

For an elastic medium with position dependent properties, we write

$$\rho(\mathbf{r})\frac{\partial^2 u_i(\mathbf{r},t)}{\partial t^2} = \frac{\partial}{\partial x_j}\left[C_{ijkl}(\mathbf{r})\varepsilon_{lm}(\mathbf{r},t)\right] \tag{5.1}$$

Take $\rho(\mathbf{r})$ and $C_{ijkl}(\mathbf{r})$ to be periodic as

$$f(\mathbf{r}+\gamma)=f(\mathbf{r})$$

where γ is the periodic translation vector, which may be specified in such a manner that the medium is one-, two-, or three-dimensionally periodic. In accordance with Floquet theory we assume a solution in the form

$$u_i(\mathbf{r},t)=\hat{U}_i(\mathbf{r},\mathbf{k})\exp[i(\mathbf{k}\cdot\mathbf{r}-\omega t)] \tag{5.2}$$

where $\hat{U}_i(\ )$ is required to be periodic and $\mathbf{k}$ is the real, vector wave number. Substituting (5.2) in (5.1) the resulting equations of motion can be put into the form

$$-\rho(\mathbf{r})\omega^2(\mathbf{k})\hat{U}_i(\mathbf{r},\mathbf{k})=\frac{\partial}{\partial x_j}\left[C_{ijlm}(\mathbf{r})\frac{\partial\hat{U}_l(\mathbf{r},\mathbf{k})}{\partial x_m}\right]+ik_m\left[\frac{\partial C_{ijlm}(\mathbf{r})}{\partial x_j}\hat{U}_l(\mathbf{r},\mathbf{k})\right.$$
$$\left.+2C_{ijlm}(\mathbf{r})\frac{\partial\hat{U}_l}{\partial x_j}(\mathbf{r},\mathbf{k})\right]-k_mk_jC_{ijlm}(\mathbf{r})\hat{U}_l(\mathbf{r},\mathbf{k}) \tag{5.3}$$

where

$$\mathbf{k}=k_1\mathbf{i}+k_2\mathbf{j}+k_3\mathbf{k}$$

where $\mathbf{i}$, $\mathbf{j}$, and $\mathbf{k}$ are unit vectors and the usual symmetry properties of C_{ijlm} have been used.

For a fixed direction of the wave number vector $\mathbf{k}$, expand $\hat{U}_i(\ )$ as

$$\hat{U}_i(\mathbf{r},\mathbf{k})=u_i^{(0)}(\mathbf{r})+(ik)u_i^{(1)}(\mathbf{r})+(ik)^2u_i^{(2)}(\mathbf{r})+\cdots \tag{5.4}$$

Consistent with (5.4) we expand frequency as a function of wave number, for a fixed direction $\mathbf{k}$:

$$\omega^2(k)=v_0^2(\theta,\phi)k^2-\beta(\theta,\phi)k^4+\cdots \tag{5.5}$$

where θ and ϕ are spherical coordinates. In general, there are three distinct acoustic velocities v_0 for use in (5.5), as discussed in Section 7.1. There also are three distinct values for β and all other coefficients that enter (5.5). We retain only the first two terms in (5.5). The zero frequency wave speed v_0 and the dispersion parameter β may be known either from analytical studies or as the result of experimental observation. We assume these parameters are known.

Write (5.3) in the form

$$\omega^2 \rho(\mathbf{r})\, \hat{U}_i(\mathbf{r},\mathbf{k}) = m_{ij}(\mathbf{k})\, \hat{U}_j(\mathbf{r},\mathbf{k}) \tag{5.6}$$

where $m_{ij}(\mathbf{k})$ is a differential operator that we write as

$$m_{ij} = m_{ij}^{(0)} + k_p m_{ijp}^{(1)} + k_p k_q m_{ijpq}^{(2)} \tag{5.7}$$

where $m_{ij}^{(0)}$, $m_{ijp}^{(1)}$, and $m_{ijpq}^{(2)}$ identify directly with the operators in (5.3). Substituting (5.4) and (5.5) into (5.3) and setting coefficients of powers of k equal to zero gives

$$m_{ij}^{(0)} u_j^{(0)}(\mathbf{r}) = 0$$

and

$$m_{ij}^{(0)} u_j^{(1)}(\mathbf{r}) + k_p m_{ijp}^{(1)} u_j^{(0)}(\mathbf{r}) = 0 \tag{5.8}$$

where these two equations result from the coefficients of $(k)^0$ and k. Now $m_{ij}^{(0)}$ is a differential operator; thus we write

$$u_j^{(0)}(\mathbf{r}) = A_j \tag{5.9}$$

where A_j may depend only on wave number. Using the form of $m_{ij}^{(0)}$ and $m_{ijp}^{(1)}$ from (5.6) and (5.3), the last of (5.8) becomes

$$\frac{\partial}{\partial x_j}\left[C_{ijlm}(\mathbf{r}) \frac{\partial}{\partial x_m} \right] u_j^{(1)}(\mathbf{r}) = -ik_m \left[\frac{\partial C_{ijlm}(\mathbf{r})}{\partial x_j} \right] A_l \tag{5.10}$$

From (5.2), (5.4), and (5.9) we now have

$$u_i(\mathbf{r},t) = \left[A_i + (ik) u_i^{(1)}(\mathbf{r}) + \cdots \right] \exp\left[i(k\cdot\mathbf{r} - \omega t) \right] \tag{5.11}$$

From (5.10) we can write

$$u_i^{(1)} = \hat{u}_{ij}^{(1)}(\mathbf{r}) A_j \tag{5.12}$$

where the explicit solution for $\hat{u}_{ij}^{(1)}(\mathbf{r})$ must come from (5.10). Using (5.12) we write (5.11) as

$$u_i(\mathbf{r},t) = \left[\delta_{ij} + v_{ijl}^{(1)}(\mathbf{r}) \frac{\partial}{\partial x_l} + \cdots \right] A_l \exp\left[i(\mathbf{k}\cdot\mathbf{r} - \omega t) \right] \tag{5.13}$$

where

$$v_{ijl}^{(1)}(\mathbf{r})\frac{\partial}{\partial x_l}\exp(i\mathbf{k}\cdot\mathbf{r})=ik\hat{u}_{ij}^{(1)}(\mathbf{r})\exp(i\mathbf{k}\cdot\mathbf{r}) \tag{5.14}$$

in the direction of propagation.

We are now nearly at the point at which we can state the fundamental results. Let

$$U_i(\mathbf{r},t)=A_i\exp[i(\mathbf{k}\cdot\mathbf{r}-\omega t)] \tag{5.15}$$

Then (5.13) has the form

$$u_i(\mathbf{r},t)=\left[\delta_{ij}+v_{ijl}^{(1)}(\mathbf{r})\frac{\partial}{\partial x_l}+\cdots\right]U_i(\mathbf{r},t) \tag{5.16}$$

Now use Fourier synthesis for the transient case by writing (5.15) as

$$U_i(\mathbf{r},t)=\frac{1}{2\pi}\int_{-\infty}^{\infty}A_i(k)\exp[i(ky-\omega t)]\,dk \tag{5.17}$$

where now y is the direction of propagation. Next we prove that the governing equation for $U_i(\mathbf{r},t)$ is given by

$$\frac{\partial^2 U_i(y,t)}{\partial t^2}=\left(v_0^2\frac{\partial^2}{\partial y^2}+\beta\frac{\partial^4}{\partial y^4}+\cdots\right)U_i(y,t) \tag{5.18}$$

To effect this proof, substitute $U_i(\mathbf{r},t)$ from (5.17) into (5.18) to get

$$\int_{-\infty}^{\infty}A_i(k)(-\omega^2)\exp[i(ky-\omega t)]\,dk$$
$$=\int_{-\infty}^{\infty}A_i(k)(-v_0^2k^2+\beta k^4+\cdots)\exp[i(ky-\omega t)]\,dk \tag{5.19}$$

Relation (5.19) is identically satisfied when we substitute (5.5) for ω^2 in (5.19).

The final results of the derivation then are relations (5.16) and (5.18). We see from (5.16) that the displacement function is governed by an envelope function $U_i(\mathbf{r},t)$ that satisfies relation (5.18), in the direction of propagation. The parameters v_0 and β, and so on, in (5.18) must be known independently for each of the three possible types of waves in any given direction of propagation. The small scale strains and stresses can be deduced from (5.16). Note that the function $v_{ijl}^{(1)}$ in (5.16) is related to $\hat{u}_{ij}^{(1)}$ in (5.14), which in turn is deduced from (5.12). However, this later equation is difficult to solve in general form in the manner we employed in the preceding section.

The macroscale effects are determined by the envelope function $U_i(\mathbf{r}, t)$. This solution, based on Fourier synthesis for long wave lengths, models wave dispersion effects through the presence of the parameter β in the governing equation (5.18). In fact, we see that the governing equation for the envelope function, $U_i(\mathbf{r}, t)$, Eq. (5.18), is identical in form to the governing equation we deduced for the special layered media problem of the preceding section. Now, however, we have deduced the result (5.18) for completely general three-dimensionally periodic media. We need only specify v_0 and β for the type of wave and direction of propagation, in order to solve transient problems by the same general method as in the preceding section. However, an essential difference exists between the general solutions here and that of wave propagation normal to the layers in layered media. In the later case, to the lowest order approximation stress was found to be governed by an equation similar to that of (5.18); that is, stress behaves as a large scale macrovariable. In the present general method even the lowest order stress solution would be found to involve rapid fluctuations, as a microscale variable. This condition results from the fact that not all components of stress are continuous across the interfaces between phases.

As seen from the derivation, the present theory, truncated after the second term in (5.18), is the lowest order theory to include the fundamental dispersive effects due to the heterogeneity of the medium. In this long wave length approach, we have not had to deal explicitly with wave reflections and refractions at interfaces. Such effects are implicitly included through the specification of the dispersive properties of the medium. However, under conditions in which the wave lengths are not long compared with the characteristic dimension of the inhomogeneity, wave reflections must be explicitly taken into account, and the present method is impractical. The study of such problems is beyond the scope intended here. It should be recognized, however, that in cases in which the inhomogeneities are very small, the present long wave length approach may still permit the inclusion of effects in the kilo- or megahertz frequency range. For that matter, in many practical problems the frequency content of the wave relative to the scale of the inhomogeneity permits the dispersive effect to be ignored altogether, and the wave problem is that for an effectively homogeneous elastic medium, Section 7.1.

7.6 ATTENUATION DUE TO RANDOM INHOMOGENEITIES

The preceding studies in this chapter have revealed a dispersion effect in the propagation of waves in heterogeneous media. In this section we see an entirely different effect, that of attenuation.

As discussed in Section 1.2, harmonic wave propagation in viscoelastic materials involves an attenuation of the wave as it propagates due to the conversion of mechanical energy into heat. However, in this section we are only concerned with wave propagation in heterogeneous elastic materials. Now, elastic materials are completely conservative; thus we may ask how there can be an attenuation effect when mechanical energy is conserved. The answer is very simple. Total mechanical energy must be conserved in elastic media; however, there can be an apparent attenuation effect from incoherent scattering due to random inhomogeneities. Thus we are concerned with a random type of heterogeneous media characterization. The random nature of the properties variation is the essential ingredient to provide the apparent attenuation effect (hereafter referred to as the attenuation effect). In contrast, there were scattering effects implicit in the preceding studies of wave motion in the periodic media. However, the scattered waves remained coherent, because of the periodicity of the medium, and thus no attenuation effect appeared.

We specifically study the behavior of long length harmonic waves in an elastic medium containing a dilute suspension of rigid spherical particles. The particles are taken to be randomly distributed in space, such that wave scattering effects off each inclusion remain incoherent, and the scattered energy must come out of the primary harmonic wave. This account follows the work of Moon and Mow [7.21], who employed a mixture theory methodology rather than directly calculating scattered energy and deducting it from the energy of the primary wave to obtain an attenuation effect.

We consider the case of a dilute suspension of perfectly rigid spherical particles in an infinite elastic medium. Furthermore, we assume the particles to be randomly distributed in space. This last requirement is necessitated by our assumption that all scattered energy is taken out of the primary wave. Even though the suspension is dilute, if it were to be of a periodic nature, the scattered waves would remain coherent with and reinforce the primary wave, eliminating the attenuation effect. Furthermore, we assume the case of very long wave length, time harmonic waves, whereby the wave length is long compared with the mean particle spacing.

Wave Scattering Effects

First we outline a solution of the scattering problem for a single spherical scatterer, as given by Pao and Mow [7.22] and Mow [7.23]. The incoming harmonic wave is taken to be a longitudinal wave. The general displace-

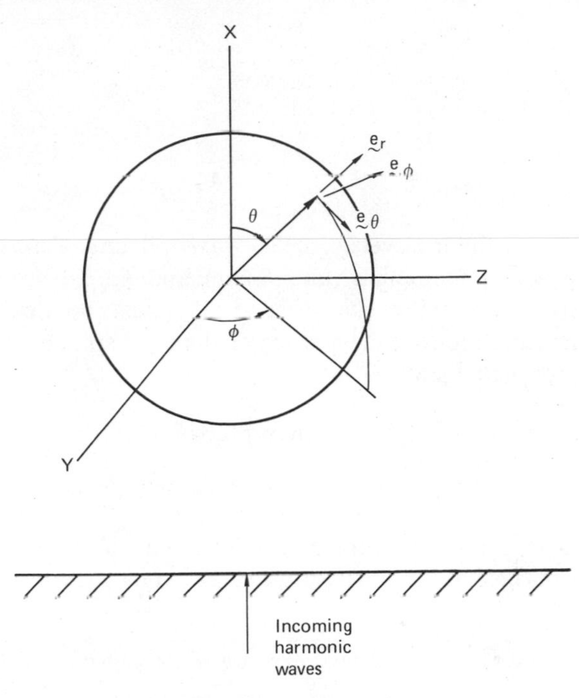

Fig. 7.8 Spherical scatterer.

ment solution for the wave motion in the elastic medium can be written as

$$\mathbf{u} = \nabla\phi + \nabla \times \mathbf{e}_\phi \frac{\partial \psi}{\partial \theta} \tag{6.1}$$

where ϕ and ψ are scalar potentials that satisfy the wave equation, and spherical coordinates are employed with $\mathbf{e}_\phi$ being the unit vector shown in Fig. 7.8. The incoming longitudinal wave is specified by

$$\phi^I = \phi_0 \exp\left[i\omega\left(\frac{x}{v_L} - t \right) \right] \tag{6.2}$$

where v_L is the longitudinal phase velocity in the elastic medium and ω is the frequency. The reflected wave solution can be written as

$$\phi^{(r)} = \exp(-i\omega t) \sum_{n=0}^{\infty} A_n h_n(\alpha r) P_n(\cos\theta)$$

$$\psi^{(r)} = \exp(-i\omega t) \sum_{n=0}^{\infty} B_n h_n(\beta r) P_n(\cos\theta) \tag{6.3}$$

where

$$\alpha=\frac{\omega}{v_L}$$

$$\beta=\frac{\omega}{v_T} \tag{6.4}$$

with v_T being the transverse (shear) wave velocity. Terms $P_n(\cos\theta)$ in (6.3) are Legendre polynomials, and $h_n(\alpha r)$ and $h_n(\beta r)$ are spherical Bessel's functions. The coefficients A_n and B_n must be determined from the boundary conditions at the surface of the rigid sphere, involving continuity of displacement such that

$$u_r=U\cos\theta$$

$$u_\theta=-U\sin\theta$$

where $U(t)$ is the translational displacement of the sphere. Also, stresses over the surface of the sphere must relate to its equation of motion through

$$\tfrac{4}{3}\rho_i\pi a^3\frac{d^2U}{dt^2}=2\pi a^2\int_0^\pi(\sigma_{rr}\cos\theta-\sigma_{r\theta}\sin\theta)\sin\theta\,d\theta=F(t) \tag{6.5}$$

where a is the radius of the sphere and ρ_i is the density of the spherical inclusion. We have later reference to the force $F(t)$ acting on the sphere. The displacement of the sphere must be harmonic, as

$$U(t)=\hat{U}\exp(i\omega t) \tag{6.6}$$

Under long wave length conditions we have $\alpha\ll 1$ and $\beta\ll 1$, in (6.4). To satisfy the boundary conditions on the rigid sphere, only the terms $n=1$ in (6.3) need be retained. Furthermore, under the long wave length conditions, we need retain only the first terms in the expansion of the spherical Bessel's functions in (6.3). It is then found that the force acting on the sphere is

$$F(\omega)=\frac{F(t)}{\exp(i\omega t)}=-\frac{\rho_m V_0}{\tau_0^2}\left\{\frac{9[\hat{U}(\omega)-\hat{u}(\omega)]}{(2\kappa^2+1)}\left(1-i\alpha\frac{2\kappa^3+1}{2\kappa^2+1}\right)\right.$$

$$\left.-\alpha^2 f_1\hat{U}(\omega)+\alpha^2 f_2\hat{u}(\omega)\right\} \tag{6.7}$$

where

$$f_1 = \frac{2+9\kappa+\kappa^2+\left[18\kappa^3(\kappa+2)\right]/\left[(2\kappa^2+1)^2\right]-\left[9\kappa(\kappa+1)(2\kappa+1)\right]/(2\kappa^2+1)}{2\kappa^2+1}$$

$$f_2 = \frac{\frac{9}{2}+9\kappa+3\kappa^2+\left[18\kappa^3(\kappa+2)\right]/\left[(2\kappa^2+1)^2\right]-\left[9\kappa(\kappa+1)(2\kappa+1)\right]/(2\kappa^2+1)}{2\kappa^2+1}$$

where ρ_m is the density of the elastic matrix material, V_0 is the volume of the inclusion, $u^I(x_i, t)$ is the matrix material motion at the origin of the inclusion if the inclusion were not present, as

$$u^I = \hat{u}\exp(i\omega t)$$

and finally

$$\kappa = \frac{\beta}{\alpha} = \frac{v_L}{v_T}$$

$$\tau_0 = \frac{a}{v_L}$$

Variable u^I is, of course, directly due to the incident wave ϕ^I in (6.2).

At this point we restrict attention to the case where $\rho_i \gg \rho_m$; thus the particles are very dense compared with the matrix material. Under this condition the last two terms in (6.7) can be neglected, leaving

$$F(\omega) = \frac{-9\rho_m V_0}{\tau_0^2}\,\frac{\left[\hat{U}(\omega)-\hat{u}(\omega)\right]}{(2\kappa^2+1)}\left[1 - i\alpha\frac{(2\kappa^3+1)}{(2\kappa^2+1)}\right] \tag{6.8}$$

Using relation (6.8) in the equation of motion (6.5) and writing the result in terms of derivatives with respect to time gives

$$\rho_i\frac{d^2U}{dt^2} + \frac{9\rho_m(2\kappa^3+1)}{\tau_0(2\kappa^2+1)^2}\left(\frac{dU}{dt} - \frac{du^I}{dt}\right) + \frac{9\rho_m}{\tau_0^2(2\kappa^2+1)}(U-u^I)=0 \tag{6.9}$$

This relation has a very interesting structure. The last term in (6.9) is of an elastic energy nature, as with a spring effect, whereas the rate terms in (6.9) provide an effective dissipation of energy effect, due to the scattering of wave energy.

Mixture Theory Formulation

We see that the structure of the equation of motion (6.9) suggests a mixture theory type of behavior, whereby we associate the motion of the suspension with that of two separate phases. Specifically, we take two continuous displacement field variables, $u_i(x,t)$ and $u_m(x,t)$ associated with inclusion phase motion and matrix phase motion, respectively. And we specifically identify the incident field variable u^I in (6.9) with the matrix material variable $u_m(x,t)$, as

$$u^I = u_m(x,t)$$

and take

$$U = u_i(x,t) \tag{6.10}$$

and index i refers to inclusion phase, not tensor index. We already have the governing equation of motion (6.9) for the inclusion phase; we must find the corresponding equation of motion for the matrix phase. We use a variational method to obtain this governing equation.

Specifically, we take a Lagrangian function

$$L = cL_i + T_m - V_m \tag{6.11}$$

where c is the volume fraction of inclusions, L_i is the Lagrangian that provides the appropriate terms in the equation of motion (6.9), and T_m and V_m are the kinetic energy and potential energy associated with the matrix phase. Lagrange's equations take the forms

$$\frac{d}{dt}\frac{\partial L}{\partial \dot{u}_i} - \frac{\partial L}{\partial u_i} + \frac{\partial R}{\partial \dot{u}_i} = 0 \tag{6.12}$$

and

$$\frac{d}{dt}\frac{\partial L}{\partial \dot{u}_m} - \frac{\partial L}{\partial u_m} + \frac{\partial R}{\partial \dot{u}_m} = 0 \tag{6.13}$$

where R is the dissipation function; see Goldstein [7.24]. With

$$L_i = T_i - V_i \tag{6.14}$$

where

$$T_i = \tfrac{1}{2}\rho_i \dot{u}_i^2$$

$$V_i = \frac{9\rho_i(u_1 - u_m)^2}{2\tau_0^2(2\kappa^2+1)}$$

$$R = -\frac{9\rho_i(2\kappa^3+1)(\dot{u}_i - \dot{u}_m)^2}{2\tau_0(2\kappa^2+1)^2} \tag{6.15}$$

we find that (6.12) written for L_i gives the equation of motion (6.9).

Next we need to construct the energy functions T_m and V_m for use in the Lagrangian (6.11). The kinetic energy of the matrix phase is taken as

$$T_m = \frac{(1-c)}{2}\rho_m \dot{u}_m^2 \tag{6.16}$$

To find V_m we use the dilute solution effective moduli for μ and k (II-2.23 and 2.26) to write the effective modulus for the dilute, rigid inclusion case as

$$\lambda + 2\mu = (\lambda_m + 2\mu_m)\left[1 + \frac{3(3-5\nu_m)c}{(4-5\nu_m)}\right] \tag{6.17}$$

Then the potential energy term is given by

$$V_m = \tfrac{1}{2}(\lambda + 2\mu)\left(\frac{\partial u_m}{\partial x}\right)^2$$

where $\lambda + 2\mu$ is as in (6.17).

The Lagrange equations (6.12) and (6.13) with the Lagrangian function and dissipation function specified earlier give the equations of motion as

$$\frac{\partial^2 u_i}{\partial t^2} + 2\eta\left(\frac{\partial u_i}{\partial t} - \frac{\partial u_m}{\partial t}\right) + \Omega^2(u_i - u_m) = 0 \tag{6.18}$$

and

$$\rho_m(1-c)\frac{\partial^2 u_m}{\partial t^2} - (\lambda + 2\mu)\frac{\partial^2 u_m}{\partial x^2} = -\rho_i c\frac{\partial^2 u_i}{\partial t^2} \tag{6.19}$$

where

$$\eta=\frac{9\rho_m(2\kappa^3+1)}{2\rho_i\tau_0(2\kappa^2+1)^2}$$

$$\Omega^2=\frac{9\rho_m}{\rho_i\tau_0^2(2\kappa^2+1)}$$

$(\lambda+2\mu)$ is given by (6.17), $\kappa=v_L/v_T$, and $\tau_0=a/v_L$. The term on the right-hand side of (6.19) provides an effective body force due to the inclusion motion, which resists the matrix motion.

Wave Dispersion and Attenuation

We now wish to determine the harmonic wave propagation characteristics for a composite medium governed by the equations of motion (6.18) and (6.19). Let

$$u_i(x,t)=U_i\exp[i(\zeta x-\omega t)]$$

$$u_m(x,t)=U_m\exp[i(\zeta x-\omega t)] \tag{6.20}$$

and parameter α must necessarily be complex as

$$\zeta=\zeta_1+i\zeta_2 \tag{6.21}$$

The substitution of (6.20) into the equations of motion (6.18) and (6.19) gives a set of homogeneous equations in U_i and U_m. The resulting characteristic equation yields the solution for ζ as a function of frequency ω. The phase velocity $v=\omega/\zeta_1$ is found to be

$$v^2=v_0^2\left[1-\left(\frac{\omega}{\Omega}\right)^2\frac{c\rho_i}{\rho_{\text{MEAN}}}\right] \tag{6.22}$$

where

$$v_0^2=\frac{\lambda+2\mu}{\rho_{\text{MEAN}}}$$

$$\rho_{\text{MEAN}}=c\rho_i+(1-c)\rho_m$$

Relation (6.22) shows the dispersive effect of a frequency dependent wave velocity.

The attenuation factor ζ_2, which governs wave attenuation through the term $\exp(-\zeta_2 x)$ in (6.20) is found to be given by

$$\zeta_2 = \frac{\rho_i}{\rho_{\mathrm{MEAN}}}\left(\frac{\rho_l}{\rho_m}\right)^2 \frac{c}{2v_0}\left(\frac{2\kappa^3+1}{9}\right)\tau_0^3\omega^4 \tag{6.23}$$

This is the fundamental result of our derivation. From this formula we can predict the rate at which waves attenuate as they travel through the suspension. The ω^4 dependence is that of the well known Rayleigh scattering.

The attenuation effect is due solely to the energy scattered from the randomly positioned inclusions. We were able to solve the problem in a completely deterministic manner even though random positioning was involved. This result suggests the possibility of treating random media in a statistical sense. A large amount of work has been done along these lines. It is found that the wave attenuation effect can be completely treated by the statistical approach. We do not pursue that line here, other than to note the well known study of that type by Karal and Keller [7.25]. Other approaches to the attenuation problem have been given by McCoy [7.26] and Christensen [7.27].

7.7 A MIXTURE THEORY APPLICATION: DYNAMIC INSTABILITY IN FLUIDIZED COLUMNS

We have just seen how a system of rigid particles dispersed in a continuous elastic phase can be idealized as a mixture of interpenetrating phases. That system provided a realistic example of mixture theory. In this section we give an example of mixture theory as applied to a fluid suspension.

Many different forms of mixture theories have been derived, but it is not our purpose to review them here. Typical general theories are those of Green and Naghdi [7.28], Bowen and Garcia [7.29], and Ingram and Eringen [7.30], and mixture theories explicitly for application to composites have been given by Bedford and Stern [7.14] and by Tiersten and Jahanmir [7.15]. Let us merely note that the general theoretical formulation of mixture theories involves a large number of properties and interaction parameters. For example, the mixture theory derived by Tiersten and Jahanmir [7.15] has a total of 171 independent material constants in the general anisotropic but linear case. There are 20 constants in the transversely isotropic case and 8 in the isotropic case. The determination of the mixture theory material properties is the central problem of the theory. Probably more progress of this type has been made in the application of mixture theory to fluid than to solids. The fluid example we now give shows the usefulness of the general approach, as does the work of the last section.

The problem of interest is the stability of fluidized beds. Many chemical operations involve the so-called fluidized bed, in which a suspension of rigid particles is dispersed in a rising fluid (liquid or gas) stream, with the particles buoyant weight balanced by upward drag forces. Stability of the dispersed particle phase is a common problem. We apply a wave propagation type of stability criterion to the fluidized bed.

The analysis to be given here directly follows that of Anderson and Jackson [7.31]. We merely state the form of the mixture theory derived by them in earlier work [7.32]. As seen in the preceding section, the mixture theory idealizes the continuum so that the multiple material phases share all material points in space. Thus each phase is idealized to be completely continuous, but the multicontinua are taken to have interaction effects in recognition of the constraining effects of their geometric combination.

We now state the mixture theory derived by Andersen and Jackson [7.32] for application to the fluidized bed problem. The continuity equation for the fluid phase has the form

$$\dot{c}_f + (c_f u_k)_{,k} = 0 \tag{7.1}$$

where c_f is the volume fraction of the fluid phase, u_k is the fluid velocity, and the superimposed dot is here taken to be the partial derivative with respect to time. The continuity equation for the particle phase is

$$\dot{c}_p + [c_p v_k]_{,k} = 0 \tag{7.2}$$

where c_p is the volume fraction of the particle phase, v_k is the velocity of the particle phase, and

$$c_f + c_p = 1 \tag{7.3}$$

The two vector equations of motion are given by

$$\rho_f(\dot{u}_i + u_j u_{i,j}) = \sigma_{ij,j}^{(f)} - \frac{f_i}{c_f} + \rho_f g_i \tag{7.4}$$

and

$$\rho_s c_p(\dot{v}_i + v_j v_{i,j}) - \rho_f c_p(\dot{u}_i + u_k u_{i,k})$$

$$= \frac{f_i}{c_f} + c_p(\rho_s - \rho_f) g_i + \sigma_{ij,j}^{(p)} \tag{7.5}$$

where g_i is the gravitational force per unit mass, ρ_f and ρ_p are the densities of fluid and solid matter, respectively, $\sigma_{ij}^{(f)}$ and $\sigma_{ij}^{(p)}$ are the partial stress tensors for the fluid phase and the particle phase, respectively, and f_i is a term related to the interaction force between the fluid and the particles.

The terms f_i, $\sigma_{ij}^{(f)}$, and $\sigma_{ij}^{(p)}$ are specified by the constitutive relations

$$f_i = c_f\beta(u_i - v_i) + c_p C\rho_f \frac{d}{dt}(u_i - v_i) \tag{7.6}$$

$$\sigma_{ij}^{(f)} = -p\delta_{ij} + \lambda u_{k,k}\delta_{ij} + \mu\left(u_{i,j} + u_{j,i} - \tfrac{2}{3}u_{k,k}\delta_{ij}\right) \tag{7.7}$$

$$\sigma_{ij}^{(p)} = p'\delta_{ij} + \lambda' v_{k,k}\delta_{ij} + \mu'\left(v_{i,j} + v_{j,i} - \tfrac{2}{3}v_{k,k}\delta_{ij}\right) \tag{7.8}$$

where β is a drag coefficient, C is a mass effect called the *virtual mass coefficient*, and λ, μ, λ', and μ' are bulk and shear viscosities associated with the two phases. All these coefficients are functions of the volume fraction of each phase. Symbols p and p' are the reactive pressures in the two phases. The viscosities associated with the particle phase, λ' and μ', should not be confused with the fact that the particles are rigid; these viscosities characterize an interactive effect of the system on the partial stress $\sigma_{ij}^{(p)}$.

The governing relations (7.4)–(7.6) have a great similarity to the governing forms in the preceding section for a solid suspension. The primary differences are twofold. First, we now must use velocities rather than displacements as the field variables, and second, the volume fraction of phases is here taken to be a basic unknown, rather than a given quantity. These governing equations can be used to solve for the particle volume fraction c_p as a function of fluid velocity in a steady state flow. We assume this has been done. In the developments given by Anderson and Jackson [7.29] the relative acceleration term in (7.6) is taken as

$$\frac{d}{dt}(u_i - v_i) = \left(\dot{u}_i + u_j u_{i,j}\right) - \left(\dot{v}_i + v_j v_{i,j}\right) \tag{7.9}$$

The stability of the steady flow system is investigated by introducing a small, spatially harmonic wave disturbance. Then it is determined whether or not the disturbance grows with time. The field variables are thus taken as

$$\mathbf{u} = \mathbf{i}u_0 + \hat{\mathbf{u}}e^{st}e^{ikx}$$

$$\mathbf{v} = \hat{\mathbf{v}}e^{st}e^{ikx}$$

$$c_f = c_0 + \hat{c}e^{st}e^{ikx}$$

$$p = p_0 + \hat{p}e^{st}e^{ikx} \tag{7.10}$$

where the x direction and unit vector $\mathbf{i}$ are taken in the upward flow direction with u_0, c_0, and p_0 being corresponding steady state values. The wave disturbance is taken as a plane wave in the x direction, with real wave number $k=2\pi/\lambda$ being inversely proportional to the wave length λ. The parameter s in (7.10) is to be determined, and its behavior determines the stability characteristics of the system. The particle phase pressure p' is taken to be a given function of the volume fraction, c_f or c_p. Relations (7.10) are substituted into the governing relations (7.1), (7.2) and (7.4), (7.5), using (7.6)–(7.8). The resulting eight equations are linearized using the smallness of the perturbation parameters, $\hat{\mathbf{u}}$, $\hat{\mathbf{v}}$, $\hat{c}$, and $\hat{p}$. This set of eight equations is homogeneous, and thereby leads to the characteristic equation to be used to determine s.

Anderson and Jackson [7.31] show that the system of eight equations can be used to eliminate variables to express a single equation in terms of $\hat{c}$, as

$$\hat{c}\left[As^2+(B+dD+i2bF)s+(eD-b^2F+ibBE)\right]=0 \tag{7.11}$$

where

$$\begin{aligned}
A&=1+\frac{\rho_p}{\rho_f}\frac{c_f}{c_p}+\frac{C}{c_f c_p}\\
B&=\left(\frac{\rho_p-\rho_f}{\rho_f}\right)\frac{c_f}{c_p}\frac{g}{\bar{u}}\\
E&=1-2c_f+c_f\frac{\beta'}{\beta}\\
D&=\frac{\rho_p}{\rho_f}\frac{c_f}{c_p}\\
F&=1+\frac{C}{c_f}\\
b&=\frac{\bar{u}k}{c_f}\\
d&=\left(\frac{\lambda'+\frac{4}{3}\mu'}{\rho_p c_p}\right)k^2\\
e&=\left(\frac{\tilde{p}}{\rho_p c_p}\right)k^2
\end{aligned} \tag{7.12}$$

and where

$$\bar{u} = c_f u_0$$

$$\beta' = \frac{-v}{n}\left(\frac{\partial \beta}{\partial c_f}\right)$$

$$\tilde{p} = \frac{-v}{n}\left(\frac{\partial p'}{\partial c_f}\right) \tag{7.13}$$

where n is the number of particles per unit volume and v is the volume of one particle. All variables in (7.12) and (7.13) are evaluated at the steady flow velocity u_0 and the corresponding volume fractions of each phase.

The roots of (7.11) are expressed as

$$s = \xi \pm i\eta \tag{7.14}$$

where the real part is given by

$$\xi = \frac{B}{2A}\left\{-\left(1+\frac{Dd}{B}\right) \pm \sqrt{\frac{\left[(1+w)^2+q^2\right]^{1/2}+(1+w)}{2}}\right.$$

and the imaginary part by (7.15)

$$\eta = \frac{B}{2A}\left\{\frac{2Fb}{B} \pm \sqrt{\frac{\left[(1+w)^2+q^2\right]^{1/2}-(1+w)}{2}}\right.$$

where

$$w = 2\left(\frac{dD}{B}\right) + \left(\frac{dD}{B}\right)^2 + 4\left(\frac{b}{B}\right)^2\left[F(A-F) - \frac{ADe}{b^2}\right]$$

$$q = 4\left(\frac{b}{B}\right)\left(AE - F - \frac{FDd}{B}\right) \tag{7.16}$$

The negative signs in (7.15) provide a negative value for ξ and thus give a time attenuating disturbance. Therefore only the positive signs in (7.15) are of interest for the stability analysis. Although ξ in (7.14) and in the perturbation forms of (7.10) determines time-wise growth, or decay of the disturbance, parameter η in (7.14) when inserted in (7.10) gives the disturbance velocity as η/k.

The stability parameter ξ and the disturbance velocity η/k from (7.15) must, in general, be evaluated numerically. However, the limiting values of ξ and η can be evaluated in the cases of zero and infinite wave length. From (7.15) along with (7.12) it is found that

$$\lim_{k\to 0}\xi \to \frac{\bar{u}^2k^2}{c_f^2B}\left[AE^2+F(1-2E)+\frac{De}{b^2}\right]=0$$

$$\lim_{k\to 0}\frac{\eta}{k}\to\frac{\bar{u}}{c_f}E \tag{7.17}$$

and

$$\lim_{k\to\infty}\xi\to\frac{\bar{u}^2\rho_p c_p}{c_f^2\left(\lambda'+\frac{4}{3}\mu'\right)D}\left(F-\frac{De}{b^2}\right)$$

$$\lim_{k\to\infty}\frac{\eta}{k}\to\frac{2\bar{u}F}{c_fA}$$

where the ratio e/b^2 from (7.12) is independent of wave number. Thus we see that for very long wave length disturbances, the rate of growth of the disturbance goes to zero and the velocity approaches a constant limiting value. For very short wave lengths the time rate of growth and the disturbance velocity both approach constant limiting values.

Anderson and Jackson [7.31] take known values or estimates for all the properties coefficients involved in the theory. They find the disturbance is always unstable, in the range studied by them. But for typical liquid fluidized systems the attenuation coefficient of the disturbance is so low that it would have to travel very far to become of significant size. For this reason the instability prediction for the liquid fluidized system is of little practical consequence when compared with the dimensions of typical chemical plant operations. The gas fluidized system, however, is another matter. It was shown in [7.31] that disturbances in an air fluidized system have growth rates about 100 times as great as in a water fluidized system. The prediction of the extreme instability of gas fluidized systems is in accordance with practical experience. This theoretical prediction of instability is a considerable achievement in the use of the mixture theory.

The mixture theory discussed here has been applied further by Medlin, Wong, and Jackson [7.33] and Medlin and Jackson [7.34] to study instabilities in fluidized beds of finite length and other effects. It is shown that convective instabilities may develop as a circulatory motion.

PROBLEMS

1. Prove that the short wave velocity v_T^M, Eq. (2.5), is greater than the long wave length velocity v_T, Eq. (2.2).
2. Using the wave transmission and reflection formulas (2.12), derive the following expression for the reflection from the second interface after passage through two interfaces in a two material periodically layered medium.

$$\frac{\sigma_R}{\sigma_I} = \frac{2\sqrt{\rho_1 G_1}\left(\sqrt{\rho_2 G_2} - \sqrt{\rho_1 G_1}\right)}{\left(\sqrt{\rho_1 G_1} + \sqrt{\rho_2 G_2}\right)^2}$$

3. In the wave velocity expansion

$$v = v_0 + \frac{\alpha}{2}\omega^2 + \cdots$$

deduce the expressions for v_0 and α from the dispersion equation (3.13).

4. Derive expression (4.31) for the term β in the expansion

$$\omega^2(k) = v_0^2 k^2 - \beta k^4 + \cdots$$

5. Discuss the experimental procedure by which one would directly determine the coefficient β in the expansion shown in Problem 4. What type of instrumentation would be required?
6. How would one proceed to estimate the very long wave length speed v_0 in elastic media with single size rigid spherical inclusions arranged in a three-dimensionally rectangular stacking pattern? Is the medium macroscopically isotropic?
7. Perform a numerical evaluation of the pulse propagation solution (4.28) to show how the pulse changes shape as it propagates. Select your own numerical values for the various terms entering the expression, as being representative of a "realistic" problem.
8. Outline the method by which one could deduce the attenuation effect studied in Section 7.6, without recourse to the mixture theory formulation given there.

REFERENCES

7.1 S. M. Rytov, "Acoustical properties of a thinly laminated medium," *Sov. Phys. Acoust.*, vol. 2, 68 (1956).

7.2 C. -T. Sun, J. D. Achenbach, and G. Herrmann, "Continuum theory for a laminated medium," *J. Appl. Mech.*, vol. 35, 467 (1968).

7.3 L. M. Brekhovskikh, *Waves in Layered Media*, Academic, New York, 1960.

7.4 C. Sve, "Time-harmonic waves traveling obliquely in a periodically laminated medium," *J. Appl. Mech.*, vol. 38, 477 (1971).

7.5 L. Brillouin, *Wave Propagation in Periodic Structures*, Dover, New York, 1963.

7.6 W. Kohn, "Propagation of low-frequency elastic disturbances in a composite material," *J. Appl. Mech.*, vol. 41, 97 (1974).

7.7 J. C. Peck and G. A. Gurtman, "Dispersive pulse propagation parallel to the interfaces of a laminated composite," *J. Appl. Mech.*, vol. 36, 479 (1969).

7.8 J. S. Whittier and J. C. Peck, "Experiments on dispersive pulse propagation in laminated composites and comparison with theory," *J. Appl. Mech.*, vol. 36, 485 (1969).

7.9 D. S. Drumheller and A. Bedford, "Wave propagation in elastic laminates using a second-order microstructure theory," *Int. J. Solids Structures*, vol. 10, 61 (1974).

7.10 A. Bedford and D. S. Drumheller, "The propagation of stress waves into a laminated half space using a second-order microstructure theory," *Int. J. Solids Structures*, vol. 11, 841 (1975).

7.11 J. D. Achenbach, *A Theory of Elasticity with Microstructure for Directionally Reinforced Composites*, Springer Verlag, New York, 1975.

7.12 G. A. Hegemeier and A. H. Nayfeh, "A continuum theory for wave propagation in laminated composites. Case 1: Propagation normal to the laminates," *J. Appl. Mech.*, vol. 40, 503 (1973).

7.13 M. Ben-Amoz, "On wave propagation in laminated composites—I. Propagation parallel to the laminates," *Int. J. Eng. Sci.*, vol. 13, 43 (1975).

7.14 A. Bedford and M. Stern, "A multi-continuum theory for composite elastic materials," *Acta Mech.*, vol. 14, 85 (1972).

7.15 H. F. Tiersten and M. Jahanmir, "A theory of composites modeled as interpenetrating solid continua," *Arch. Ration. Mech. Anal.*, vol. 65, 154 (1977).

7.16 S. Nemat-Nasser and S. Minagawa, "Harmonic waves in layered composites: comparison among several schemes," *J. Appl. Mech.*, vol. 42, 699 (1975).

7.17 S. Mukherjee and E. H. Lee, "Dispersion relations and mode shapes for waves in laminated viscoelastic composites by variational methods," *Int. J. Solids Structures*, vol. 14, 1 (1978).

7.18 J. D. Achenbach, "Waves and vibrations in directionally reinforced composites," in *Composite Materials*, vol. 2, G. P. Sendeckyj, Ed., Academic, 1974.

7.19 F. C. Moon, "Wave propagation and impact in composite materials," in *Composite Materials*, vol. 7, C. C. Chamis, Ed., Academic, New York, 1974.

7.20 W. Kohn, "Propagation of low frequency elastic disturbances in a three-dimensional composite material," *J. Appl. Mech.*, vol. 42, 159 (1975).

7.21 F. C. Moon and C. C. Mow, "Wave propagation in a composite material containing dispersed rigid spherical inclusions," Rand Corporation Report RM-6139-PR, 1970.

7.22 Y. H. Pao and C. C. Mow, "Scattering of plane compressional waves by a spherical obstacle," *J. Appl. Phys.*, vol. 34, 493 (1963).

7.23 C. C. Mow, "Transient response of a rigid spherical inclusion in an elastic medium," *J. Appl. Mech.*, vol. 32, 637 (1965).

7.24 H. Goldstein, *Classical Mechanics*, Addison-Wesley, Reading, Mass., 1959.

7.25 F. C. Karal, Jr., and J. B. Keller, "Elastic, electromagnetic and other waves in a random medium," *J. Math. Phys.*, vol. 5, 537 (1964).

7.26 J. J. McCoy, "On the dynamic response of disordered composites," *J. Appl. Mech.*, vol. 40, 511 (1973).

7.27 R. M. Christensen, "Wave propagation in layered elastic media," *J. Appl. Mech.*, vol. 42, 153 (1975).

7.28 A. E. Green and P. M. Naghdi, "On basic equations for mixtures," *Quart. J. Mech. Appl. Math.*, vol. 22, 427 (1969).

7.29 R. M. Bowen and D. J. Garcia, "On the thermodynamics of mixtures with several temperatures," *Int. J. Eng. Sci.*, vol. 8, 63 (1970).

7.30 J. D. Ingram and A. C. Eringen, "A continuum theory of chemically reacting media—II. Constitutive equations of reacting fluid mixtures," *Int. J. Eng. Sci.*, vol. 5, 289 (1967).

7.31 T. B. Anderson and R. Jackson, "Fluid mechanical description of fluidized beds," *Ind. Eng. Chem. Fundam.*, vol. 7, 12 (1968).

7.32 T. B. Anderson and R. Jackson, "A fluid mechanical description of fluidized beds," *Ind. Eng. Chem. Fundam.*, vol. 6, 527 (1967).

7.33 J. Medlin, H-W. Wong, and R. Jackson, "Fluid mechanical description of fluidized beds. Convective instabilities in bounded beds," *Ind. Eng. Chem. Fundam.*, vol. 13, 247 (1974).

7.34 J. Medlin and R. Jackson, "Fluid mechanical description of fluidized beds. The effect of distribution thickness on convective instabilities," *Ind. Eng. Chem. Fundaml., vol. 14, 315 (1975).*

CHAPTER VIII
INELASTIC AND NONLINEAR EFFECTS

In our work up to this point we have been concerned with characterizations based on linear elasticity theory. In fact, with the exception of the last section of Chapter VII, dealing with mixture theory, we have exclusively employed linear elasticity formulations. The emphasis on elasticity reflects its usefulness and importance, not only as a basic theory but also for providing the means to develop practical design methods. By no means, however, can all practical systems be idealized as behaving according to linear elasticity theory. Composite materials are often used in situations that involve material and/or kinematic nonlinearity. Furthermore, many types of materials involve constitutive behavior that is distinctly inelastic. Practical cases of these types are considered in this chapter.

The basic material constitutive types in common use are those of elasticity, viscoelasticity, and inviscid plasticity. Typically, inviscid plasticity modeling is the most useful and applicable type for metal behavior above the yield limit. There are many different types of composite materials involving one or more metallic phases; thus we are concerned here with problems of this type. Specifically, we deal with the deformation of a fiber system containing a metal matrix that deforms according to plasticity theory. Also, we consider the problem of the collapse of a metal matrix porous material.

Viscoelasticity theory is used to characterize polymer response in the ranges of temperature above that of glassy behavior. All polymers exhibit viscoelastic effects, and such effects can be of importance in many practical situations. Since polymers are commonly employed as one or more phases in composite materials, we do indeed wish to cover certain aspects of viscoelastic behavior here. Finally, we treat a class of problems that involve arbitrarily large deformation conditions.

8.1 PLASTIC DEFORMATION OF POROUS MEDIA

We are here concerned with the behavior of a porous, foam type material that is subjected to hydrostatic pressure conditions causing the pore space to decrease in volume. If the material is elastic, we take the solution for the effective bulk modulus k of the composite spheres model as governing the deformation. In the case of voids, from (II-3.17) the solution for k is given by

$$k = k_m - \frac{ck_m}{1-(1-c)k_m/\left(k_m + \frac{4}{3}\mu_m\right)} \tag{1.1}$$

where c is the volume fraction of the pore space and k_m and μ_m are the elastic properties of the material phase of the foam. This result, which governs the compressibility of the foam, is realistic only under small deformation conditions with regard to the initial state of the material.

Our interest here is in large deformation states in which a significant amount of void collapse has been experienced and where material nonlinearity is expected to be of importance. We specifically assume the material to be governed by the inviscid theory of plasticity, assuming elastic-perfectly plastic behavior. We appeal to the composite spheres model of Section 2.3 to idealize the behavior of the entire void containing material by that of a single hollow sphere. The analysis here follows that of Carroll and Holt [8.1].

Take a hollow sphere of initial internal and external radii a_0 and b_0 as shown in Fig. 8.1. A monotonically increasing pressure p is applied at the outer boundary. The material is isotropic, and elastic-perfectly plastic. Furthermore, the material is taken to be completely incompressible, both elastically and plastically. The material has an elastic shear modulus μ and it satisfies either the Mises or Tresca yield criteria.

The assumption of incompressibility greatly simplifies the attack on the problem. Since the problem is spherically symmetric, the deformation can be specified as

$$r^3 = r_0^3 - B \tag{1.2}$$

where r_0 is the initial position of any material point and r is its position

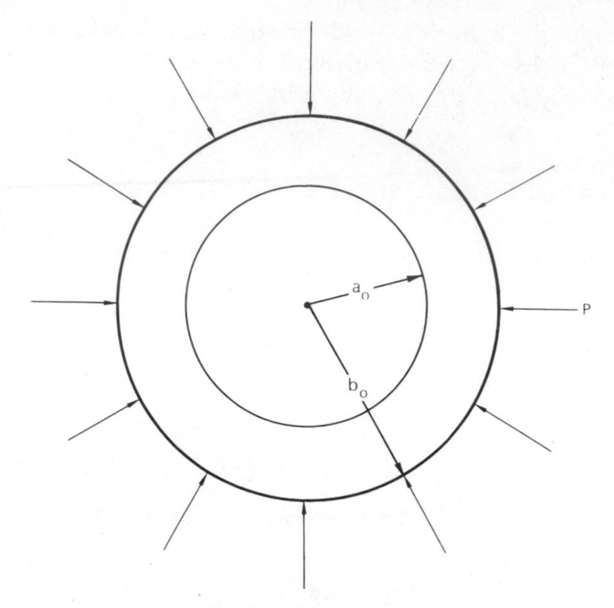

Fig. 8.1 Spherical cell.

after the volume preserving deformation. The parameter B can be specified in terms of the porosity. The porosity is defined by

$$\alpha = \frac{b^3}{(b^3 - a^3)} \tag{1.3}$$

where a and b are the inner and outer radii after deformation and the initial porosity is given by

$$\alpha_0 = \frac{b_0^3}{(b_0^3 - a_0^3)} \tag{1.4}$$

Combining (1.2)–(1.4) gives

$$a^3 = a_0^3 \frac{(\alpha - 1)}{(\alpha_0 - 1)}$$

$$b^3 = \frac{a_0^3 \alpha}{\alpha_0 - 1}$$

and

$$B = a_0^3 \frac{(\alpha_0 - \alpha)}{(\alpha_0 - 1)} \tag{1.5}$$

Initially, the entire region is elastic, and we need the corresponding analysis of the elastic initial region. The deviatoric stresses are given by

$$s_{rr} = 2\mu \frac{\partial u}{\partial r} = \frac{4\mu B}{3r^3}$$

and

$$s_{\theta\theta} = s_{\phi\phi} = 2\mu \frac{u}{r} = -\frac{2\mu B}{3r^3} \tag{1.6}$$

where u is the infinitesimal radial displacement. The total stresses are then

$$\sigma_{rr} = \sigma(r) + \frac{4\mu B}{3r^3}$$

$$\sigma_{\theta\theta} = \sigma_{\phi\phi} = \sigma(r) - \frac{2\mu B}{3r^3} \tag{1.7}$$

where σ is the reactive hydrostatic stress term.

The governing equation of equilibrium is

$$\frac{\partial \sigma_{rr}}{\partial r} + \frac{2}{r}(\sigma_{rr} - \sigma_{\theta\theta}) = 0 \tag{1.8}$$

and the boundary conditions are

$$\text{at } r = a, \qquad \sigma_{rr} = 0$$

$$\text{at } r = b, \qquad \sigma_{rr} = -p \tag{1.9}$$

Substituting (1.7) into (1.8), integrating, satisfying the boundary conditions, and finally using the last of (1.5) gives

$$p = \frac{4\mu(\alpha_0 - \alpha)}{3\alpha(\alpha - 1)} \tag{1.10}$$

Although the linear elasticity stress-strain relations are used, the boundary conditions are satisfied on the deformed boundaries (1.9).

The solution (1.10) applies until the yield limit is reached at some point in the medium. The yield function is written as

$$\sigma_{rr}-\sigma_{\theta\theta}=2k \tag{1.11}$$

where k is the yield stress in simple shear. It can be shown that the yield stress is first reached on the inner boundary at a value of porosity given by

$$\alpha_1=\frac{\mu\alpha_0+k}{\mu+k} \tag{1.12}$$

With continued loading beyond the level at which first yielding commences, there will be an elastic region and a plastic region. The elastic-plastic interface moves outward with increased loading. Taking the elastic-plastic interface to be at a radius d, the equation of equilibrium in the plastic region is

$$\frac{\partial\sigma_{rr}}{\partial_r}+\frac{4k}{r}=0 \tag{1.13}$$

This equation is integrated, and the constant is evaluated to satisfy the inner boundary condition. At the interface there results

$$\sigma_{rr}|_{r=d}^{\text{plastic}}=-4k\ln\frac{d}{a} \tag{1.14}$$

In the elastic region the solution that satisfies the outer boundary condition gives

$$\sigma_{rr}|_{r=d}^{\text{elastic}}=-p+\frac{4\mu B}{3}\left(\frac{1}{d^3}-\frac{1}{b^3}\right) \tag{1.15}$$

Also, at the elastic-plastic boundary $r=d$, the elastic solution must satisfy the yield condition, which leads to

$$\frac{\mu B}{d^3}=k \tag{1.16}$$

Eliminating $\sigma_{rr}|_{r=d}$ and d between (1.14)–(1.16) and using (1.5) gives the solution

$$p=\frac{4}{3}\left[\mu+k-\frac{\mu\alpha_0}{\alpha}+k\ln\frac{\mu(\alpha_0-\alpha)}{k(\alpha-1)}\right] \tag{1.17}$$

This result applies until the elastic-plastic interface reaches the outer boundary, as $d=b$. From (1.5) and (1.16) this occurs at a value of porosity

$$\alpha_2 = \frac{\mu\alpha_0}{\mu+k} \tag{1.18}$$

The sphere will be in a fully plastic state for loading beyond the level at which the elastic-plastic boundary reaches the outer boundary. We now obtain the solution in the fully plastic state. The solution of the equilibrium equation (1.13) and the satisfaction of the boundary conditions (1.9) gives

$$p = 4k \ln\frac{b}{a} \tag{1.19}$$

Using (1.5) this becomes

$$p = \frac{4k}{3}\ln\frac{\alpha}{\alpha-1} \tag{1.20}$$

This solution is seen to be independent of the initial porosity α_0.

The complete solution for the pressure-porosity relation is given by (1.10), (1.17), and (1.20), with the transitions between solutions occurring at values of porosity given by (1.12) and (1.18). The corresponding transition pressures are given by

$$p_1 = \frac{k}{3\alpha_1}$$

and

$$p_2 = \frac{k}{3}\ln\frac{\alpha_2}{\alpha_2-1} \tag{1.21}$$

In Fig. 8.2 the porosity-pressure curve for aluminum is shown, for the case of an initial porosity of about 1.30. It is seen that nearly the entire collapse occurs during the fully plastic state. The solution given here is expected to be valid up to a point where the collapse has compacted the cell structure sufficiently that compressibility effects in the material would be of importance. Butcher, Carroll, and Holt [8.2] have given a generalization of this work to include compressibility and dynamic effects. The earliest work on this problem was that of Chadwick [8.3]. See also Chu and Hashin [8.4].

This method of analysis using inviscid plasticity theory is well suited to the problem of pore collapse under hydrostatic pressure. However, under

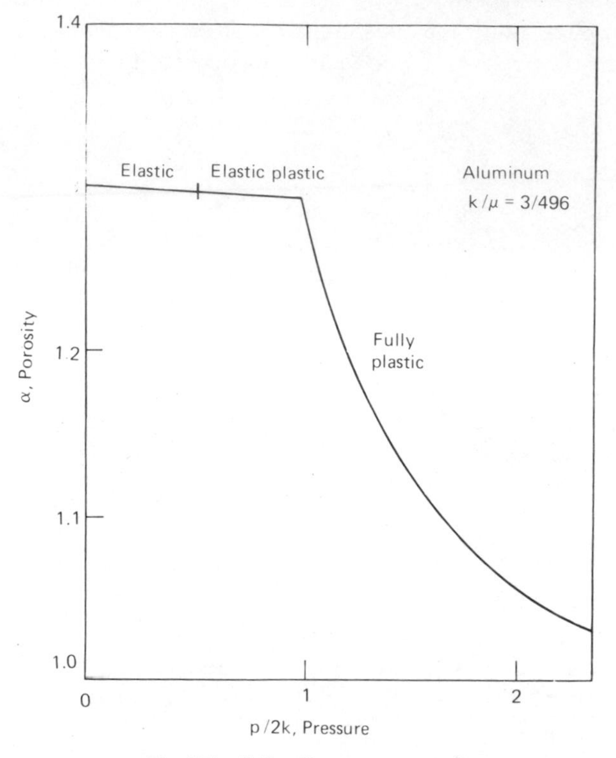

Fig. 8.2 Cell collapse response diagram.

other states of stress, such as shear, there could be bending and buckling of the cell walls, particularly for highly porous media, and special methods of analysis would be needed to attack such problems. Furthermore, even in the present problem, plastic instability could be an important effect in thin walled cell geometries.

8.2 PLASTIC DEFORMATION OF FIBER SYSTEMS

Fiber composite systems are often used with metal matrices. There is a considerable weight penalty involved in using a metal rather than a polymeric matrix phase; however, metals have other advantages. The primary advantage of the metal matrix is its high temperature capability compared with that of most polymers. Metal matrix composites have a rather different type of mechanical behavior from that of polymers, and we now study a fundamental problem that highlights these differences.

The problem of interest is that of the longitudinal behavior of an aligned fiber system containing a metal matrix. We appeal to the composite cylinders model of Section 3.2 to model the entire fiber system by that of a single composite cylinder. Thus we study a single composite cylinder composed of an elastic fiber embedded in a concentric cylinder of material governed by the theory of inviscid plasticity. Only axisymmetric deformation states are considered; thus we allow axial extension, as well as lateral contraction, or a state of laterally applied pressure. We specifically wish to relate overall composite behavior to the properties of the constituent phases. This is the identical objective that guided much of our work with elastic composites. We shall see that the problem is more complicated when we allow plasticity behavior. Nevertheless, we obtain some very meaningful results. The approach taken here is from Dvorak and Rao [8.5].

We could proceed along the same lines we followed in the previous section: solving boundary value problems and keeping track of the separate regions of elastic behavior and plastic behavior. However, we proceed by a somewhat more general method, and we obtain much more general results. The matrix phase is taken to be that of elastic-perfectly plastic behavior, with the Mises criteria specifying the matrix material yield function. We shall see that this leads to a work hardening type response for the overall behavior of the composite.

Yield Function

As usual, let coordinate x_1 be in the fiber direction. Relative to a Cartesian coordinate system the average stresses acting on the composite are specified by σ_{ij}. The invariants of stress appropriate to transverse isotropy are given by

$$I_1=\sigma_{11}$$

$$I_2=\tfrac{1}{2}(\sigma_{22}+\sigma_{33})$$

$$I_3=\sigma_{12}^2+\sigma_{13}^2$$

$$I_4=\tfrac{1}{2}(\sigma_{22}-\sigma_{33})^2+2\sigma_{23}^2$$

$$I_5=\tfrac{1}{2}(\sigma_{22}-\sigma_{33})(\sigma_{12}^2-\sigma_{13}^2)+2\sigma_{12}\sigma_{23}\sigma_{31} \tag{2.1}$$

The initial yield function of the composite cylinder is then specified as

$$f(I_1,I_2,I_3,I_4,I_5)=0 \tag{2.2}$$

In loading from the undeformed state the matrix phase deforms in a completely elastic manner, until the yield stress is first reached. This yielding typically occurs first at the fiber-matrix interface, just as in the problem of the preceding section, where yielding was first reached on the inner surface of the spherical shell. It is assumed that the plastic region in the matrix phase is a cylindrical region having an inner boundary at the fiber-matrix interface. Relative to cylindrical coordinates, the only nonzero stresses are σ_{rr}, $\sigma_{\theta\theta}$, and σ_{zz}. The interface stresses in the matrix phase in the elastic range can be written as

$$\begin{bmatrix} \sigma_{rr} \\ \sigma_{\theta\theta} \\ \sigma_{zz} \end{bmatrix} = \mathbf{AI} \tag{2.3}$$

where

$$\mathbf{I} = \begin{bmatrix} I_1 \\ I_2 \end{bmatrix} \tag{2.4}$$

and

$$\mathbf{A} = \begin{bmatrix} A_{11} A_{12} \\ A_{21} A_{22} \\ A_{31} A_{32} \end{bmatrix} \tag{2.5}$$

with A_{ij} being functions of the elastic properties of the fiber and matrix phases and the volume fraction of each phase. The equation for the Mises criteria (I-3.6) can be written as

$$\mathbf{I}^T\mathbf{A}^T\mathbf{CAI} - Y^2 = 0 \tag{2.6}$$

where (2.3) has been used, Y designates the yield stress of the matrix material in uniaxial deformation, and $\mathbf{C}$ is the matrix

$$\mathbf{C} = \frac{1}{2}\begin{bmatrix} 2 & -1 & -1 \\ -1 & 2 & -1 \\ -1 & -1 & 2 \end{bmatrix}$$

The superscript T designates transpose. The yield function (2.6) specifies an ellipse in the I_1, I_2 plane. We must be very careful not to confuse the yield function, (2.6), for the matrix material at the interface, with that of the overall composite (2.2).

The deviatoric stress is given by

$$\mathbf{s} = \begin{bmatrix} s_{rr} \\ s_{\theta\theta} \\ s_{zz} \end{bmatrix} = \tfrac{2}{3}\mathbf{CAI} \tag{2.7}$$

Using (2.7), Eq. (2.6) can be written as

$$\tfrac{3}{2}\mathbf{s}^T\mathbf{s} - Y^2 = 0 \tag{2.8}$$

Relation (2.7) also can be written as

$$\mathbf{s} = \begin{bmatrix} s_{\theta\theta} \\ s_{zz} \end{bmatrix} = \mathbf{PI} \tag{2.9}$$

where

$$\mathbf{P} = \frac{1}{3}\begin{bmatrix} -1 & 2 & -1 \\ -1 & -1 & 2 \end{bmatrix}\mathbf{A} \tag{2.10}$$

Relation (2.9) also defines s_{rr} since $s_{kk} = 0$.

Take the composite cylinder as being loaded to some state such that $I_1 = I_1^L, I_2 = I_2^L$, with deviatoric stresses $\mathbf{s}^L$ at the interface. Upon unloading there will be residual stresses given by $\mathbf{s}^R$, which from (2.9) is

$$\mathbf{s}^R = \mathbf{s}^L - \mathbf{PI}^L \tag{2.11}$$

where the term $\mathbf{PI}^L$ is due to elastic unloading. Now find an unloading path such that the residual stresses $\mathbf{s}^R = \mathbf{0}$. This unloading state is given by

$$\mathbf{PI}^R = -(\mathbf{s}^L - \mathbf{PI}^L) \tag{2.12}$$

where (2.9) and (2.11) are used to specify (2.12) and $\mathbf{I}^R$ are the residual composite stresses. Solving (2.12) for $\mathbf{I}^R$ gives

$$\mathbf{I}^R = \mathbf{I}^L - \mathbf{P}^{-1}\mathbf{s}^L \tag{2.13}$$

Finally, solving (2.13) for $\mathbf{s}^L$ gives

$$\mathbf{s}^L = \mathbf{P}(\mathbf{I}^L - \boldsymbol{\alpha}) \tag{2.14}$$

where

$$\boldsymbol{\alpha} = \mathbf{I}^R \tag{2.15}$$

The interpretation of (2.14) is very simple. For deviatoric stresses $\mathbf{s}^L$ in the plastic state, the corresponding residual composite cylinder stresses $\boldsymbol{\alpha}$ are required to return the interface stresses to the null state. It now follows that relative to this state the yield function (2.6) is written as

$$(\mathbf{I}-\boldsymbol{\alpha})^T\mathbf{A}^T\mathbf{C}\mathbf{A}(\mathbf{I}-\boldsymbol{\alpha})-Y^2=0 \tag{2.16}$$

We have now established a major result. The original yield function (2.6), for the matrix material, takes the form (2.16) when expressed in terms of the average stresses on the composite cylinder. In terms of these average stresses the yield function is seen to behave as though it undergoes kinematic work hardening (see Section 1.3 for definition). It follows that the yield function (2.2) for the composite cylinder must undergo kinematic strain hardening with loading; thus (2.2) has the general form

$$f(I_1-\alpha_1, I_2-\alpha_2, I_3, I_4, I_5)=0$$

Hardening Rule

The next step is to determine the hardening rule that specifies $\boldsymbol{\alpha}$, the kinematic work hardening parameter, as a function of loading history. The load increment can be written as the sum of two parts, as

$$\begin{bmatrix} dI_1 \\ \\ dI_2 \end{bmatrix} = \begin{bmatrix} d\alpha_1 \\ \\ d\alpha_2 \end{bmatrix} + d\mu \begin{bmatrix} -\dfrac{\partial f}{\partial I_2} \\ \dfrac{\partial f}{\partial I_1} \end{bmatrix} \tag{2.17}$$

where the last term involves neutral loading on the loading surface, and we must find the term $d\mu$. The general character of the loading surface is shown in Fig. 8.3. It is assumed that the vector $d\boldsymbol{\alpha}$ is directed radially from the center of the elliptical loading function, as shown in Fig. 8.3. The justification for this step must await comparison with experimental data or with a numerical solution. The construction shown in Fig. 8.3 is that of the hardening rule (I-3.11) given in Section 1.3. Note that we are here employing differential increments, rather than the notation involving rates given in Section 1.3. The term $d\mu$ in (2.17) can be found by requiring that $d\boldsymbol{\alpha}$ have the radial direction shown in Fig. 8.3. This is accomplished by

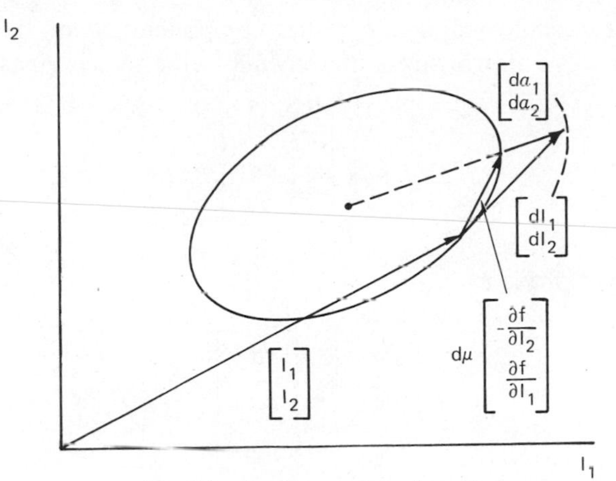

Fig. 8.3 Loading surface character.

requiring that the two vectors

$$\begin{bmatrix} I_1-\alpha_1+dI_1 \\ \\ I_2-\alpha_2+dI_2 \end{bmatrix} \quad \text{and} \quad \begin{bmatrix} I_1-\alpha_1-d\mu\dfrac{\partial f}{\partial I_2} \\ \\ I_2-\alpha_2+d\mu\dfrac{\partial f}{\partial I_1} \end{bmatrix}$$

be coaxial. Setting the cross product of these vectors equal to zero gives

$$d\mu=\frac{(I_1-\alpha_1)\,dI_2-(I_2-\alpha_2)\,dI_2}{(I_1-\alpha_1)\dfrac{\partial f}{\partial I_1}+(I_2-\alpha_2)\dfrac{\partial f}{\partial I_2}} \tag{2.18}$$

With $d\boldsymbol{\alpha}$ solved from (2.17) using (2.18), we now know the motion of the loading surface as specified by the kinematic work hardening parameter.

Flow Rule

The last item we need to complete the solution is the complete specification of the flow rule. The flow rule (I-3.7) here has the form

$$d\varepsilon^P=d\lambda\begin{bmatrix} \dfrac{\partial f}{\partial I_1} \\ \dfrac{\partial f}{\partial I_2} \end{bmatrix} \tag{2.19}$$

where $\boldsymbol{\varepsilon}^P$ is the plastic component of strain, with terms $\varepsilon_1 = \varepsilon_{zz}, \varepsilon_2 = 2\varepsilon_{rr}$. It is assumed that during loading the radial component $d\boldsymbol{\alpha}$ of the change in loading $d\mathbf{I}$ causes uniform stress increments in the matrix phase as

$$\begin{bmatrix} d\sigma_{rr}^m \\ d\sigma_{\theta\theta}^m \\ d\sigma_{zz}^m \end{bmatrix} = \begin{bmatrix} d\alpha_2 \\ d\alpha_2 \\ d\alpha_2 \end{bmatrix} \tag{2.20}$$

and in the fiber phase as

$$\begin{bmatrix} d\sigma_{rr}^f \\ d\sigma_{\theta\theta}^f \\ d\sigma_{zz}^f \end{bmatrix} = \begin{bmatrix} d\alpha_2 \\ d\alpha_2 \\ d\sigma_{zz}^f \end{bmatrix} \tag{2.21}$$

Relations (2.20) and (2.21) are simply a recognition of the fact that longitudinal stress in the fiber phase may be very large compared with the other components of stress in both phases. All these other components of stress are of the same order of magnitude relative to the fiber stress, σ_{zz}^f.

The equilibrium of the composite cylinder in the longitudinal direction during increment $d\boldsymbol{\alpha}$ requires the balance of forces

$$\left(d\sigma_{zz}^f\right) + (1-c)((d\sigma_{zz}^m) = d\alpha_1 \tag{2.22}$$

where c is the volume fraction of the fiber phase. From (2.20) and (2.22) there follows

$$c\left(d\sigma_{zz}^f\right) = d\alpha_1 - (1-c)(d\alpha_2) \tag{2.23}$$

Solving (2.23) explicitly for $d\sigma_{zz}^f$ and converting the result to longitudinal strain increment gives

$$d\varepsilon_1 = \frac{1}{E_f}\left[\frac{d\alpha_1}{c} - \left(\frac{1-c}{c} + 2\nu_f\right) d\alpha_2\right] \tag{2.24}$$

where E_f and ν_f are the elastic properties of the fiber phase and the last term in (2.24) follows from the lateral stress effect.

The plastic strain increment can be written as

$$d\varepsilon_1^P = d\varepsilon_1 - \left[\kappa_{11}\kappa_{12}\right] d\boldsymbol{\alpha} \tag{2.25}$$

where $[\kappa_{11}\kappa_{12}]$ are the elastic compliances of the composite cylinder. Substituting (2.24) into (2.25), and then eliminating $d\varepsilon_1^P$ between (2.19) and (2.25) gives the solution for the scalar multiplier $d\lambda$

$$d\lambda = \frac{1}{\dfrac{\partial f}{\partial I_1}}\left[\left(-\frac{2\nu_f+(1-c/c)}{E_f}-\kappa_{11}\right)d\alpha_2+\left(\frac{1}{cE_f}-\kappa_{12}\right)d\alpha_1\right] \tag{2.26}$$

We now have the complete formulation for the plastic deformation of the composite cylinder. The flow rule is given by (2.19) with (2.26), and the kinematic hardening parameter follows from (2.17) with (2.18). The complete mathematical machinery is now assembled by which plastic deformation solutions can be obtained.

It is important to note that nowhere have we restricted the plastic volume change to vanish, and, in general, it does not vanish, contrary to common assumption. Following through the details of the analysis we see that the spreading plastic zone in the elastic-perfectly plastic matrix phase is the physical mechanism that induces the kinematic hardening behavior in the composite. As pointed out by Dvorak and Rao [8.5], it is entirely plausible that strain hardening in engineering materials could be due to spreading zones of plastic behavior in materials that actually are elastic-perfectly plastic on a microscale. Such materials would be heterogeneous on the microscale, as with the grains in a metal.

The results of this rather simple theoretical formulation have been compared by Dvorak and Rao [8.5] with the results of a digital computer solution that evaluates all local effects. The comparison is completely acceptable, even in rather complicated deformation or loading history programs. This characterization of the macroscopic behavior of the elastic-plastic composite material is somewhat unusual. Most solutions merely examine some local aspects of the plastic deformation in a heterogeneous material system. The present result relates the macroscopic behavior of the composite to the properties and local behavior of the individual phases.

There have been many studies of yielding and possible fracture in composites, for example, Drucker [8.6], and Mulhern, Rogers, and Spencer [8.7]. See [8.5] for further revelant references to the plastic behavior in composite materials.

8.3 VISCOELASTIC PROPERTIES OF COMPOSITES

Composite materials typically contain one or more phases of polymeric constitution. Such polymer phases impart a viscoelastic type of behavior to the composite. Nevertheless, for many purposes the viscoelastic behavior is simply idealized as being elastic. There are other situations in which the viscoelastic effects cannot be idealized as elastic, and due account must be taken of the time and memory dependent effects. This section and the next one focus expressly on viscoelastic effects in heterogeneous media.

Correspondence Principle Application

With regard to effective properties, we have developed many elasticity results. The question confronting us now is whether these results can be converted to viscoelastic results. First we consider the results for exact elasticity solutions for the effective properties, and we defer until later the consideration of bounds. The effective elastic properties can be directly converted to viscoelastic properties, using the correspondence principle of Section 1.2. That is, an exact solution for an elastic effective property can be reinterpreted as a viscoelastic solution in the transform domain by replacing the elastic properties by the transform parameter multiplied Laplace transforms of the corresponding viscoelastic properties. See Hashin [8.8] for a general discussion of the topic.

A simple example illustrates this procedure. The effective bulk modulus for the composite spheres model is given by (II-3.17), repeated here as

$$\frac{k-k_m}{k_i-k_m}=\frac{c}{1+\left[(1-c)(k_i-k_m)\right]/\left(k_m+\frac{4}{3}\mu_m\right)} \tag{3.1}$$

Let s be the Laplace transform parameter. Replace each modulus in (3.1) by s times the Laplace transform of the corresponding viscoelastic relaxation function, to obtain

$$\frac{\bar{k}-\bar{k}_m}{\bar{k}_i-\bar{k}_m}=\frac{c}{1+\left[(1-c)(\bar{k}_i-\bar{k}_m)\right]/\left(\bar{k}_m+\frac{4}{3}\bar{\mu}_m\right)} \tag{3.2}$$

Given the relaxation functions $k_i(t)$, $k_m(t)$, and $\mu_m(t)$, take the Laplace transforms of each, substitute into (3.2), and invert $\bar{k}(s)$, to obtain the effective volumetric relaxation function $k(t)$.

In the case of complex moduli the correspondence principle process of Section 1.2 is even more simple; replace each modulus in (3.1) by the corresponding complex modulus, to obtain

$$\frac{k^*-k_m^*}{k_i^*-k_m^*}=\frac{c}{1+\left[(1-c)(k_i^*-k_m^*)\right]/\left(k_m^*+\frac{4}{3}\mu_m^*\right)} \tag{3.3}$$

With the frequency dependent specification of the two phase complex moduli, as

$$\mu_m = \mu_m'(\omega) + i\mu_m''(\omega)$$

$$k_m^* = k_m'(\omega) + ik_m''(\omega)$$

$$k_i^* = k_i'(\omega) + ik_i''(\omega) \tag{3.4}$$

then (3.3) gives the effective complex modulus $k^* = k'(\omega) + ik''(\omega)$. These simple procedures apply to all the rigorously derived results for effective properties.

A useful special form for the complex effective properties can be obtained in the case of small loss tangents for all phases. This procedure was pointed out by Hashin [8.8] and Schapery [8.9]. Let the effective elastic property be written as

$$\mathbf{C} = \mathbf{C}(\boldsymbol{\kappa}_i, c_i) \tag{3.5}$$

where there are N phases of elastic moduli $\boldsymbol{\kappa}_i$ and volume fractions c_i. Henceforth the dependence on the volume fractions c_i is understood. The corresponding viscoelastic result is then

$$\mathbf{C}^* = \mathbf{C}(\boldsymbol{\kappa}_i^*) \tag{3.6}$$

which we write explicitly as

$$\mathbf{C}^* = \mathbf{C}(\boldsymbol{\kappa}_i' + i\boldsymbol{\kappa}_i'') \tag{3.7}$$

We assume that we can expand (3.7) as a multiple complex power series. We specifically take, for each tensor component,

$$\boldsymbol{\kappa}_i'' \ll \boldsymbol{\kappa}_i'$$

and expand about $\boldsymbol{\kappa}_i'$ to obtain

$$\mathbf{C}^* = \mathbf{C}(\boldsymbol{\kappa}_i') + \sum_{i=1}^{N} (\boldsymbol{\kappa}_i^* - \boldsymbol{\kappa}_i') \frac{\partial \mathbf{C}(\boldsymbol{\kappa}_i')}{\partial \boldsymbol{\kappa}_i'} + \cdots \tag{3.8}$$

which we truncate at the level shown to give

$$\mathbf{C}^* \simeq \mathbf{C}(\boldsymbol{\kappa}_i') + i \sum_{i=1}^{N} \boldsymbol{\kappa}_i'' \frac{\partial \mathbf{C}(\boldsymbol{\kappa}_i')}{\partial \boldsymbol{\kappa}_i'} \tag{3.9}$$

The derivative operations shown in (3.8) and (3.9) must be carried out in terms of components. This result is a very simple form to use, since only the elastic function $\mathbf{C}(\boldsymbol{\kappa}_i)$ and its derivatives are involved, and the end result is already separated into real and imaginary parts.

Viscoelastic Bounds and Other Results

Next we turn to the question of bounds. Do the elasticity theory bounds on effective properties imply corresponding bounds in the viscoelastic case? In general the answer is no, or at least it is not known how to effect the transition from elastic bounds to viscoelastic bounds. It is only in very special situations that viscoelastic bounds on effective properties can be obtained, namely, those cases in which a continuous phase of homogeneous viscoelastic material contains either voids or perfectly rigid inclusions. Furthermore, the viscoelastic material must be isotropic, with Poisson's ratio taken as a real constant, which means the relaxation functions in shear and dilatation must be proportional. It is only in these highly restrictive cases that the appropriate viscoelastic minimum theorems are available for use. The results, taken from Christensen [8.10], are outlined here.

Under the conditions mentioned, there are two viscoelastic minimum theorems, one expressed in terms of the relaxation function in shear, and one in terms of the creep function in shear. The viscoelastic minimum theorems have a parallel structure to those of elasticity and it follows that the elastic bound results can be reinterpreted as a viscoelastic result. For example, let an elastic bound be given by

$$\frac{\mu}{\mu_m} \leqslant F_1(c, \nu_m) \tag{3.10}$$

where c is the volume fraction of the void or perfectly rigid phase and ν_m and μ_m are the properties of the other phase. The viscoelastic minimum theorem allows the elastic result (3.10) to be converted to the viscoelastic result

$$\mu(t) \leqslant F_1(c, \nu_m)\mu_m(t) \tag{3.11}$$

Similarly, an elastic result in terms of shear compliances may have the form

$$\frac{J}{J_m} \leqslant F_2(c, \nu_m) \tag{3.12}$$

and the corresponding viscoelastic result is then

$$J(t) \leqslant F_2(c, \nu_m) J_m(t) \tag{3.13}$$

We see that we obtain upper bounds on the relaxation function and creep function. Neither of these results can be converted to provide a lower bound on the other since $\mu(t)$ and $J(t)$ are not reciprocals, as in elasticity where $\mu = 1/J$. It has been claimed that lower bounds on $\mu(t)$ and $J(t)$ can be obtained, but that is incorrect, or at least it has not yet been done rigorously. Corresponding bounds on viscoelastic complex moduli and compliances are discussed in [8.10].

Next we discuss a special aspect of viscoelastic behavior in the case of a homogeneous isotropic medium that contains voids or rigid inclusions. Also, we take the medium as having a constant, real Poisson's ratio. The solution for the effective elastic property can be written as

$$\frac{\mu}{\mu_m} = f(c, \nu_m) \tag{3.14}$$

where again c is the volume fraction of voids or rigid inclusions. Using the correspondence principle, the solution for the effective viscoelastic complex modulus is given by

$$\frac{\mu^*}{\mu_m^*} = f(c, \nu_m) \tag{3.15}$$

Writing the loss tangents as

$$\phi = \tan^{-1}\frac{\mu''}{\mu'}$$

$$\phi_m = \tan^{-1}\frac{\mu_m''}{\mu_m'} \tag{3.16}$$

it follows from (3.15) and (3.16) that

$$\phi = \phi_m \tag{3.17}$$

Thus in this type of viscoelastic composite material containing voids or rigid inclusions, the effective loss tangent phase angle for the composite is identical to that for the matrix material. This result was first noted by Hashin [8.8]. Similarly, results can be obtained for anisotropic composites. A useful review article on viscoelastic composites was given by Schapery [8.9].

Wave Behavior Example

We now show a particular viscoelastic composite system and calculate its effective viscoelastic properties, as an illustration of these results. Small amounts of rubber are added to glassy polymers such as polystyrene to "toughen" these materials. That is, when the rubber is added, the materials are much less likely to experience brittle fracture. There are many different explanations for this effect; we discuss one such possible mechanism in the present context. Figure 8.4 shows a schematic diagram of a typical dispersion of rubber in the glassy polymer. It is seen that the rubber particle itself contains a high volume concentration of inclusions of the glassy polymer within it. Thus the configuration is that of a composite within a composite. We wish to model the viscoelastic complex modulus property of the entire composite. The glassy polymer is taken to be perfectly elastic, and the rubber phase is modeled by a viscoelastic complex modulus, which is consistent with interest in high frequency conditions appropriate to a wave propagation condition. For simplicity, to illustrate the results, both phases are taken to be incompressible.

Begin by recalling the appropriate effective elastic shear modulus formula (II-2.23) for very low volume concentration composites. For incompressibility this is rewritten as

$$\frac{\mu}{\mu_m} = 1 - \frac{5(1 - \mu_i/\mu_m)}{3 + 2(\mu_i/\mu_m)}c, \qquad c \ll 1 \tag{3.18}$$

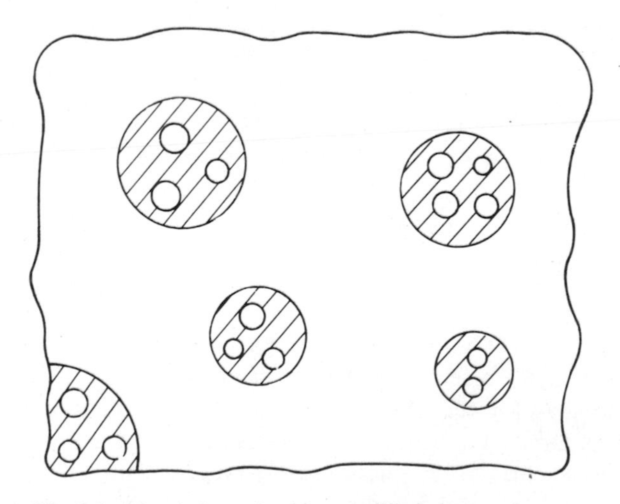

Fig. 8.4 Morphology of rubber modified glassy polymers.

Let the glassy polymer modulus be given by μ_G and the rubber material modulus by μ_R; then

$$\frac{\mu}{\mu_G} = 1 - \frac{5(1-\mu_R/\mu_G)}{3+2(\mu_R/\mu_G)}c \tag{3.19}$$

where c is the volume fraction of rubber particles. We can directly convert this form to the viscoelastic result using the correspondence principle, but first we obtain elastic results suggested by the morphology of Fig. 8.4.

Consistent with the geometry of Fig. 8.4, take the rubber particles as having particles of the glassy phase embedded in them. The rubber phase is taken to be sufficiently compliant that the glassy particles are rigid in comparison. Then from (3.18) the rubber phase effective property μ_{RP} is given by

$$\frac{\mu_{RP}}{\mu_R} = 1 + \tfrac{5}{2}\hat{c} \tag{3.20}$$

where $\hat{c}$ is the volume fraction of rigid, glassy inclusions in the rubber phase. Next these rubber phase particles containing rigid inclusions are to be embedded in the glassy polymer. With c still designating the volumetric ratio of homogeneous rubber to glassy polymer plus rubber, it can be shown that the volume fraction of the rubber phase with inclusions is given by

$$\frac{\begin{pmatrix}\text{volume of rubber phase}\\ \text{with inclusions}\end{pmatrix}}{(\text{total volume})} = \frac{c}{1-\hat{c}(1-c)}$$

where $\hat{c}$ is the volume fraction of glassy polymer particles in the rubber phase. From (3.18), the effective shear modulus in this situation is given by

$$\frac{\mu}{\mu_G} = 1 - \frac{5(1-\mu_{RP}/\mu_G)}{3+2(\mu_{RP}/\mu_G)}\left[\frac{c}{1-\hat{c}(1-c)}\right] \tag{3.21}$$

where μ_{RP} is given by (3.20).

Under the restriction of a very compliant rubber phase, $\mu_R/\mu_G \ll 1$, formulas (3.19) and (3.21) become

$$\frac{\mu}{\mu_G} = 1 - \tfrac{5}{3}\left(1 - \tfrac{5}{3}\frac{\mu_R}{\mu_G}\right)c \qquad \text{(no inclusions in rubber)}$$

and

$$\frac{\mu}{\mu_G}=1-\tfrac{5}{3}\left[1-\tfrac{5}{3}\left(1+\tfrac{5}{2}\hat{c}\right)\frac{\mu_R}{\mu_G}\right]\left[\frac{c}{1-\hat{c}(1-c)}\right] \qquad \left(\begin{matrix}\text{rigid inclusions}\\ \text{in rubber}\end{matrix}\right) \tag{3.22}$$

Relations (3.22) may be directly converted to viscoelastic forms in the manner already discussed. Taking the glassy polymer as elastic, the rubber phase is taken with a complex modulus

$$\mu_R^*=\mu_R'+i\mu_R''$$

Utilizing μ_R^* in (3.22) gives the results

$$\frac{\mu^*}{\mu_G}=1+i\frac{25}{9}\frac{\mu_R''}{\mu_G}c \qquad \text{(no inclusions in rubber)}$$

and

$$\frac{\mu^*}{\mu_G}=1+i\frac{25}{9}\frac{\mu_R''}{\mu_G}c\left[\frac{\left(1+\frac{5}{2}\hat{c}\right)}{1-\hat{c}(1-c)}\right] \qquad \left(\begin{matrix}\text{rigid inclusions}\\ \text{in rubber}\end{matrix}\right) \tag{3.23}$$

where approximations consistent with those already introduced have been used in going from (3.22) to (3.23).

The loss tangents corresponding to the results in (3.23) are given by

$$\tan\phi=\frac{25}{9}\frac{\mu_R''}{\mu_G}c \qquad \text{(no inclusions in rubber)}$$

and

$$\tan\phi=\frac{25}{9}\frac{\mu_R''}{\mu_G}c\left[\frac{1+\frac{5}{2}\hat{c}}{1-\hat{c}(1-c)}\right] \qquad \text{(rigid inclusions in rubber)} \tag{3.24}$$

We see that the loss tangent is greatly enhanced by the presence of the rigid glassy inclusions in the rubber phase. For realistic volume fractions of $c=0.10$ and $\hat{c}=0.50$ the enhancement factor shown in brackets in (3.24) has a value of 4. The use of the value of $\hat{c}=0.5$ in the present dilute suspension forms simply means that the basic effects are being underestimated.

The practical significance of these results is as follows. Under dynamic conditions the initiation and growth of cracks appears to have an intimate relationship to the character of wave propagation in the material. The loss tangent is the fundamental determining characteristic of wave propagation in viscoelastic materials. It is the loss tangent that determines the attenuation rate of the traveling waves, Eqs. (I-2.19, 2.20). High loss tangent materials can quickly attenuate dynamic effects. We see, therefore, that the results shown in (3.24) imply that the material with the inclusions in the rubber phase will be much more effective at attenuating wave effects. Thus even though the same amount of rubber and the same amount of glassy polymer is used in the two cases of (3.24), the material involving the glassy inclusions in the rubber provides a greatly more efficient use of the material. These results are consistent with actual observations.

We see that the composite material behavior results (3.23) and (3.24) are simple, meaningful results, which provide the basic materials properties to be used in analyzing wave propagation in viscoelastic materials. Comparable forms easily can be obtained with less restrictive assumptions than were invoked here for illustrative purposes.

8.4 A VISCOELASTIC FLUID SUSPENSION MODEL

In the preceding section we saw an example of a heterogeneous viscoelastic solid. Viscoelastic characterization is just as appropriate to the behavior of flowing fluids as it is to solids. Of course, at the level of infinitesimal deformation there need be no distinction between solids and fluids. It is only under large deformation conditions that the distinction occurs. We study the fluid suspension problem of interest here first at the infinitesimal deformation level, and after that we consider the extension of the results to model a flowing fluid.

Small Deformation Characterization

The fluid suspension of interest here is the simplest type of suspension that exhibits viscoelastic effects. Specifically, we consider a dilute suspension of elastic spherical particles in a Newtonian viscous matrix. Both phases are taken to be incompressible. Note that neither phase of the suspension is itself viscoelastic; nevertheless the effective property of the suspension will be found to be viscoelastic.

The starting point of the method is the elastic suspension result (II-2.23):

$$\frac{\mu}{\mu_m} = 1 - \frac{5(1-\mu_i/\mu_m)}{3+2(\mu_i/\mu_m)}c, \qquad c \ll 1 \tag{4.1}$$

for incompressible materials. We use the correspondence principle to convert (4.1) to the proper viscoelastic form. For elastic particles, modulus μ_i directly characterizes the behavior of the inclusion. For the matrix phase, here taken to be a Newtonian viscous fluid, as discussed in Section 1.2, we have

$$\mu_m \rightarrow i\omega\eta$$

where η is the coefficient of viscosity and ω is the frequency. The generalization of (4.1) to model the suspension, is then

$$\frac{\mu^*(\omega)}{i\omega\eta} = 1 - \frac{5(1 + i\mu_i/\omega\eta)}{3 - 2i(\mu_i/\omega\eta)}c \tag{4.2}$$

The complex compliance follows immediately from (4.2) as

$$J^*(\omega) = \frac{1}{\mu^*(\omega)} \tag{4.3}$$

Our first objective is to obtain the corresponding creep function and relaxation function. The creep function is decomposed into two parts, as in Section 1.2, as

$$J(t) = \tilde{J}(t) + \frac{t}{\hat{\eta}} \tag{4.4}$$

As shown by Gross [8.11] the creep function $J(t)$ is related to the complex compliance through

$$\frac{d\tilde{J}(t)}{dt} = \frac{2}{\pi}\int_0^\infty J'(\omega)\cos\omega t\, d\omega \tag{4.5}$$

where

$$J^*(\omega) = J'(\omega) + iJ''(\omega)$$

Using $J'(\omega)$ from (4.2) and (4.3) in (4.5) gives

$$\tilde{J}(t) = \frac{25c}{4\mu_i}(1 - e^{-(t/\xi)}) \tag{4.6}$$

where

$$\xi = \frac{3}{2}\frac{\eta}{\mu_i} \tag{4.7}$$

To obtain $\hat{\eta}$ in (4.4) take

$$\left.\frac{dJ(t)}{dt}\right|_{t\to\infty}=\frac{1}{\eta^*(\omega)|_{\omega\to 0}} \tag{4.8}$$

where

$$\eta^*(\omega)=\frac{\mu^*(\omega)}{i\omega}$$

It is found that

$$\hat{\eta}=\left(1+\tfrac{5}{2}c\right)\eta \tag{4.9}$$

With (4.6) and (4.9), the creep function (4.4) for the suspension is completely determined in terms of the viscosity of the matrix phase and the modulus and volume fraction of the elastic spherical particles.

The relaxation function for the suspension is given by

$$\bar{\mu}(s)=\frac{1}{s^2\bar{J}(s)} \tag{4.10}$$

This relation is obtained from (I-2.6), where Laplace transform forms are used with transform variable s. Using $J(t)$ from (4.4), (4.6), and (4.9) in (4.10), and after performing all transform operations, it is found that

$$\mu(t)=\left(1-\tfrac{5}{3}c\right)\eta\delta(t)+\tfrac{25}{9}c\mu_i e^{-(t/\tau)} \tag{4.11}$$

where

$$\tau=\frac{3}{2}\frac{\eta}{\mu_i} \tag{4.12}$$

with $\delta(t)$ being the Dirac delta function. The appearance of the delta function in the relaxation function (4.11) cannot be surprising. The matrix phase, being a viscous fluid, has no instantaneous elasticity effect, and this leads to the delta function type of behavior. We now have the complete characterization of the effective viscoelastic properties of the suspension under small deformation conditions. It is indeed revealing that viscoelastic effects appear in the macroscopic behavior even though neither phase itself is viscoelastic. Clearly, the main ingredients necessary for a material to exhibit viscoelastic memory effects are the capacity to dissipate energy and the simultaneous capacity to store energy. The present type of suspension has both of these capabilities; thus viscoelastic effects appear.

Non-Newtonian Flow Behavior

It is logical to inquire whether the small deformation results just found can be generalized to model the large deformation kinematics inherent in a flowing suspension. In fact, this generalization can be accomplished, and we now proceed to outline the method and results, which are taken from Christensen [8.12]. The key to the generalization lies with the capacity of the suspension to store energy, as alluded to earlier. Materials that store energy admit the existence of a free energy function and a thermodynamical derivation of constitutive relations. The method was explicitly given by Coleman [8.13] for viscoelastic materials.

Our interest is restricted to slow flow conditions, and we begin by obtaining certain basic results for the flow of a homogeneous viscoelastic fluid. After that, we use the result to model the equivalent homogeneous behavior of the suspension. We take an expansion of the free energy as

$$\rho A = \int_{-\infty}^{t}\int_{\infty}^{t} \Delta(t-\tau_1, t-\tau_2)\dot{G}_{ij}(\tau_1)\dot{G}_{ij}(\tau_2)\,d\tau_1\,d\tau_2$$

$$+\int_{-\infty}^{t}\int_{-\infty}^{t} \gamma(t-\tau_1, t-\tau_2)\dot{G}_{ii}(\tau_1)\dot{G}_{jj}(\tau_2)\,d\tau_1\,d\tau_2 + \cdots \tag{4.13}$$

where

$$G_{ij}(\tau) = C_{ij}(\tau) - \delta_{ij}$$

$$C_{ij}(\tau) = \frac{\partial x_k(\tau)}{\partial x_i(t)}\frac{\partial x_k(\tau)}{\partial x_j(t)} \tag{4.14}$$

where the "dot" denotes derivatives with respect to the argument shown and rectangular Cartesian coordinates are employed with $x_i(t)$ being the particle position. Symbol t refers to current time and τ to past times. The higher-order terms in (4.13) are neglected, consistent with the slow flow assumption. Following the thermodynamical method of Coleman [8.13], the free energy form can be shown to lead to the constitutive relation

$$\sigma_{ij}(t) = -p\delta_{ij} + 4\int_{-\infty}^{t} \Delta(t-\tau, 0)\dot{G}_{ij}(\tau)\,d\tau$$

$$-4\int_{-\infty}^{t}\int_{-\infty}^{t} \Delta(t-\tau_1, t-\tau_2)\dot{G}_{ik}(\tau_1)\dot{G}_{kj}(\tau_2)\,d\tau_1\,d\tau_2$$

$$-4\int_{\infty}^{t}\int_{\infty}^{t} \gamma(t-\tau_1, t-\tau_2)\dot{G}_{ij}(\tau_1)\dot{G}_{kk}(\tau_2)\,d\tau_1\,d\tau_2 \tag{4.15}$$

where symmetry of $\Delta(\tau_1,\tau_2)$ and $\gamma(\tau_1,\tau_2)$ in their arguments have been used.

Now the constitutive relation (4.15), being of a general nonlinear form, must admit small deformation conditions as a special case. Relation (4.15) when restricted to small deformation conditions has a linear viscoelastic form that comes from the single integral term in (4.15). For that result to model the suspension of interest here, the relaxation function $4\Delta(t-\tau,0)$ in (4.15) must be identified with the relaxation function (4.11); thus we write

$$4\Delta(t,0)=\alpha\delta(t)+\beta e^{-(t/\tau)} \tag{4.16}$$

where

$$\begin{aligned}
\alpha &= \left(1-\tfrac{5}{3}c\right)\eta \\
\beta &= \tfrac{25}{9}c\mu_i \\
\tau &= \frac{3\eta}{2\mu_i}
\end{aligned} \tag{4.17}$$

Our interest is in the flow problem involving simple shear deformation. This is specified by

$$\begin{aligned}
x_1(t) &= X_1+k(t)X_2 \\
x_2(t) &= X_2 \\
x_3(t) &= X_3
\end{aligned} \tag{4.18}$$

where $k(t)$ is used to specify the rate of flow and X_i are fixed coordinates. It can be shown that the deformation state, (4.14), corresponding to (4.18) is given by

$$[G_{ij}(\tau)]=\begin{bmatrix} 0 & [k(\tau)-k(t)] & 0 \\ [k(\tau)-k(t)] & [k(\tau)-k(t)]^2 & 0 \\ 0 & 0 & 0 \end{bmatrix} \tag{4.19}$$

Our interest is further restricted to the steady state flow conditions, and we assume that all initial transients have decayed away. In this state of steady simple shearing flow, it can be shown that only the integrals in (4.15) involving $\Delta(\ \ )$ contribute to the lowest-order terms. The terms due to $\gamma(\ \ )$ in (4.15) are of higher order and therefore are neglected. Furthermore, in view of the exponential character of $\Delta(t,0)$, Eq. (4.16), the arguments of

$\Delta(t-\tau_1, t-\tau_2)$ in (4.15) are taken in additive form $\Delta(2t-\tau_1-\tau_2)$, with $\Delta(t)$ given by (4.16).

The steady state solution is found to be

$$\begin{aligned} \sigma_{11} &= -p - \beta\tau^2\kappa^2 \\ \sigma_{22} &= -p - 3\beta\tau^2\kappa^2 \\ \sigma_{33} &= -p \\ \sigma_{12} &= (\alpha + \beta\tau)\kappa \end{aligned} \tag{4.20}$$

where κ is the shear rate. Using α, β, and τ from (4.17), then (4.20) has the explicit form

$$\begin{aligned} \sigma_{11} &= -p - \tfrac{25}{4} c \frac{\eta}{\mu_i} \kappa^2 \\ \sigma_{22} &= -p - \tfrac{75}{4} c \frac{\eta}{\mu_i} \kappa^2 \\ \sigma_{33} &= -p \\ \sigma_{12} &= \left(1 + \tfrac{5}{2} c\right)\eta\kappa \end{aligned} \tag{4.21}$$

We see that the shearing flow of the suspension gives rise not only to shearing stress, but also to normal stresses. We see from the derivation that these normal stresses are directly due to the capacity of the suspension to store energy by deforming the elastic particles. If the particles are perfectly rigid, $\mu_i \rightarrow \infty$, then from (4.21) the normal stress effects disappear.

Although it is possible to retain higher-order terms in κ in the various terms in (4.21), because of the relaxation function $\Delta(t)$, it is not consistent to do so. This is because other higher-order terms in the original expansion step of the derivation, (4.13), also contribute terms of order κ^4 and higher in the normal stress terms of (4.21), and of order κ^3 in the shear stress term. Thus the method only can be used to deduce the slow flow characteristics of the suspension as it asymptotically approaches a zero shear rate. Nevertheless the knowledge of the mere existence and, more definitively, the magnitude of the normal stress terms, in terms of the suspension properties, is valuable information. The characteristic relaxation time for the suspension, $\tau = 3\eta/2\mu_i$, is also of practical interest.

The flow properties of this type of suspension have been studied by many authors, from the point of view of integrating the Navier-Stokes equations under creeping flow conditions and satisfying the boundary conditions on the surface of the deformed sphere. Perhaps the most complete work of that type is that of Goddard and Miller [8.14]. Their normal stress terms are of the same form and magnitude as those in (4.21) but with different coefficients. Leal [8.15] has criticized the approach of using thermodynamics to aid in the characterization of suspension behavior. He argues that suspension problems can only be approached by such methods as direct integration of the fundamental balance of momentum relations. Of course, there is no truly fundamental level at which to approach these problems. Assumptions and hypotheses abound at all levels. Furthermore our understanding of the mechanical behavior of viscoelastic materials would be extremely limited were it not for the underlying thermodynamical derivations. Our point of view is that there are many different mathematical approaches to modeling material behavior. No one approach is uniquely best. The important aspect of all derivations is that they be performed with consistent assumptions and with reasonable mathematical rigor.

8.5 LARGE DEFORMATION OF FIBER SYSTEMS

Problems involving the large deformation of heterogeneous material systems must be expected to be much more complicated than those for small deformation. The difficulty with the more general problem is not only that of the nonlinear kinematics; the general and proper forms of constitutive relations for use under nonlinear conditions are uncertain. To be sure, there are a great many different nonlinear constitutive relations designed to model various effects; however, agreement on the most general and most versatile forms has not been reached. There are, however, a few nonlinear problems for which general solutions can be found independent of the constitutive relations. In fact, there is a class of problems, of relevance to fiber reinforced materials, for which general solutions can be obtained independent of constitutive relations.

The general types of problems that involve us here are those of the behavior of materials reinforced with inextensible fibers. In Section 6.1 we developed a boundary layer theory appropriate to a condition of very stiff fibers. Here we take the idealization one step further and assume that the fibers are infinitely rigid in the axial direction. The general field of reinforcement with inextensible chords was developed by Adkins and Rivlin [8.16] beginning in 1955. The theory is summarized by Green and Adkins [8.17]. The two-dimensional theory to be covered here was developed in part by Mulhern, Rogers, and Spencer [8.18] in the context of plasticity theory, and finally in general form by Pipkin and Rogers [8.19]. General accounts of the work have been given by Pipkin [8.20] and Rogers [8.21], the former of which we follow here.

General Theory

Our concern here is with a three-dimensional medium reinforced with a single family of initially parallel, straight, inextensible fibers. As in much of our work, we are not concerned with the behavior of the individual phases, but rather with the macroscopic behavior of the composite material. In terms of the average field variables, the effect of inextensibility in one (initial) direction is a type of constraint on the response of the material. In fact, this type of one-dimensional constraint is the counterpart of the three-dimensional constraint imposed by an assumption of incompressibility. That brings us to our second major assumption, which is that of incompressibility. The third major assumption is that of two-dimensional, plane strain behavior such that

$$
\begin{aligned}
x &= x(X,Y)\\
y &= y(X,Y)\\
z &= Z
\end{aligned}
\tag{5.1}
$$

where the initial material positions are given by $\mathbf{X}$ and the positions after deformation by $\mathbf{x}=\mathbf{x}(\mathbf{X})$. The preceding assumptions impose severe constraints on the types of deformation that can occur. Nevertheless we find that an important class of deformations are possible.

We refer to the set of material points in the direction of inextensibility as a fiber. We take the fibers to be initially straight and parallel to the X direction. After deformation, then, the fiber is a curve of the material points. The field of fiber angular orientations from the X direction is specified by $\theta(x,y)$, where for the initial direction, $\theta_0=0$.

Take unit vectors $\mathbf{a}$ and $\mathbf{n}$ such that

$$
\begin{aligned}
\mathbf{a} &= \mathbf{i}\cos\theta+\mathbf{j}\sin\theta\\
\mathbf{n} &= -\mathbf{i}\sin\theta+\mathbf{j}\cos\theta
\end{aligned}
\tag{5.2}
$$

where $\mathbf{i}$ and $\mathbf{j}$ are unit vectors in the X and Y directions, respectively. Note that

$$\frac{d\mathbf{a}(\theta)}{d\theta}=\mathbf{n}(\theta)$$

and

$$\frac{d\mathbf{n}(\theta)}{d\theta}=-\mathbf{a}(\theta) \tag{5.3}$$

The directions specified by $\mathbf{n}$ are called *normal lines*, and these lines are normal to the fibers.

Let ∇ be the gradient operator with respect to $\mathbf{x}$. The curvature of the fibers and normal lines are defined by

$$\frac{1}{r_a} = \mathbf{a} \cdot \nabla \theta$$

$$\frac{1}{r_n} = \mathbf{n} \cdot \nabla \theta \tag{5.4}$$

where r_a and r_n are the radii of curvature. Taking various multiplicative combinations of (5.3) and (5.4) and using the chain rule gives

$$(\mathbf{a} \cdot \nabla)\mathbf{a} = \frac{\mathbf{n}}{r_a}, \qquad (\mathbf{a} \cdot \nabla)\mathbf{n} = -\frac{\mathbf{a}}{r_a}$$

$$(\mathbf{n} \cdot \nabla)\mathbf{a} = \frac{\mathbf{n}}{r_n}, \qquad (\mathbf{n} \cdot \nabla)\mathbf{n} = -\frac{\mathbf{a}}{r_n} \tag{5.5}$$

These results, well known as the Serrat-Frenet formulas, relate the curvatures to the derivatives of the unit vectors.

Next we consider the deformation gradients defined by

$$F_{ik} = \frac{\partial x_i}{\partial X_k}$$

Obviously, the constraints of the problems impose restrictions on the components of the deformation gradient tensor. The fiber cannot extend or contract, as we already know. However, the assumption of incompressibility requires that area elements normal to the fibers be conserved. Further, then, the requirement of no deformation in the Z direction means that the distance between fibers remain unchanged. Thus there can be no change in length along the fiber, or normal to it. This leaves the only possible mode of deformations as that of simple shear in the fiber direction. Thus we interrelate the deformed and undeformed coordinates by

$$\mathbf{x} = (X + kY)\mathbf{a} + Y\mathbf{n} + Z\mathbf{k} \tag{5.6}$$

where k specifies the shear distortion. The directional derivatives of (5.6) with respect to the undeformed coordinates are

$$(\mathbf{a}_0 \cdot \nabla_0)\mathbf{x} = \mathbf{a}$$

$$(\mathbf{n}_0 \cdot \nabla_0)\mathbf{x} = \mathbf{n} + k\mathbf{a}$$

$$(\mathbf{k}_0 \cdot \nabla_0)\mathbf{x} = \mathbf{k} \tag{5.7}$$

where ∇_0 is the gradient operator with respect to the undeformed coordinates, with $\mathbf{a}_0$, $\mathbf{n}_0$, and $\mathbf{k}_0$ being unit vectors in the undeformed coordinate directions. Relations (5.7) then give the components of the deformation gradient as

$$F_{i,k} = \begin{bmatrix} \cos\theta & -\sin\theta + k\cos\theta & 0 \\ \sin\theta & \cos\theta + k\sin\theta & 0 \\ 0 & 0 & 1 \end{bmatrix} \tag{5.8}$$

The functions k and θ specify the deformation gradient of the material. These are a function of position, as $k = k(\mathbf{x})$ and $\theta = \theta(\mathbf{x})$, or $k = k(\mathbf{X})$ and $\theta = \theta(\mathbf{X})$. But k and θ cannot be independent functions of position since both are derived from $\mathbf{x} = \mathbf{x}(\mathbf{X})$. Thus there are compatibility conditions to be satisfied, which may be derived from the identity

$$F_{iK,L} = F_{iL,K} \tag{5.9}$$

From this and (5.7) it follows that

$$\frac{\partial \mathbf{a}}{\partial Y} = \frac{\partial(\mathbf{n} + k\mathbf{a})}{\partial X} \tag{5.10}$$

Using the components from (5.8) in (5.10) gives

$$\frac{\partial(\cos\theta)}{\partial Y} = \frac{\partial}{\partial X}(-\sin\theta + k\cos\theta)$$

and

$$\frac{\partial(\sin\theta)}{\partial Y} = \frac{\partial}{\partial X}(\cos\theta + k\sin\theta) \tag{5.11}$$

Now using the chain rule, these can be used to obtain

$$\frac{\partial(\theta - k)}{\partial X} = 0$$

and

$$\frac{\partial\theta}{\partial Y} - k\frac{\partial\theta}{\partial X} = 0 \tag{5.12}$$

The first part of (5.12) gives

$$k = \theta + f(Y) \tag{5.13}$$

Thus the shear deformation is dependent on angle θ and a term that is constant along fiber directions. Note that the second part of (5.12) can be written as

$$\mathbf{n}\cdot\nabla\theta=0 \tag{5.14}$$

where from (5.6)

$$\mathbf{n}=\frac{\partial\mathbf{x}}{\partial Y}-k\frac{\partial\mathbf{x}}{\partial X}$$

thus

$$\mathbf{n}\cdot\nabla=\frac{\partial}{\partial Y}-k\frac{\partial}{\partial X}$$

Since angle θ also specifies the orientation of the normal lines, (5.14) shows that the normal lines are straight. This result is of fundamental use in constructing deformation solutions.

Finally, to complete the theory we must specify the equilibrium equations. First, stress must be characterized in accordance with the constraints. In tensor notation the stress is given by

$$\sigma_{ij}=Ta_ia_j-p\delta_{ij}+\tau_{ij} \tag{5.15}$$

where p is the reactive hydrostatic pressure, T is the reactive fiber stress, and the extra stress, τ_{ij}, is that part due to the shear deformation of the medium and the constraint in the X_3 direction. In dyadic notation we write this as

$$\boldsymbol{\sigma}=T\mathbf{aa}-p\mathbf{I}+\boldsymbol{\tau} \tag{5.16}$$

It is convenient to rewrite (5.16) in the form

$$\boldsymbol{\sigma}=T\mathbf{aa}-p(\mathbf{I}-\mathbf{aa})+S(\mathbf{an}+\mathbf{na})+S_3\mathbf{kk} \tag{5.17}$$

where now T is the total stress in the fiber direction. In this form, S_3-p is the constraint stress in the X_3 direction and S is the amplitude of the shear stress term. Both S and S_3 are functions or functionals of the shear deformation. Thus the constitutive relation for the shear deformation is needed to specify S and S_3. At the present level of generality there is no restriction on the type of constitutive relation; it could be elastic, viscoelastic, plastic, or of any other form. The general form of the solution can be obtained in many cases without actually specifying the constitutive relation.

The stress (5.17) when substituted into the stress equilibrium equations,

$$\nabla \cdot \boldsymbol{\sigma} = 0$$

gives the determining equations for p and T, the relative stresses. Using the Serret-Frenet formulas (5.5) and the divergence formulas, $\nabla \cdot \mathbf{a} = 1/r_n$ and $\nabla \cdot \mathbf{n} = -1/r_a$, and recalling from (5.14) that the curvature $1/r_n$ vanishes, we obtain the two determining equilibrium equations as

$$\mathbf{a} \cdot \nabla T = \frac{2S}{r_a} - \mathbf{n} \cdot \nabla S \tag{5.18}$$

and

$$\mathbf{n} \cdot \nabla p - \frac{p}{r_a} = \frac{T}{r_a} + \mathbf{a} \cdot \nabla S \tag{5.19}$$

Thus if we can consider the kinematics of the problem as known, then (5.18) can be integrated to give the field of T, where values at one point on each fiber must be specified. With T known, p can be found by integrating (5.19) along normal lines, with the value of p specified at one point along each normal line.

The key element in this procedure involving integration to get stresses is the knowledge of the kinematics of the deformed state. In many examples, as we shall see, it is possible to determine the state of deformation, independent of the full equilibrium solution. This is in sharp contrast to the situation in the usual theories, where, in general, one cannot determine the state of deformation without explicitly satisfying equilibrium. The special situation here evolves as the direct result of the high degree of constraints on the problem. We see next, from an example involving both displacement and stress boundary conditions, that the state of deformation can be determined from consideration of equilibrium in terms of force resultants.

Deformation of a Cantilever Plate

We seek to determine the state of deformation in the cantilever plate in Fig. 8.5a, loaded by a distribution of end force. The end at $X=0$ is constrained by the rigid attachment. Now the line of attachment is a normal line before deformation, and it must remain as one afterward. The fiber lines, being perpendicular to the normal lines, must remain normal to the rigid support after deformation. Thus, along the support line there is no shear deformation, and $\theta = 0$; it follows from (5.13) that $f(Y) = 0$, leaving

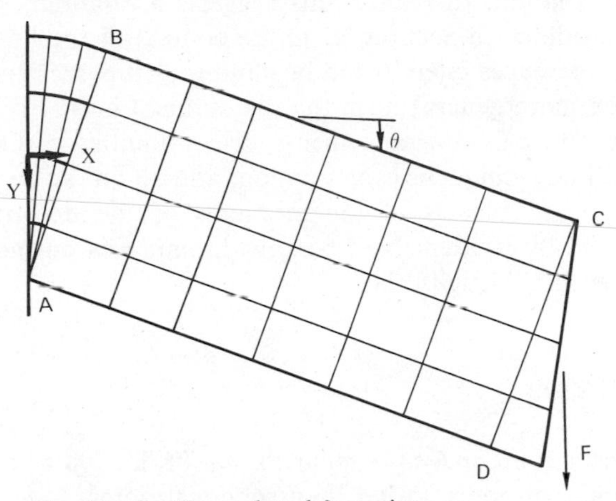

Fig. 8.5 Cantilever plate deformation.

the condition that the shear k is everywhere given by just θ. Along a normal line, which must be straight, therefore the shear deformation is constant, and correspondingly the shearing stress is constant along normal lines. The total shearing force across the normal line cross section is then $hS(\theta)$, where h is the thickness. For equilibrium this must be equal to the appropriate component of the end load; thus

$$hS(\theta) = F\cos\theta \tag{5.20}$$

Only one value of θ emerges as the solution of (5.20), once the constitutive relation $S(\theta)$ is specified; thus the part of the plate shown by $ABCD$ in Fig. 8.5 is in a state of uniform shear. For the normal lines and the fiber lines to remain perpendicular, the deformation region near the support must be fan shaped as shown in Fig. 8.5. We do not have enough information to discuss the deformation in the region of the end load, Fig. 8.5. The general deformation solution of this problem contrasts dramatically with the bending solution of elementary theory. In fact, from the point of view of the elementary theory, the inextensibility of the fibers would dictate that no deformation could occur in the loaded plate.

The stresses could now be found from (5.18) and (5.19), but the solution is practically self-apparent. The shear stresses are uniform through most of the plate. However, there are boundary conditions of zero tractions on the top and bottom surfaces. Accordingly, the top and bottom fiber surfaces must carry a singular stress level. This is seen as follows: infinite normal stresses are required for equilibrium across the fiber layer, which is of infinitesimal thickness but has shear stress on opposite sides that varies by a finite amount. In reality, this suggests a boundary layer effect such as was considered in Section 5.1 in the context of the linear theory.

The references cited at the beginning of this section develop the theory in much more general form than is outlined here. Formulations are given for concentric fiber arrangements, discontinuities in the slope of radial and normal lines, uniqueness of solution, and so on. A book by Spencer [8.22] covers many aspects of the problems of the deformation of materials reinforced by inextensible fibers, with emphasis on modes of deformation governed by plasticity theory.

PROBLEMS

1. Carry out an analysis similar to that of Section 8.1, but take the case of cylindrical pores, rather than spherical pores. Use the composite cylinders model in this connection.
2. Consider the shear deformation of a porous material and idealize the material as that of rigid-perfectly plastic. Investigate the determination of a plastic collapse load for the material.
3. Take an idealized fiber material as being elastic-perfectly plastic. In an aligned fiber system, take the matrix phase as elastic. What would be the macroscopic stress-strain characteristics of the composite, regarding a state of uniaxial tension in the direction of the fibers?
4. Consider a system of aligned elastic fibers in an elastic-perfectly plastic matrix, with the fibers in a cubical stacking arrangement. Concerning transverse states of shear deformation, will unlimited flow of the composite occur at stress levels above, below, or the same as that of the matrix material? Consider dilute and nondilute fiber concentration cases. What would be the effect of strain hardening in the matrix phase? See [8.6].
5. Relations (3.11) and (3.13) give upper bounds on effective relaxation and creep functions, respectively. Write the corresponding upper bound forms for complex modulus and compliance functions, and use these to obtain upper and lower bound results on a single property.
6. Obtain the real and imaginary parts of the effective complex shear modulus under dilute suspension conditions involving spherical particles. Take both phases as being viscoelastic, and begin with the corresponding elastic formula (3.18).
7. Specialize the result obtained in Problem 6 to the case of spherical inclusions of a viscous material in an elastic matrix. Compare the

resulting complex modulus with that obtained in Section 8.4 involving elastic inclusions in a viscous matrix.

8. Consider the deformation of a system of aligned inextensible fibers, in which one surface is brought into contact with a rigid cylindrical surface. Obtain the solution for the stress state in the deformed body. See [8.20].

REFERENCES

8.1 M. M. Carroll and A. C. Holt, "Static and dynamic pore-collapse relations for ductile porous materials," *J. Appl. Phys.*, vol. 43, 1626 (1972).

8.2 B. M. Butcher, M. M. Carroll, and A. C. Holt, "Shock-wave compaction of porous aluminum," *Int. J. Mech. Sci.*, vol. 45, 3864 (1974).

8.3 P. Chadwick, "Compression of a spherical shell of work hardening material," *Int. J. Mech. Sci.*, vol. 5, 165 (1963).

8.4 T. Y. Chu and Z. Hashin, "Plastic behavior of composites and porous media under isotropic stress," *Int. J. Eng. Sci.*, vol. 9, 971 (1971).

8.5 G. J. Dvorak and M. S. M. Rao, "Axisymmetric plasticity theory of fibrous composites," *Int. J. Eng. Sci.*, vol. 14, 361 (1976).

8.6 D. C. Drucker, "Engineering and continuum aspects of high-strength materials," in *High Strength Materials*, V. F. Zakay, Ed., Wiley, New York, 1965.

8.7 J. F. Mulhern, T. G. Rogers, and A. J. M. Spencer, "A continuum theory of a plastic-elastic fibre-reinforced material," *Int. J. Eng. Sci.*, vol. 7, 129 (1969).

8.8 Z. Hashin, "Complex moduli of viscoelastic composites—I. General theory and application to particulate composites," *Int. J. Solids Structures*, vol. 6, 539 (1970).

8.9 R. A. Schapery, "Viscoelastic behavior and analysis of composite materials," in *Composite Materials*, vol. 2, G. P. Sendeckyj, Ed., Academic, 1974.

8.10 R. M. Christensen, "Viscoelastic properties of heterogeneous media," *J. Mech. Phys. Solids*, vol. 17, 23 (1969).

8.11 B. Gross, *Mathematical Structure of the Theories of Viscoelasticity*, Hermann, Paris, 1953.

8.12 R. M. Christensen, "A special theory of viscoelastic fluids for application to suspensions," *Acta Mech.*, vol. 16, 183 (1973).

8.13 B. D. Coleman, "Thermodynamics of materials with memory," *Arch. Ration. Mech. Anal.*, vol. 17, 1 (1964).

8.14 J. D. Goddard and C. Miller, "Nonlinear effects in the rheology of dilute suspensions," *J. Fluid Mech.*, vol. 28, 657 (1967).

8.15 G. L. Leal, *Appl. Mech. Rev.*, vol. 27, review no. 8537 (1974).

8.16 J. E. Adkins and R. S. Rivlin, "Large elastic deformations of isotropic materials—X. Reinforcement by inextensible cords," *Phil. Trans. R. Soc. Lond.*, vol. A248, 201 (1955).

8.17 A. E. Green and J. E. Adkins, *Large Elastic Deformations*, 2nd ed., Oxford, 1970.

8.18 J. F. Mulhern, T. G. Rogers, and A. J. M. Spencer, "A continuum model for fibre-reinforced plastic materials," *Proc. R. Soc. Lond.*, vol. A301, 473 (1967).

8.19 A. C. Pipkin and T. G. Rogers, "Plane deformations of incompressible fiber-reinforced materials," *J. Appl. Mech.*, vol. 38, 634 (1971).

8.20 A. C. Pipkin, "Finite deformations of ideal fiber-reinforced composites," in *Composite Materials*, vol. 2, G. P. Sendeckyj, Ed., Academic, New York, 1974.

8.21 T. G. Rogers, "Finite deformations of strongly anisotropic materials," in *Theoretical Rheology*, J. F. Hutton, J. R. A. Pearson, and K. Walters, Eds., Wiley, 1975.

8.22 A. J. M. Spencer, *Deformations of Fibre-Reinforced Materials*, Oxford University Press, New York, 1972.

CHAPTER IX
EFFECTIVE THERMAL PROPERTIES

In this final chapter we return to the problem of the determination of effective properties for heterogeneous media. In the first few chapters we determined effective elastic properties. It is quite clear that a similar situation can be posed for other field theory properties such as the usual thermal, electrical, and magnetic properties. Because of the close interaction between mechanical and thermal effects, we are here concerned explicitly with thermal properties.

Specifically, we shall determine the effective thermal conductivity, coefficient of thermal expansion, and specific heats for certain common types of heterogeneous media. The averaging procedures used to determine the effective properties are the same as those introduced in Section 2.1.

Our starting point necessarily is a derivation of the proper form of a theory of thermoelasticity in which the coupling between mechanical and thermal properties is displayed.

9.1 THERMOELASTICITY THEORY

Our interest in thermal properties motivates us to examine the defining forms for these properties. The term *defining forms* is here taken to mean the proper, consistent forms in which these properties appear in a rigorously derived theory of behavior. We are thus naturally led to a thermodynamical derivation of the theory of thermoelasticity. At the present level we are assuming homogeneity. In the following sections we examine the heterogeneous material aspects of thermal behavior.

The thermodynamical method and results are widely known and used, and we repeat these results here in the interests of completeness. Reference should be made to Truesdell and Toupin [9.1] and Truesdell and Noll [9.2] for general derivations. We began with the statements of the local balance of energy equation as

$$\rho r - \rho \dot{U} + \sigma_{ij} d_{ij} - q_{i,i} = 0 \tag{1.1}$$

where ρ is the mass density, r is the heat supply or removal function per unit mass, U is the internal energy per unit mass, σ_{ij} is the stress tensor, d_{ij} is the rate of deformation tensor, and q_i denotes the heat flux vector. As usual, the superimposed dot denotes the time rate of change. The entropy production inequality (Clasius-Duhem inequality) is taken in the form

$$T\rho\dot{S} - \rho r + q_{i,i} - q_i\left(\frac{T_{,i}}{T}\right) \geqslant 0 \tag{1.2}$$

where T is absolute temperature and S is entropy per unit mass.

Helmholtz Free Energy Derivation

We have reason to use an alternate form of the energy balance equation (1.1). Specifically, we introduce the Legendre transformation

$$\rho U = \rho A + T\rho S \tag{1.3}$$

where A is the Helmholtz free energy per unit mass. The use of (1.3) in (1.1) gives

$$\rho r - \rho\left[\dot{A} + \dot{T}S + T\dot{S}\right] + \sigma_{ij} d_{ij} - q_{i,i} = 0 \tag{1.4}$$

The heat supply function ρr can be eliminated between (1.2) and (1.4) to leave

$$-\rho S\dot{T} - \rho\dot{A} + \sigma_{ij} d_{ij} - q_i \frac{T_{,i}}{T} \geqslant 0 \tag{1.5}$$

At this point we restrict attention to small deformation conditions. We also take

$$T = T_0 + \theta$$

where T_0 is a given base temperature and θ / T_0 is an infinitesimal of the same order, O(ε), as the infinitesimal strain tensor ε_{ij}. With these restrictions (1.5) becomes

$$-\rho S\dot{\theta}-\rho\dot{A}+\sigma_{ij}\dot{\varepsilon}_{ij}-q_i\frac{\theta_{,i}}{T_0}\geqslant 0 \tag{1.6}$$

Under the conditions of infinitesimal change from a given state, the mass density ρ now could be taken to be ρ_0, that of the initial configuration. For simplicity, however, we retain the notation ρ.

Next we expand the free energy ρA as a power series in the field variables ε_{ij} and θ. Thus we write

$$\begin{aligned}\rho A=\rho A_0&+D_{ij}\varepsilon_{ij}-\beta\theta\\&+\tfrac{1}{2}C_{ijkl}\varepsilon_{ij}\varepsilon_{kl}\\&-\phi_{ij}\varepsilon_{ij}\theta\\&-\tfrac{1}{2}C_v\theta^2+O(\varepsilon^3)\end{aligned} \tag{1.7}$$

where ρA_0, D_{ij}, β, C_{ijkl}, ϕ_{ij}, and C_v are constants that we shall interpret later. Consistent with the conditions of an infinitesimal change of state, we truncate the expansion (1.7) at the order $O(\varepsilon^2)$ explicitly shown.

Substituting (1.7) into (1.6) gives

$$\begin{aligned}&(-D_{ij}-C_{ijkl}\varepsilon_{kl}+\phi_{ij}\theta+\sigma_{ij})\dot{\varepsilon}_{ij}\\&+(\beta+C_v\theta+\phi_{ij}\varepsilon_{ij}-\rho S)\dot{\theta}-q_i\frac{\theta_{,i}}{T_0}\geqslant 0\end{aligned} \tag{1.8}$$

where $C_{ijkl}=C_{klij}$ has been used. For the inequality (1.8) to be true for all values of $\dot{\varepsilon}_{ij}$ and $\dot{\theta}$ it is necessary that their respective coefficients vanish, giving the following constitutive relations:

$$\sigma_{ij}=D_{ij}+C_{ijkl}\varepsilon_{kl}-\phi_{ij}\theta$$

and

$$\rho S=\beta+\phi_{ij}\varepsilon_{ij}+C_v\theta \tag{1.9}$$

With (1.9), then (1.8) is left as

$$-q_i\frac{\theta_{,i}}{T_0}\geqslant 0 \tag{1.10}$$

We see that the first order terms in the free energy expansion (1.7) merely lead to constant values, D_{ij} and β, for stress and entropy in (1.9). These terms are dropped in subsequent developments. The tensor C_{ijkl} is, of course, that of the elastic moduli. It is now convenient to take an alternate form for ϕ_{ij}. Let

$$\phi_{ij} = C_{ijkl}\alpha_{kl} \tag{1.11}$$

Then from (1.9) we have

$$\sigma_{ij} = C_{ijkl}(\varepsilon_{kl} - \alpha_{kl}\theta) \tag{1.12}$$

where now we see α_{kl} as the thermal expansion tensor.

Combining the results just found with the suitably linearized energy equation (1.4), it now takes the form

$$\rho r - T_0\phi_{ij}\dot{\varepsilon}_{ij} - T_0 C_v\dot{\theta} - q_{i,i} = 0 \tag{1.13}$$

It remains to obtain the constitutive equation for the heat flux vector q_i. Consistent with the conditions of infinitesimal change we take the Fourier law of heat conduction

$$q_i = -k_{ij}\theta_{,j} \tag{1.14}$$

where k_{ij} is the thermal conductivity tensor. Substituting (1.14) into (1.10) gives

$$\frac{k_{ij}\theta_{,i}\theta_{,j}}{T_0} \geqslant 0 \tag{1.15}$$

where we see that k_{ij} must be a symmetric positive definite tensor.

Finally, combining the heat conduction law (1.14) into (1.13) gives

$$\rho r - T_0\phi_{ij}\dot{\varepsilon}_{ij} - T_0 C_v\dot{\theta} + k_{ij}\theta_{,ij} = 0 \tag{1.16}$$

This we recognize as the standard heat conduction equation, appropriate to homogeneous but anisotropic materials. Taking the case of no heat conduction and $\dot{\varepsilon}_{ij} = 0$ leaves the equation as $\rho r - C_v\dot{\theta} = 0$; thus C_v is the constant of proportionality that relates the temperature change with the heat energy change, and $T_0 C_v/\rho$ is the specific heat at constant deformation. The term in (1.16) involving ϕ_{ij} provides the coupling between thermal and mechanical effects.

For isotropic materials the preceding results reduce to the following forms. The stress constitutive relations become

$$s_{ij} = 2u\varepsilon_{ij}$$

$$\sigma_{kk} = 3k(\varepsilon_{kk} - 3\alpha\theta) \tag{1.17}$$

where u and k are the shear modulus and bulk modulus, respectively. The entropy form is given by

$$\rho S = \phi\delta_{ij}\varepsilon_{ij} + C_v\theta \tag{1.18}$$

whereas the heat flux vector has the form

$$q_i = -k\theta_{,i} \tag{1.19}$$

where the thermal conductivity k must not be confused with the bulk modulus k. Finally, the heat conduction equation takes the familiar form

$$\rho r - T_0\phi\dot{\varepsilon}_{kk} - T_0 C_v\dot{\theta} + k\theta_{,ii} = 0 \tag{1.20}$$

where $\phi = k\alpha$, with k in this expression being the bulk modulus, as in (1.17.)

Gibbs Free Energy Derivation

There is an alternative approach to that just given based on the use of the Helmholtz free energy. Rather than using the energy transformation (1.3), we take

$$\rho U = \rho G + T\rho S + \sigma_{ij}d_{ij} \tag{1.21}$$

where G is the Gibbs free energy per unit mass. This transformation serves the purpose of allowing stress and temperature to be the governing variables of the theory, rather than strain and temperature, as in the preceding derivation. We substitute (1.21) into the energy equation (1.1), and thereafter follow exactly the same procedure as that just outlined. Relative to some base level, we take an expansion for ρG similar to that of (1.7), but in terms of stress and temperature, as

$$\rho G = -\tfrac{1}{2}S_{ijkl}\sigma_{ij}\sigma_{kl}$$

$$-\alpha_{ij}\sigma_{ij}\theta - \tfrac{1}{2}C_p\theta^2 + O(\varepsilon^3) \tag{1.22}$$

we truncate at the explicit level shown. We have not retained terms of lower order than quadratic, for as explained previously these terms contribute nothing to the theory being sought.

From (1.22) and the governing thermodynamical relations (1.1), (1.2), and (1.21) we find the governing constitutive relations as

$$\varepsilon_{ij} = S_{ijkl}\sigma_{kl} + \alpha_{ij}\theta \tag{1.23}$$

where S_{ijlk} is the tensor of elastic compliances. The entropy constitutive relation is given by

$$\rho S = C_p\theta + \alpha_{ij}\sigma_{ij} \tag{1.24}$$

The heat conduction equation takes the form

$$\rho r - T_0\alpha_{ij}\dot{\sigma}_{ij} - T_0 C_p\dot{\theta} + k_{ij}\theta_{,ij} = 0 \tag{1.25}$$

where the Fourier law of heat condition (1.14) has again been employed. We note that α_{ij} in (1.22) is identical with that involved in (1.11). With (1.11), (1.23) can be written as

$$\varepsilon_{ij} = S_{ijkl}(\sigma_{kl} + \phi_{kl}\theta) \tag{1.26}$$

Finally, we note that $T_0 C_p/\rho$ in (1.25) is the specific heat at constant stress. The corresponding isotropic forms of these results may be easily stated.

We now have assembled the basic results we need to proceed with our developments for heterogeneous media. Our objective is to deduce the proper forms for the effective thermal conductivity k_{ij}, the effective thermal expansion coefficients α_{ij}, and the effective specific heats C_p and C_v for some practical types of composite materials.

9.2 THERMAL CONDUCTIVITY

As introduced in the last section, the Fourier law of heat conduction is given by

$$q_i = -k_{ij}\theta_{,j} \tag{2.1}$$

Under conditions of no coupling between mechanical and thermal effects, and under steady state conditions, the heat conduction equation, (1.16), gives

$$k_{ij}\theta_{,ij} = 0 \tag{2.2}$$

In a heterogeneous material we define the effective thermal conductivity coefficient through the relation

$$\bar{q}_i = -k_{ij}\bar{\theta}_{,j} \tag{2.3}$$

where $\bar{q}_j$ and $\bar{\theta}_{,j}$ are the volumetric averages of q_j and $\theta_{,j}$ in the representative volume element. As it stands (2.3) applies to a macroscopically anisotropic material, with k_{ij} depending on the direction of heat conduction. We now restrict attention to a two phase composite. In this case we can write

$$\bar{q}_j = \bar{q}_j^{(i)}c_{(i)} + \bar{q}_j^{(m)}c_{(m)} \tag{2.4}$$

where $c_{(i)}$ and $c_{(m)}$ are the volume fractions in the two phases, and the volume averages shown in (2.4) are those appropriate to the individual phases. Using (2.3) in (2.4) gives

$$\bar{q}_j = -c_{(i)}k_{(i)}\bar{\theta}_{,j}^{(i)} - c_{(m)}k_{(m)}\bar{\theta}_{,j}^{(m)} \tag{2.5}$$

We substitute $\bar{q}_j$ from (2.5) in (2.3) to get

$$-c_{(i)}k_{(i)}\bar{\theta}_{,j}^{(i)} - c_{(m)}k_{(m)}\bar{\theta}_{,j}^{(m)} = -k_{ji}\bar{\theta}_{,i} \tag{2.6}$$

Now similarly to relation (2.5) we can also write

$$\bar{\theta}_{,j} = c_{(i)}\bar{\theta}_{,j}^{(i)} + c_{(m)}\bar{\theta}_{,j}^{(m)} \tag{2.7}$$

solving for $\bar{\theta}_{,j}^{(m)}$ from (2.7) and substituting into (2.6) gives

$$k_{ji}\bar{\theta}_{,i} - k_{(m)}\bar{\theta}_{,j} = c_{(i)}(k_{(i)} - k_{(m)})\bar{\theta}_{,j}^{(i)} \tag{2.8}$$

Thus with a given state $\bar{\theta}_{,j}$ we need only solve for the average temperature gradient in the (i) phase, which we take as the inclusion phase; then we have the solution for k_{ij}.

Spherical Inclusion Model

We first solve this problem for the three phase spherical model of Section 2.4. The model is as shown in Fig. 9.1. At infinite distances from the origin, impose the condition

$$\theta\,|_{r\to\infty} \to \beta x_3 \tag{2.9}$$

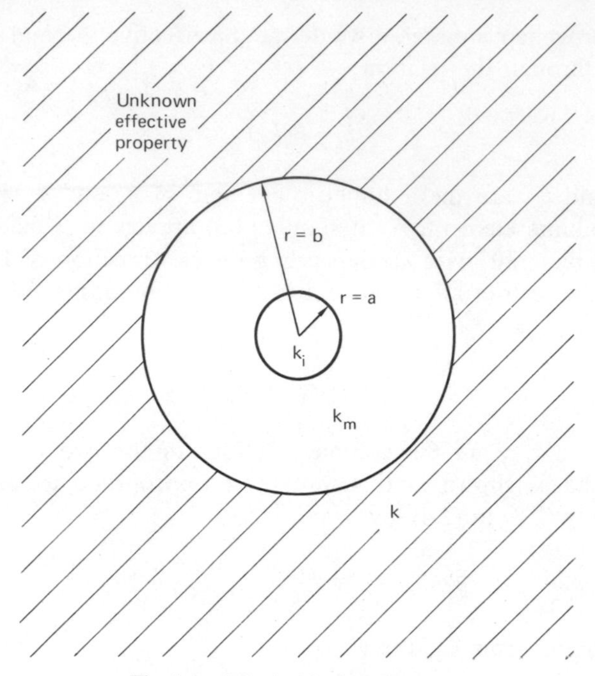

Fig. 9.1 Spherical inclusion model.

where x_j are Cartesian coordinates and β is the temperature gradient. We note that in the present context

$$\bar{\theta}_{,j}=\beta\delta_{3j}$$

for use in (2.8).

As governing conditions we have

$$\begin{aligned} \nabla^2\theta^{(i)}&=0, && 0\leqslant r\leqslant a \\ \nabla^2\theta^{(m)}&=0, && a\leqslant r\leqslant b \\ \nabla^2\theta&=0, && b\leqslant r\leqslant \infty \end{aligned} \tag{2.10}$$

Using spherical coordinates with axial symmetry about x_3 we have

$$\nabla^2=\frac{1}{r^2}\frac{\partial}{\partial r}\left(r^2\frac{\partial}{\partial r}\right)+\frac{1}{r^2\sin\theta}\frac{\partial}{\partial\theta}\left(\sin\theta\frac{\partial}{\partial\theta}\right)$$

Thus

$$\text{as } r\to\infty, \qquad \theta(r,\theta)\to\beta r\cos\theta$$

We then can easily show that the appropriate forms of solutions are

$$\begin{aligned} \theta^{(i)} &= A_i r \cos\theta, && 0 \leqslant r \leqslant a \\ \theta^{(m)} &= \left(A_m r + \frac{B_m}{r^2}\right)\cos\theta, && a \leqslant r \leqslant b \\ \theta &= \left(\alpha r + \frac{B}{r^2}\right)\cos\theta, && b \leqslant r \leqslant \infty \end{aligned} \tag{2.11}$$

The continuity conditions to be satisfied are

$$\begin{aligned} &\text{at } r=a, \quad \theta^{(i)} = \theta^{(m)}, \qquad k_{(i)}\frac{\partial \theta^{(i)}}{\partial r} = k_{(m)}\frac{\partial \theta^{(m)}}{\partial r} \\ &\text{at } r=b, \quad \theta^{(m)} = \theta, \qquad k_{(m)}\frac{\partial \theta^{(m)}}{\partial r} = k\frac{\partial \theta}{\partial r} \end{aligned} \tag{2.12}$$

Using the preceding solution (2.11) and conditions (2.12), the following solution for k is obtained

$$k = k_m\left[1 + \frac{c}{(1-c)/3 + [k_m/(k_i - k_m)]}\right] \tag{2.13}$$

where now c is the volume fraction of the inclusion phase. The result (2.13) for the effective conductivity coefficient was obtained by Hashin and Shtrikman [9.3]. Although this result was obtained here in the context of the three phase model of Fig. 9.1, Hashin and Shtrikman show that it is also the exact result for the composite spheres model of Section 2.3. This result is consistent with our expectations from the effective bulk modulus. Furthermore, Hashin and Shtrikman [9.3] show that the result (2.13) has an interpretation as a lower bound on k for arbitrary (macroscopically isotropic) phase geometry with $k_i > k_m$, and as an upper bound for $k_m > k_i$. Again this result is consistent with our observations in the mechanical properties case, Chapter IV. The explicit result (2.13) was first obtained by Kerner [9.4], in the context of electrical conductivity. The same results are also applicable to magnetic permeability and the dielectric constant.

Anisotropic Results

Now we turn to the case of a heterogeneous medium that is macroscopically anisotropic. The only restriction we make is that the medium is of the axial type, whereby the structure of the heterogeneity can be specified in a plane, normal to the axial direction. Take the temperature gradient in the axial direction

$$\theta=\beta x_1 \tag{2.14}$$

in all phases. The governing equations $\nabla^2\theta=0$ for all phases are trivially satisfied. From (2.3) we have

$$\bar{q}_1=-k_{11}\bar{\theta}_{,1} \tag{2.15}$$

where k_{11} is the effective axial conductivity. Obviously we have

$$\bar{q}_1=\sum_{n=1}^{N} c_n q_1^{(n)} \tag{2.16}$$

where there are taken to be N phases, with volume fractions c_n. With $q_1=-k\beta$ in each phase, we have by combining (2.15) and (2.16) the solution

$$k_{11}=\sum_{n=1}^{N} c_n k_n \tag{2.17}$$

and the axial conductivity is just given by the rule of mixtures.

Finally we consider the thermal conductivity relative to the composite cylinders model of Section 3.2. Since the composite cylinders model is transversely isotropic, we have

$$q_i=-k_{ij}T_{,j} \tag{2.18}$$

where because of the transverse isotropy there are only two independent terms in k_{ij} as

$$\begin{bmatrix} q_1 \\ q_2 \\ q_3 \end{bmatrix}=\begin{bmatrix} -k_{11} & 0 & 0 \\ 0 & -k_{22} & 0 \\ 0 & 0 & -k_{22} \end{bmatrix}\begin{bmatrix} \theta_{,1} \\ \theta_{,2} \\ \theta_{,3} \end{bmatrix} \tag{2.19}$$

where x_1 is in the axial direction. We already have solved for k_{11}, given by (2.17). The conductivity k_{22} in the transverse direction can be obtained in exactly the same manner we used in the spherical inclusion case. It is found that

$$k_{22}=k_m\left[1+\frac{c}{k_m/(k_i-k_m)+(1-c)/2}\right] \tag{2.20}$$

where c is the volume fraction of the cylindrical inclusion phase. Hashin [9.5] derives (2.20) explicitly for the case of the composite cylinders model.

9.3 THERMAL EXPANSION COEFFICIENT

We derive the effective thermal expansion coefficient for two phase composites in this section. The situation of more than two phases is considered in the next section in connection with the specific heat properties. The work to be given here is that of Levin [9.6] as reviewed by Rosen and Hashin [9.7].

Consider two problems, one of prescribed stress and one of prescribed temperature on the representative volume element. For the body with surface S, we write for prescribed stress

$$\left.\begin{aligned} \sigma_i &= \sigma_{ij}^0 n_j \\ \theta &= 0 \end{aligned}\right\} \quad \text{on } S \tag{3.1}$$

The volumetrically averaged stresses are given by

$$\begin{aligned} \bar{\sigma}'_{ij} &= \sigma_{ij}^0 \\ \bar{\theta} &= 0 \end{aligned} \tag{3.2}$$

where σ_{ij}^0 are conditions of uniform stress and the prime is introduced to distinguish the stress problem from the temperature problem to be stated later. Now the relationship between average strain and stress is given by

$$\varepsilon'_{ij} = S_{ijkl}\sigma_{kl}^0 \tag{3.3}$$

where S_{ijkl} are the effective compliances for the heterogeneous medium, considered to be known.

Next we state the prescribed temperature problem as

$$\left.\begin{aligned} \sigma_i &= 0 \\ \theta &= \theta_0 \end{aligned}\right\} \quad \text{on } S \tag{3.4}$$

and from this specification we have the volumetric averages

$$\begin{aligned} \bar{\sigma}_{ij} &= 0 \\ \bar{\theta} &= \theta_0 \end{aligned} \tag{3.5}$$

The corresponding average strain is then given by

$$\bar{\varepsilon}_{ij}=\alpha_{ij}\theta_0 \tag{3.6}$$

where α_{ij} are the unknown effective coefficients of thermal expansion.

Now we form the product of σ'_{ij} from the stress problem and ε_{ij} from the temperature problem to write

$$\int_V \sigma'_{ij}\varepsilon_{ij}\,dv=\int_V (\sigma'_{ij}u_i)_{,j}\,dv \tag{3.7}$$

where V is the volume and $\sigma'_{ij,j}=0$ has been used. Using the divergence theorem we get

$$\int_V \sigma'_{ij}\varepsilon_{ij}\,dv=\int_S \sigma'_{ij}u_i n_j\,ds \tag{3.8}$$

where n_j denotes the unit vector normal to the surface. But on surface S, $\sigma'_{ij}=\sigma^0_{ij}$; thus (3.8) becomes

$$\int_V \sigma'_{ij}\varepsilon_{ij}\,dv=\sigma^0_{ij}\bar{\varepsilon}_{ij}V \tag{3.9}$$

Using (3.6) this then takes the form

$$\int_V \sigma'_{ij}\varepsilon_{ij}\,dv=\sigma^0_{ij}\alpha_{ij}\theta_0 V \tag{3.10}$$

By definition we have

$$\bar{\sigma}'_{ij}=\sum_{n=1}^{N} c_n\bar{\sigma}'^{(n)}_{ij} \tag{3.11}$$

where we temporarily consider the case of N phases with c_n being the phase volume fractions. We next write the linear transformation between $\bar{\sigma}'^{(n)}_{ij}$ and σ^0_{kl} as

$$\bar{\sigma}'^{(n)}_{ij}=B^{(n)}_{ijkl}\sigma^0_{kl} \tag{3.12}$$

where $B^{(n)}_{ijkl}$ can be determined from the solution of the stress problem. Combining (3.11) and (3.12) gives us

$$\bar{\sigma}'_{ij}=\sum_{n=1}^{N} c_n B^{(n)}_{ijkl}\sigma^0_{kl} \tag{3.13}$$

But from (3.2), $\bar{\sigma}'_{ij} = \sigma^0_{ij}$; thus (3.13) gives

$$\sum_{n=1}^{N} c_n B^{(n)}_{ijkl} = I_{ijkl} = \tfrac{1}{2}(\delta_{ik}\delta_{jl} + \delta_{il}\delta_{jk}) \tag{3.14}$$

Now we rewrite (3.10) in terms of the phase contributions as

$$\sum_{n=1}^{N} B^{(n)}_{ijkl}\sigma^0_{kl}\,\alpha^{(n)}_{ij}\theta_0 V^{(n)} = \sigma^0_{ij}\alpha_{ij}\theta_0 V \tag{3.15}$$

where (3.13) and (3.6) have been used. Solving for α_{ij}, we have

$$\alpha_{ij} = \sum_{n=1}^{N} c_n \alpha^{(n)}_{kl} B^{(n)}_{klij} \tag{3.16}$$

where $c_n = V^{(n)}/V$.

By definition we have

$$\bar{\varepsilon}'_{ij} = \sum_{n=1}^{N} c_n \bar{\varepsilon}'^{(n)}_{ij} \tag{3.17}$$

and using (3.3) this becomes

$$S_{ijkl}\sigma^0_{kl} = \sum_{n=1}^{N} c_n S^{(n)}_{ijkl}\bar{\sigma}'^{(n)}_{kl} \tag{3.18}$$

Substituting for $\bar{\sigma}'^{(n)}_{kl}$ from (3.12) in (3.18), we get

$$S_{ijkl} = \sum_{n=1}^{N} c_n S^{(n)}_{ijmn} B^{(n)}_{mnkl} \tag{3.19}$$

Now eliminate $B^{(n)}_{mnkl}$ between (3.16) and (3.19) to have a solution for α_{ij}, also using (3.14). We have three determining equations and thus only enough conditions to consider a two phase composite with unknowns α, $B^{(1)}_{ijkl}$ and $B^{(2)}_{ijkl}$. From (3.14) we have

$$c_2 B^{(2)}_{ijkl} = \tfrac{1}{2}(\delta_{ik}\delta_{jl} + \delta_{il}\delta_{jk}) - c_1 B^{(1)}_{ijkl} \tag{3.20}$$

Substitute this into (3.19), giving

$$S_{ijkl} - S^{(2)}_{ijkl} = c_1\left[S^{(1)}_{ijmn} - S^{(2)}_{ijmn}\right]B^{(1)}_{mnkl} \tag{3.21}$$

Let

$$P_{klmn}\left(S^{(1)}_{mnrs}-S^{(2)}_{mnrs}\right)=I_{klrs} \tag{3.22}$$

Then from (3.20)

$$c_1B^{(1)}_{klij}=P_{klmn}\left(S_{mnij}-S^{(2)}_{mnij}\right) \tag{3.23}$$

From (3.16) and (3.20) we have

$$\alpha_{ij}=\left(\alpha^{(1)}_{kl}-\alpha^{(2)}_{kl}\right)c_1B^{(1)}_{klij}+\alpha^{(2)}_{ij} \tag{3.24}$$

where $\alpha^{(2)}_{ij}=\alpha^{(2)}_{ji}$ has been used. Substitute $c_1B^{(1)}_{klij}$ from (3.23) into (3.24) to get the explicit result

$$\alpha_{ij}=\left(\alpha^{(1)}_{kl}-\alpha^{(2)}_{kl}\right)P_{klmn}\left(S_{mnij}-S^{(2)}_{mnij}\right)+\alpha^{(2)}_{ij} \tag{3.25}$$

Now we specialize the result to the case of isotropic phases. In this case

$$\alpha^{(n)}_{ij}=\alpha_n\delta_{ij}$$

$$S^{(n)}_{ijkk}=\frac{\sigma_{ij}}{3k_n} \tag{3.26}$$

With these forms, α_{ij} in (3.25) takes the reduced form

$$\alpha_{ij}=(\alpha_1-\alpha_2)\delta_{kl}P_{klmn}\left[S_{mnij}-S^{(2)}_{mnij}\right]+\alpha_2\delta_{ij} \tag{3.27}$$

From (3.22) we have

$$P_{kkmn}=\frac{3\delta_{mn}}{1/k_1-1/k_2}$$

which when used in (3.27) gives

$$\alpha_{ij}=(\alpha_1-\alpha_2)\frac{\delta_{mn}}{1/k_1-1/k_2}\left(3S_{mnij}-\frac{\delta_{ij}}{k_2}\right)+\alpha_2\delta_{ij} \tag{3.28}$$

where (3.26) has been used.

In the case of a macroscopically isotropic composite, we have from (3.28)

$$\alpha=\alpha_2+\frac{(\alpha_1-\alpha_2)}{(1/k_1-1/k_2)}\left[\frac{1}{k}-\frac{1}{k_2}\right] \tag{3.29}$$

which we can write in equivalent form

$$\alpha = \bar{\alpha} + \frac{(\alpha_1 - \alpha_2)}{(1/k_1 - 1/k_2)}\left[\frac{1}{k} - \left(\overline{\frac{1}{k}}\right)\right] \tag{3.30}$$

where

$$\bar{\alpha} = c_1\alpha_1 + c_2\alpha_2$$

$$\left(\overline{\frac{1}{k}}\right) = \frac{c_1}{k_1} + \frac{c_2}{k_2} \tag{3.31}$$

These very simple results give the effective coefficient of thermal expansion, assuming the theoretical or experimental knowledge of the effective bulk modulus k.

It is left as an exercise to show that for transversely isotropic media the result (3.28) reduces to

$$\alpha_{\text{AXIAL}} = \bar{\alpha} + \frac{\alpha_1 - \alpha_2}{(1/k_1 - 1/k_2)}\left[\frac{3(1-2\nu_{12})}{E_{11}} - \left(\overline{\frac{1}{k}}\right)\right] \tag{3.32}$$

and

$$\alpha_{\text{TRANSVERSE}} = \bar{\alpha} + \frac{\alpha_1 - \alpha_2}{(1/k_1 - 1/k_2)}\left[\frac{3}{2K_{23}} - \frac{3\nu_{12}(1-2\nu_{12})}{E_{11}} - \left(\overline{\frac{1}{k}}\right)\right] \tag{3.33}$$

where $\bar{\alpha}$ and $(\overline{1/k})$ are still given by (3.31) and E_{11}, K_{23}, and ν_{12} are the effective uniaxial modulus, effective plane strain bulk modulus, and effective axial Poisson's ratio for the medium.

It is probably obvious that one can use the results for α to get bounds on the effective coefficient(s) of thermal expansion when bounds on the effective mechanical properties are specified.

9.4 SPECIFIC HEATS

We now approach the problem of determining the effective specific heats for heterogeneous media. In the process we also obtain results appropriate to the coefficient of thermal expansion, more generally than were found in the preceding section. Again the method follows that of Rosen and Hashin

[9.7]. We proceed through the use of an energy method, which was first given by Schapery [9.8] in connection with the derivation of the effective coefficient of thermal expansion.

General Formulation

Define the thermoelastic potential energy as

$$U_P = \frac{1}{V}\left[\int_V A\,dv - \int_V F_i u_i\,dv - \int_{S_\sigma} \sigma_i u_i\,ds\right] \tag{4.1}$$

where V is the volume of the region, F_i are the body force components, S_σ is that part of the surface on which tractions are prescribed, A is the Helmholtz free energy (per unit volume), specified in Section 9.1 and repeated here with only minor modification as

$$A = \tfrac{1}{2}C_{ijkl}(x_j)\varepsilon_{ij}\varepsilon_{kl} - \phi_{ij}(x_j)\varepsilon_{ij}\theta - \tfrac{1}{2}c_v(x_j)\frac{\theta^2}{T_0}{}^{*} \tag{4.2}$$

where $c_v = T_0 C_v$, with C_v being the term used in Section 9.1. The term c_v/ρ is the specific heat at constant deformation. Coupling coefficient ϕ_{ij} is related to the coefficient of thermal expansion through (1.11).

Similarly we define the thermoelastic complementary energy as

$$U_C = -\frac{1}{V}\left[\int_V G\,dv + \int_{S_u} \sigma_i u_i\,ds\right] \tag{4.3}$$

where S_u is that part of the surface on which displacements are prescribed and G is the Gibbs free energy defined in Section 9.1 and repeated here as

$$G = -\tfrac{1}{2}S_{ijkl}(x_j)\sigma_{ij}\sigma_{kl} - \alpha_{ij}(x_j)\sigma_{ij}\theta - \tfrac{1}{2}c_p(x_j)\frac{\theta^2}{T_0} \tag{4.4}$$

where c_p/ρ is the specific heat at constant stress.

It is left as an exercise to use

$$A = \sigma_{ij}\varepsilon_{ij} + G$$

*Note that herein all terms involving c_v and c_p have a sign reversal from those in [9.7], the latter of which are not consistent with the thermodynamical derivation of Section 9.1.

and the thermoelastic constitutive relations to prove that

$$c_p - c_v = T_0 C_{ijkl} \alpha_{ij} \alpha_{kl} \tag{4.5}$$

The thermoelastic energies U_P and U_C can be proved to possess a minimum character for the displacement and stress fields from the solution compared with any other admissible fields. The admissible fields are defined in Section 1.1, and of course the thermoelastic minimum principles are the counterparts of the minimum principles discussed in Section 1.1. The volumetric averages of A and G are given by

$$\frac{1}{V}\int_V A\,dv = \tfrac{1}{2}\bar{\sigma}_{ij}\bar{\varepsilon}_{ij} - \tfrac{1}{2}\phi_{ij}\bar{\varepsilon}_{ij}\theta - \tfrac{1}{2}c_v\frac{\theta^2}{T_0} \tag{4.6}$$

and

$$\frac{1}{V}\int_V G\,dv = -\tfrac{1}{2}\bar{\sigma}_{ij}\bar{\varepsilon}_{ij} - \tfrac{1}{2}\alpha_{ij}\bar{\sigma}_{ij}\theta - \tfrac{1}{2}c_p\frac{\theta^2}{T_0} \tag{4.7}$$

where now ϕ_{ij}, α_{ij}, c_p, and c_v are effective properties of the heterogeneous medium and V is taken to be the volume of the representative volume element.

Let us prescribe a stress problem wherein

$$\sigma_{ij} = \sigma_{ij}^0 n_j \qquad \text{on } S \tag{4.8}$$

where σ_{ij}^0 is a uniform stress state and n_j are the components of the unit normal vector. With no body forces, $F_i = 0$, we then have

$$U_P = -\tfrac{1}{2}\sigma_{ij}^0\bar{\varepsilon}_{kj} - \tfrac{1}{2}\phi_{ij}\bar{\varepsilon}_{ij}\theta - \tfrac{1}{2}c_v\frac{\theta^2}{T_0} \tag{4.9}$$

and

$$U_C = \tfrac{1}{2}\sigma_{ij}^0\bar{\varepsilon}_{ij} + \tfrac{1}{2}\alpha_{ij}\sigma_{ij}^0\theta + \tfrac{1}{2}c_p\frac{\theta^2}{T_0}$$

Similarly for a problem with prescribed displacements

$$u_i = \varepsilon_{ij}^0 x_j \qquad \text{on } S \tag{4.10}$$

where ε_{ij}^0 are uniform, then

$$U_P = \tfrac{1}{2}\bar{\sigma}_{ij}\varepsilon_{ij}^0 - \tfrac{1}{2}\phi_{ij}\varepsilon_{ij}^0\theta - \tfrac{1}{2}c_v\frac{\theta^2}{T_0} \tag{4.11}$$

and

$$U_C = -\tfrac{1}{2}\bar{\sigma}_{ij}\varepsilon_{ij}^0 + \tfrac{1}{2}\alpha_{ij}\bar{\sigma}_{ij}\theta + \tfrac{1}{2}c_p\frac{\theta^2}{T_0}$$

The statement of the minimum principles is that

$$\begin{aligned} U_P &\leqslant \tilde{U}_P \\ U_C &\leqslant \tilde{U}_C \end{aligned} \tag{4.12}$$

where $\tilde{U}_P$ and $\tilde{U}_C$ are the energies calculated for the admissible fields. These minimum principles were proved by Rosen [9.9].

Now for the stress problem (4.8) relation (4.5) can be used to show that

$$U_C = -U_P \tag{4.13}$$

Combining (4.12) and (4.13) gives

$$-\tilde{U}_C \leqslant -U_C \leqslant \tilde{U}_P$$

Using (4.1) and (4.3) these conditions become

$$\begin{aligned} -\frac{1}{V}\int_V \left\{ \tfrac{1}{2}S_{ijkl}\sigma_{ij}\sigma_{kl} + \alpha_{ij}\sigma_{ij}\theta + \tfrac{1}{2}c_p\frac{\theta^2}{T_0} \right\} dv \\ \leqslant -\tfrac{1}{2}S_{ijkl}\sigma_{ij}^0\sigma_{kl}^0 - \alpha_{ij}\sigma_{ij}^0\theta - \tfrac{1}{2}c_p\frac{\theta^2}{T_0} \\ \leqslant \frac{1}{V}\int_V \left\{ \tfrac{1}{2}C_{ijkl}\varepsilon_{ij}\varepsilon_{kl} - \phi_{ij}\varepsilon_{ij}\theta - \tfrac{1}{2}c_v\frac{\theta^2}{T_0} \right\} dv \\ -\frac{1}{V}\sigma_{ij}^0\int_V \varepsilon_{ij}\,dv \end{aligned} \tag{4.14}$$

where admissible stresses are to be used in the left-hand side of the inequalities and admissible displacements in the right-hand side. S_{ijkl}, α_{ij}, and c_p are the effective properties.

Similarly for the displacement problem of (4.10) we find

$$-\frac{1}{V}\int_V\left\{\tfrac{1}{2}S_{ijkl}\sigma_{ij}\sigma_{kl}+\alpha_{ij}\sigma_{ij}\theta+\tfrac{1}{2}c_p\frac{\theta^2}{T_0}\right\}dv+\frac{1}{V}\varepsilon_{ij}^0\int_V\sigma_{ij}\,dv$$
$$\leqslant \tfrac{1}{2}C_{ijkl}\varepsilon_{ij}^0\varepsilon_{kl}^0-\phi_{ij}\varepsilon_{ij}^0\theta-\tfrac{1}{2}c_v\frac{\theta^2}{T_0}$$
$$\leqslant \frac{1}{V}\int_V\left\{\tfrac{1}{2}C_{ijkl}\varepsilon_{ij}\varepsilon_{kl}-\phi_{ij}\varepsilon_{ij}\theta-\tfrac{1}{2}c_v\frac{\theta^2}{T_0}\right\}dv \tag{4.15}$$

Of course, with $\theta=0$ these relations, (4.14) and (4.15), are just the results of the classical theory. In determining c_p and c_v we only need to use one or the other of (4.14) and (4.15) since we have a second independent relation from (4.5).

Macroscopically Isotropic Case

At this point we restrict attention to macroscopically isotropic composites; we return to the anisotropic case later. Take applied stresses in the stress problem as

$$\sigma_{ij}^0=\sigma_0\delta_{ij}$$

and as admissible stresses take

$$\sigma_{ij}=\sigma_0\delta_{ij} \tag{4.16}$$

In the displacement problem take the surface displacements as

$$u_i^0=\varepsilon_0\delta_{ij}x_j=\varepsilon_0x_i$$

and as admissible displacements take

$$u_i=\varepsilon_0x_i \tag{4.17}$$

Substituting (4.16) and (4.17) into (4.14) gives

$$-\tfrac{1}{2}\bar{S}_{iijj}\sigma_0^2-\bar{\alpha}_{ii}\sigma_0\theta-\tfrac{1}{2}\bar{c}_p\frac{\theta^2}{T_0}$$

$$\leqslant -\frac{\sigma_0^2}{2k}-3\alpha\sigma_0\theta-\tfrac{1}{2}c_p\frac{\theta^2}{T_0}$$

$$\leqslant \tfrac{1}{2}\bar{C}_{iijj}\varepsilon_0^2-\bar{\phi}_{ii}\varepsilon_0\theta-\tfrac{1}{2}\bar{c}_v\frac{\theta^2}{T_0}-3\sigma_0\varepsilon_0 \tag{4.18}$$

where volumetric averages of the properties have been taken, such as

$$\bar{\alpha}_{ii}=\int_V \alpha_{ii}(x_i)\,dv$$

Minimize the upper bound in (4.18) with respect to ε_0. Then the upper bound character of (4.18) can be written as

$$-\frac{\sigma_0^2}{2k}-3\alpha\sigma_0\theta-\tfrac{1}{2}c_p\frac{\theta^2}{T_0}$$

$$\leqslant -\frac{\left(3\sigma_0+\bar{\phi}_{ii}\theta\right)^2}{2\bar{C}_{jjkk}}-\tfrac{1}{2}\bar{c}_v\frac{\theta^2}{T_0} \tag{4.19}$$

Collecting coefficients of powers of (σ_0/θ) gives (4.19) as

$$A\left(\frac{\sigma_0}{\theta}\right)^2+(\alpha-B)\left(\frac{\sigma_0}{\theta}\right)+(\gamma-C)\geqslant 0 \tag{4.20}$$

where

$$\gamma=\frac{c_p-\bar{c}_v}{6T_0}$$

$$A=\frac{1}{6}\left(\frac{1}{k}-\frac{9}{\bar{C}_{iijj}}\right)$$

$$B=\frac{\phi_{ii}}{\bar{C}_{jjkk}}$$

$$C=\frac{\bar{\phi}_{ii}\bar{\phi}_{jj}}{6\bar{C}_{mmmm}} \tag{4.21}$$

Similarly from the lower bound in (4.18) it is found that

$$A'\left(\frac{\sigma_0}{\theta}\right)^2-(\alpha-B')\left(\frac{\sigma_0}{\theta}\right)-(\gamma-C')\geqslant 0 \tag{4.22}$$

where

$$A'=\tfrac{1}{6}\left(\bar{S}_{iijj}-\frac{1}{k}\right)$$

$$B'=\tfrac{1}{3}\bar{\alpha}_{ii}$$

$$C'=\tfrac{1}{6}\left(\overline{C_{ijkl}\alpha_{ij}\alpha_{kl}}\right) \tag{4.23}$$

For $\theta/\sigma_0=0$, (4.20) and (4.22) give the Reuss and Voigt bounds

$$\frac{9}{\bar{C}_{iijj}}\leqslant\frac{1}{k}\leqslant\bar{S}_{iijj}$$

For $\sigma_0/\theta=0$ we get the elementary bounds

$$\frac{\bar{\phi}_{ii}\bar{\phi}_{jj}}{\bar{C}_{mmmm}}\leqslant\frac{c_p-\bar{c}_v}{T_0}\leqslant\left(\overline{C_{ijkl}\alpha_{ij}\alpha_{kl}}\right)$$

It also follows from (4.20) and (4.22) that

$$(\alpha-B)^2-4A(\gamma-C)\leqslant 0$$

$$(\alpha-B')^2+4A'(\gamma-C')\leqslant 0 \tag{4.24}$$

These are two inequalities in two unknowns α and γ, with γ related to c_p through (4.21). The behavior of relations (4.24) are as shown in Fig. 9.2. The intersection of the parabolas is given by

$$\gamma=C'-\frac{1}{4A'}(\alpha-B')^2 \tag{4.25}$$

Substituting this value of γ into the equality form of the first relation in (4.24) gives

$$\alpha^2(A+A')-2\alpha(AB'+A'B)+\left[A(B')^2+A'B^2+4AA'(C-C')\right]=0$$

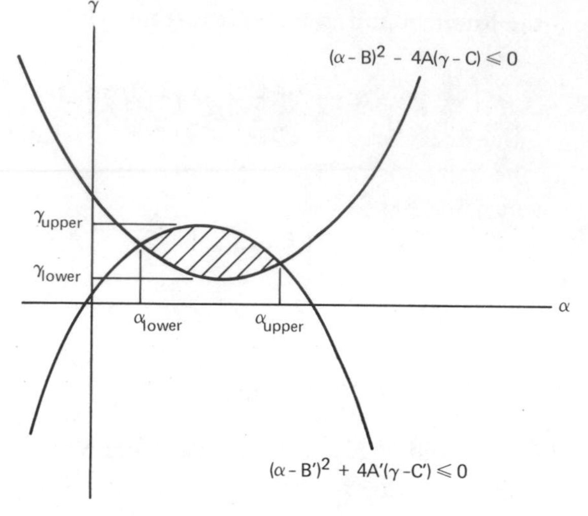

Fig. 9.2 Relations (4.24).

This relation then gives the bounds on α as

$$\begin{aligned}\frac{\alpha^{(+)}}{\alpha^{(-)}} &= \frac{AB'+A'B}{A+A'}\\ &\pm\frac{1}{A+A'}\left\{AA'\left[4(A+A')(C'-C)\right]-(B-B')^2\right\}^{1/2}\end{aligned} \tag{4.26}$$

For isotropic phases (4.26) reduces to

$$\begin{aligned}\frac{\alpha^{(+)}}{\alpha^{(-)}} &= \frac{1}{(\overline{1/k})-1/k}\left\{\frac{(\overline{k\alpha})}{\bar{k}}\left[\left(\overline{\frac{1}{k}}\right)-\frac{1}{k}\right]+\bar{\alpha}\left(\frac{1}{k}-\frac{1}{\bar{k}}\right)\right.\\ &\left.\pm\psi\left[\left(\overline{\frac{1}{k}}\right)-\frac{1}{k}\right]^{1/2}\left(\frac{1}{k}-\frac{1}{\bar{k}}\right)^{1/2}\right\}\end{aligned} \tag{4.27}$$

where

$$\psi=\left\{\left[\left(\overline{\frac{1}{k}}\right)-\frac{1}{\bar{k}}\right]\left[(\overline{k\alpha^2})-\frac{(\overline{k\alpha})^2}{\bar{k}}\right]-\left[\bar{\alpha}-\frac{(\overline{k\alpha})}{\bar{k}}\right]^2\right\}^{1/2} \tag{4.28}$$

For two phases it can be shown that $\psi=0$ and bounds coalesce with the result from (4.27) identical with that from (3.29). Note also that $\psi=0$

whenever $k_n\alpha_n$ = constant for all phases, leaving the result in this case as

$$k\alpha = k_n\alpha_n$$

We continue now to find the specific heats for a two phase composite material. Using the solution for α (3.29) in expression (4.25), and thence solving for c_p, using γ from (4.21), gives

$$\frac{c_p - \bar{c}_p}{T_0} = 9\left(\frac{\alpha_2 - \alpha_1}{1/k_2 - 1/k_1}\right)^2 \left[\frac{1}{k} - \left(\overline{\frac{1}{k}}\right)\right] \tag{4.29}$$

With the solution for c_p (4.29), c_v follows from (4.5), as

$$c_p - c_v = 3k\alpha^2 T_0$$

This completes the development of the effective specific heats for macroscopically isotropic, two phase composites. Now we turn to the case of a two phase transversely isotropic medium.

Transversely Isotropic Case

As admissible fields take

$$u_i = \varepsilon_{ij}^0 x_j$$

$$\sigma_{ij} = \sigma_{ij}^0 \tag{4.30}$$

Substituting (4.30) into (4.14) gives the bounding forms as

$$-\tfrac{1}{2}\bar{S}_{ijkl}\sigma_{ij}^0\sigma_{kl}^0 - \bar{\alpha}_{ij}\sigma_{ij}^0\theta - \tfrac{1}{2}\bar{c}_p\frac{\theta^2}{T_0}$$

$$\leqslant -\tfrac{1}{2}S_{ijkl}\sigma_{ij}^0\sigma_{kl}^0 - \alpha_{ij}\sigma_{ij}^0\theta - \tfrac{1}{2}c_p\frac{\theta^2}{T_0}$$

$$\leqslant \tfrac{1}{2}\bar{C}_{ijkl}\varepsilon_{ij}^0\varepsilon_{kl}^0 - \bar{\phi}_{ij}\varepsilon_{ij}^0\theta - \tfrac{1}{2}\bar{c}_v\frac{\theta^2}{T_0} - \sigma_{ij}^0\varepsilon_{ij}^0$$

(4.31)

As before, we determine ε_{ij}^0 such that the upper bound in (4.31) is minimized. Then the upper bound structure of (4.31) gives

$$\tfrac{1}{2}\left(S_{ijkl}-\overline{\Lambda}_{ijkl}\right)\sigma_{ij}^{0}\sigma_{kl}^{0}+\left(-\overline{\Lambda}_{ijkl}\overline{\phi}_{kl}+\alpha_{ij}\right)\sigma_{ij}^{0}\theta + \left(\frac{c_p-\bar{c}_v}{2T_0}-\tfrac{1}{2}\overline{\Lambda}_{ijkl}\overline{\phi}_{ij}\overline{\phi}_{kl}\right)\theta^2 \geqslant 0 \tag{4.32}$$

where

$$\overline{\Lambda}_{ijmn}\overline{C}_{mnkl}=I_{ijkl} \tag{4.33}$$

Taking only a single component $\sigma_{\beta\alpha}$ of σ_{ij}^{0} as being nonzero, and suspending the summation convention, gives (4.32) as

$$A\left(\frac{\sigma_{\beta\gamma}}{\theta}\right)^2+(\alpha_{\beta\gamma}-B)\left(\frac{\sigma_{\beta\gamma}}{\theta}\right)+(\gamma-C)\geqslant 0 \tag{4.34}$$

where

$$\begin{aligned} A&=\tfrac{1}{2}\left(S_{\beta\gamma\beta\gamma}-\overline{\Lambda}_{\beta\gamma\beta\gamma}\right)\\ B&=\overline{\Lambda}_{\beta\gamma ij}\overline{\phi}_{ij}\\ C&=\tfrac{1}{6}\overline{\Lambda}_{ijkl}\overline{\phi}_{ij}\overline{\phi}_{kl} \end{aligned} \tag{4.35}$$

and γ is as in (4.21). Similarly from the lower bound structure in (4.31) we find

$$A'\left(\frac{\sigma_{\beta\gamma}}{\theta}\right)^2-(\alpha_{\beta\gamma}-B')\left(\frac{\sigma_{\beta\gamma}}{\theta}\right)-(\gamma-C')\geqslant 0 \tag{4.36}$$

where

$$\begin{aligned} A'&=\tfrac{1}{2}\left(S_{\beta\gamma\beta\gamma}-\overline{\Lambda}_{\beta\gamma\beta\gamma}\right)\\ B'&=\bar{\alpha}_{\beta\gamma}\\ C'&=\tfrac{1}{6}\,\overline{C_{ijkl}\alpha_{ij}\alpha_{kl}} \end{aligned} \tag{4.37}$$

From the inequalities in (4.34) and (4.36) we find

$$(\alpha_{\beta\gamma}-B)^2-4A(\gamma-C)\leqslant 0$$

and

$$(\alpha_{\beta\gamma}-B')^2+4A'(\gamma-C')\leqslant 0 \tag{4.38}$$

Following the same procedure as that used to deduce (4.25), we now solve for γ from the equality form of either statement of (4.38); both lead to the same result. This gives

$$\gamma=C'-\frac{1}{4A'}(\alpha_{\beta\gamma}-B')^2 \tag{4.39}$$

Now using γ from (4.21) and solving for c_p gives

$$\frac{c_p-\bar{c}_v}{T_0}=\overline{C_{ijkl}\alpha_{ij}\alpha_{kl}} - \frac{3\left[\alpha_{\beta\gamma}-\bar{\alpha}_{\beta\gamma}\right]^2}{\left(S_{\beta\gamma\beta\gamma}-\overline{\Lambda}_{\beta\gamma\beta\gamma}\right)} \tag{4.40}$$

We can use any component $\alpha_{\beta\gamma}$ to solve for c_p. Taking x_1 in the axial direction of the transversely isotropic material, then with component α_{11}, (4.40) becomes

$$\frac{c_p-\bar{c}_v}{T_0}=\overline{c_{ijkl}\alpha_{ij}\alpha_{kl}} - \frac{3\left[\alpha_{11}-\bar{\alpha}_{11}\right]^2}{\left(S_{1111}\quad\overline{\Lambda}_{1111}\right)} \tag{4.41}$$

From (4.33) we have

$$\overline{\Lambda}_{11mm}\overline{C}_{mm11}=I_{1111}=1 \tag{4.42}$$

and we must have $\overline{\Lambda}_{1111}$ for use in (4.41). Now in general for isotropic materials

$$S_{ijmn}C_{mnkl}=I_{ijkl}$$

and we have for $i=j=k=l=1$

$$S_{11mn}C_{mn11}=1 \tag{4.43}$$

Thus (4.42) and (4.43) have exactly the same forms. Now we know that for an isotropic material

$$S_{1111}=\frac{1}{E}$$

Thus we see that for two phase isotropic materials

$$\bar{\Lambda}_{1111}=\frac{c_1}{E_1}+\frac{c_2}{E_2} \tag{4.44}$$

Under these conditions we also have

$$\begin{aligned} C_{ijkl}\alpha_{ij}\alpha_{kl} &= C_{iijj}\alpha^2 \\ &= 3k\alpha^2 \end{aligned} \tag{4.45}$$

Using (4.44) and (4.45) in (4.41) we obtain the final result

$$\begin{aligned} \frac{c_p-\bar{c}_v}{3T_0} &= c_1k_1\alpha_1^2+c_2k_2\alpha_2^2 \\ &\quad -\frac{(\alpha_{11}-c_1\alpha_1-c_2\alpha_2)^2}{\left(\frac{1}{E_{11}}-\frac{c_1}{E_1}-\frac{c_2}{E_2}\right)} \end{aligned} \tag{4.46}$$

where E_{11} is the effective axial modulus for the two phase transversely isotropic material, and α_{11} is the axial effective coefficient of thermal expansion, given by expression (3.32). With c_p determined by (4.46), the specific heat at constant deformation follows from (4.5).

9.5 DISCUSSION OF RESULTS

In this chapter on effective thermal properties, we have given only exact results, within the scope of the linear theory. There is a large amount of literature available that is concerned with approximations and empirical formulations. In fact, there is a long history of development in this area. Both Maxwell [9.10] and Rayleigh [9.11] obtained approximate solutions for the electrical conductivity through a continuous phase containing cylindrical and spherical inclusions of another phase.

We have been concerned only with spherical inclusions and aligned cylindrical type inclusions, in the case of thermal conductivity. It is easy to see that more general types of geometry could easily be considered, using ellipsoidal inclusions, as in Section 3.5 or randomly oriented fiber or platelet models as in Sections 4.4 and 4.5. Such work has been given by Polder and Van Santen [9.12] for the dielectric constant of a dilute suspension of ellipsoids of random orientation. A general approach to the dilute thermal conductivity problem has been given by Rocha and Acrivos [9.13, 9.14].

As we mentioned earlier, it is possible to derive bounds on the thermal conductivities in a similar and much more simple manner than that which we did in Chapter IV for elastic effects. References to these works may be found in the very complete reference article on effective "physical" properties by Hale [9.15]. In this regard there are bounding procedures that employ n point correlation functions to provide more information on phase structure than is possible simply by specifying volume fractions. Works of this type are those of Beran [9.16], Miller [9.17], who introduces a special cell type geometric idealization, and Elsayed and McCoy [9.18].

Interestingly, the two phase, effective coefficients of thermal expansion and specific heats do not require the introduction of a specific geometric model. They depend only on the symmetry characteristics of the equivalent homogeneous medium. However, they do also have a dependence on the effective elastic properties, which must be known either from experimental data or from the theoretical prediction based on a particular geometric model.

It should also be mentioned that the self-consistent scheme discussed in Section 2.5, has been applied to the determination of effective thermal properties. See, for example, Budiansky [9.19] and Laws [9.20].

It is important to observe that the curing cycle for polymeric composites can cause shrinkage effects that are analogous to thermal stress effects. Furthermore moisture or vapor environments can diffuse into composite materials and cause swelling effects. The moisture diffusion problem is analogous to that of heat diffusion, whereas the induced swelling effect causes a stress state similar to that induced by thermal effects.

PROBLEMS

1. Obtain the complete derivation of the constitutive equations for strain and entropy starting with the representation given in Section 9.1 for the Gibbs free energy. Also obtain the corresponding form of the energy equation.

2. Perform the derivation of the transverse conductivity formula (2.20), using the three phase model of Section 3.3.
3. Derive the coefficients of thermal expansion formulas (3.32) and (3.33), for transversely isotropic media, from the general result (3.28).
4. Derive the result

$$c_p - c_v = T_0 C_{ijkl} \alpha_{ij} \alpha_{kl}$$

relating the specific heats at constant pressure and volumetric stiffness tensor and the coefficients of thermal expansion.

5. A two material composite medium has one continuous phase and the other phases are of discrete inclusions. Which shape of inclusions—spheres, fibers, or platelets—would give the greatest and the least values for the effective thermal conductivity and thermal expansion coefficient? Take the composite material as being macroscopically isotropic.
6. The problem of moisture diffusion into fiber composites can provide a design limiting condition. Formulate the complete set of governing equations by using an analogy between moisture diffusion and thermal diffusion, including the effects of induced states of stress.
7. Consider a dilute suspension of spherical inclusions in a continuous matrix phase. Using the solution for the effective bulk modulus, under dilute conditions, obtain the solution for the effective coefficient of thermal expansion in the form

$$\alpha = \alpha_m + f(\quad)c$$

where $f(\quad)$ depends on the properties of both phases.

8. Many types of fibers have anisotropic properties. Outline a method by which the coefficients of thermal expansion can be obtained for an aligned fiber system of transversely isotropic fibers in an isotropic matrix phase.
9. A composite material is subjected to a uniform increase in temperature. How would one obtain an estimate of the interfacial, debonding stress in the case of a suspension of spherical inclusions in a continuous matrix phase?

REFERENCES

9.1 C. Truesdell and R. A. Toupin, in *Handbuch der Physik*, S. Flügge, Ed., vol. 3, no. 1, Springer, Berlin, 1960.

9.2 C. Truesdell and W. Noll, in *Handbuch der Physik*, S. Flügge, Ed., vol. 3, no. 3, Springer, Berlin, 1965.

9.3 Z. Hashin and S. Shtrikman, "A variational approach to the theory of the effective magnetic permeability of multiphase materials," *J. Appl. Phys.*, vol. 33, 3125 (1962).

9.4 E. H. Kerner, "The electrical conductivity of composite materials," *Proc. Phys. Soc.*, vol. B69, 802 (1956).

9.5 Z. Hashin, "Theory of fiber-reinforced materials," NASA CR-1974, 1972.

9.6 V. M. Levin, "On the coefficients of thermal expansion of heterogeneous materials," *Mekh. Tverd. Tela* (in Russian), vol. 88 (1968).

9.7 B. W. Rosen and Z. Hashin, "Effective thermal expansion coefficients and specific heats of composite materials," *Int. J. Eng. Sci.*, vol. 8, 157 (1970).

9.8 R. A. Schapery, "Thermal expansion coefficients of composite materials based on energy principles," *J. Comp. Mater.*, vol. 2, 380 (1968).

9.9 B. W. Rosen, "Thermoelastic energy function and minimum energy principles for composite materials," *Int. J. Eng. Sci.*, vol. 8, 5 (1970).

9.10 J. C. Maxwell, *A Treatise on Electricity and Magnetism*, Dover, New York, 1954.

9.11 L. Rayleigh, "On the influence of obstacles arranged in rectangular order upon the properties of a medium," *Phil. Mag.*, vol. 34, 481 (1892).

9.12 D. Polder and J. H. Van Santen, "The effective permeability of mixtures of solids," *Physica*, vol. 12, 257 (1946).

9.13 A. Rocha and A. Acrivos, "On the effective thermal conductivity of dilute dispersions: General theory for inclusions of arbitrary shape," *Quant. J. Mech. Appl. Math.*, vol. 26, 217 (1973).

9.14 A. Rocha and A. Acrivos, "On the effective thermal conductivity of dilute dispensions: Highly conducting inclusions of arbitrary shape," *Quant. J. Mech. Appl. Math.*, vol. 26, 441 (1973).

9.15 D. K. Hale, "The physical properties of composite materials," *J. Mater. Sci.*, vol. 11, 2105 (1976).

9.16 M. J. Beran, "Use of the variational approach to determine bounds for the effective permittivity in random media," *Nuovo Cim.*, vol. 38, 771 (1965).

9.17 M. N. Miller, "Bounds on the effective electrical, thermal and magnetic properties of heterogeneous materials," *J. Math. Phys.*, vol. 10, 1988 (1969).

9.18 M. A. Elsayed and J. J. McCoy, "Effective properties of three phase materials," *Fibre Sci. Tech.*, vol. 7, 281 (1974)

9.19 B. Budiansky, "Thermal and thermoelastic properties of isotropic composites," *J. Comp. Mater.*, vol. 4, 286 (1970).

9.20 N. Laws, "The overall thermoelastic moduli of transversely isotropic composites according to the self-consistent method," *Int. J. Eng. Sci.*, vol. 12, 79 (1974).

AUTHOR INDEX

Achenbach, J. D., 232, 241, 242, 252, 253, 271, 272
Acrivos, A., 37, 61, 67, 69, 71, 337, 339
Adkins, J. E., 301, 309
Anderson, T. B., 266, 267, 268, 270, 273

Batchelor, G. K., 10, 30, 63, 69
Bedford, A., 253, 265, 272
Behrens, E., 122, 151
Ben-Amoz, M., 253, 272
Beran, M. J., 149, 151, 337, 339
Bert, C. W., 199, 223
Bose, S. K., 149, 151
Boucher, S., 100, 103, 105
Bowen, R. M., 265, 273
Brandmaier, H. E., 210, 223
Brekhovskikh, L. M., 232, 272
Brenner, H., 61, 72
Brillouin, L., 243, 272
Budiansky, B., 59, 71, 337, 339
Butcher, B. M., 279, 309

Carne, T. G., 99, 105
Carroll, M. M., 275, 279, 309
Chadwick, P., 279, 309
Chamis, C. C., 272
Chen, C. H., 123, 151
Chen, H.-S., 69, 72
Cheng, S., 123, 151
Christensen, R. M., 11, 30, 58, 71, 89, 105, 125, 132, 139, 151, 174, 178, 184, 187, 214, 224, 265, 273, 290, 298, 309
Chu, T. Y., 279, 309
Coleman, B. D., 207, 224, 298, 309
Cox, H. L., 135, 151

Daniels, H. E., 207, 224
Dewey, J. M., 46, 71
Drucker, D. C., 287, 309
Drumheller, D. S., 253, 272
Dvorak, G. J., 281, 287, 309

Elsayed, M. A., 337, 339
Eringen, A. C., 265, 273
Eshelby, J. D., 23, 24, 29, 30, 38, 90, 91, 93, 94, 99, 101, 102, 105, 107
Everstine, G. C., 190, 199, 223

Frankel, N. A., 61, 67, 72

Garcia, D. J., 265, 273
Goddard, J. D., 301, 309
Goldstein, H., 262, 272
Goodier, J. N., 46, 167, 187
Green, A. E., 5, 30, 265, 273, 309
Green, J. T., 69, 72
Gross, B., 11, 30, 296, 309
Gurtman, G. A., 251, 272

Hale, D. K., 337, 339
Halpin, J. C., 137, 150, 151
Hamstad, M. A., 212

Happel, J., 61, 72
Harlow, D. G., 207, 223
Hashin, Z., 47, 51, 71, 81, 84, 105, 108, 118, 120, 151, 279, 288, 289, 291, 309, 319, 321, 325, 339
Hearmon, R. F. S., 158, 187, 189, 223
Hegemeier, G. A., 253, 272
Hermans, J. J., 89, 105, 121
Herrmann, G., 232, 241, 242, 252, 271
Hershey, A. V., 59, 71
Hill, R., 19, 30, 59, 71, 84, 105, 108, 118, 151, 200, 202, 210, 223
Holt, A. C., 275, 279, 309
Hutton, J. F., 310

Ingram, J. D., 265, 273

Jackson, R., 266, 267, 268, 270, 273
Jahanmir, M., 253, 265, 272
Jerina, K., 137, 151

Kachanov, L. M., 19, 30
Karal, F. C., Jr., 265, 272
Keller, J. B., 265, 272
Kerner, E. H., 54, 58, 71, 120, 121, 319, 339
Kohn, W., 243, 253, 272
Krieger, I. M., 68, 70, 72
Kröner, E., 59, 71, 149, 151

Landau, L. D., 64, 72
Laws, N., 337, 339
Leal, G. L., 301, 309
Lee, E. H., 253, 272
Lekhnitskii, S. G., 189, 223
Levin, V. M., 99, 105, 321, 339
Lifshitz, E. M., 64, 72
Lo, K. H., 58, 71, 89, 105, 174, 178, 184, 187
Love, A. E. H., 112, 151

McCoy, J. J., 265, 272, 337, 339
MacDonald, D., 205, 223
McLaughlin, R., 60, 72
Mal, A. K., 149, 151
Maxwell, J. C., 336, 339
Medlin, J., 270, 273
Mendelson, A., 19, 30

Miller, C., 301, 309
Miller, M. N., 337, 339
Minagawa, S., 253, 272
Moon, F. C., 253, 258, 272
Mow, C. C., 258, 272
Mukherjee, S., 253, 272
Muki, R., 99, 105
Mulhern, J. F., 287, 301, 309

Naghdi, P. M., 19, 30, 265, 273
Nanyaro, A. P., 205, 223
Nayfeh, A. H., 253, 272
Nemat-Nasser, S., 253, 272
Noll, W., 312, 328

Pagano, 157, 167, 173, 179, 185, 187
Pao, Y. H., 258, 272
Paul, B., 108, 151
Pearson, J. R. A., 310
Peck, J. C., 251, 252, 272
Phoenix, S. L., 207, 223
Pipes, R. B., 185, 187
Pipkin, A. C., 11, 30, 190, 199, 223, 301, 310
Polder, D., 337, 339
Postma, G. W., 140, 151
Prager, W., 19, 30

Rao, M. S. M., 281, 287, 309
Rayleigh, Lord, 336, 339
Richard, T. G., 57, 71
Rivlin, R. S., 301, 309
Rocha, A., 337, 339
Rogers, T. G., 287, 301, 309, 310
Rosen, B. W., 81, 84, 105, 120, 151, 207, 224, 321, 325, 328, 339
Russel, W. B., 37, 71, 90, 99, 105
Rytov, S. M., 232, 271

Schapery, R.A., 289, 291, 309, 326, 339
Seitz, F., 30
Sendeckyj, G. P., 223, 272, 309, 310
Shtrikman, S., 108, 151, 319, 339
Smith, J. C., 58, 71
Sokolnikoff, I. S., 2, 3, 4, 7, 30
Spencer, A. J. M., 193, 196, 199, 223, 287, 301, 308, 309, 310
Stern, M., 253, 265, 272

Sternberg, E., 99, 105
Sun, C.-T., 232, 241, 242, 252, 271
Sve, C., 241, 243, 272

Tennyson, R. C., 205, 223
Tiersten, H. F., 253, 265, 272
Timoshenko, S. P., 167, 187
Toupin, R. A., 312, 338
Truesdell, C., 312, 338
Tsai, S. W., 150, 157, 187, 200, 223
Turnbull, D., 30

van der Pol, C., 54, 58, 71

Van Santen, J. H., 337, 339

Waals, F. M., 125, 151
Walpole, L. J., 108, 149, 151
Walters, W., 310
Whitney, J. M., 137, 151
Whittier, J. S., 252, 272
Wong, H.-W., 270, 273
Wu, E. M., 174, 178, 184, 187, 200, 205, 212, 214, 223, 224

Zakay, V. F., 309
Zerna, W., 5, 30
Zweben, C., 207, 224

SUBJECT INDEX

Admissible displacement field, 8
Agglomeration, effect of, 70, 148
Aspect ratio, 99
  finite, 89, 104
Attenuation:
  apparent, 258
  factor, 265
  viscoelastic, 18

Balance of momentum, 3
Bauschinger effect, 22, 202
Biharmonic stress function, 189
Bond failure, 223
Boundary layer, 190, 308
  method, 199
Bounds, 147, 290
  on Poisson's ratios, 79
  Reuss and Voigt, 108, 117, 148, 331
Bulk modulus, 7, 119
  effective, 47, 48
  plane strain, 76, 83

Chopped fibers, 89, 152
Clasius-Duhem inequality, 312
Comparison material, 109
Compatibility:
  conditions, 3
  equation, 169
Complex compliance, 15
Complex modulus, 13, 14, 15, 288
Compliance tensor, 9
Composite cylinders model, 81, 84, 147
Composite spheres model, 47, 53, 59, 67, 70, 147
Constitutive assumptions in plasticity, 19
Continuity equation, 10
Continuum hypothesis, 33
Contracted notation, 5, 153, 190
Correspondence principle, *see* Elastic-viscoelastic correspondence principle
Crack growth, 207, 294
Creep functions, 13
  for fluid, 15
Creeping flow, 10
Cubical packing, 61, 71, 148
Cutoff frequency, 241

Debonding stress, 338
Deviatoric components, 7
Dielectric behavior, 150
Dilatational components, 7
Dispersion, 227, 264
  curves:
    acoustic modes, 241
    branches, 241
  relations, 232, 235, 271
Displacement vector, 10
Drucker's postulate, 22

Effective homogeneity, *see* Equivalent homogeneity
Effective modulus, 32

Effective stiffness property, 32
Effective viscosity, 16, 47
Einstein formula, 47, 69
Elasticity:
  anisotropic, 189
  linear, 2
Elastic moduli, 3
Elastic-perfectly plastic behavior, 21
Elastic-plastic behavior, 2
Elastic-plastic interface, 278
Elastic-viscoelastic correspondence principle, 17, 288
Electromagnetic behavior, 150
Ellipsoidal inclusions, prolate and oblate, 74
Entropy, 312
  production inequality, 312
Envelope function, 247, 257
Environmental effects, 207, 337
  moisture, 208
    diffusion, 338
Equivalent homogeneity, 33, 34
Eshelby's formula, 40, 44, 55, 57, 87, 109

Failure criteria, 199, 215
  generalized Mises, 200, 202
  maximum strain, 206
  maximum stress, 205
  Mises and Tresca, 200
  tensor polynomial, 200
Fatigue, 186
Fiber lines, infinitely stiff, 189
Fibers, 302
  finite length, 104
  perfectly rigid (inextensible), 104, 301
  randomly oriented, 124
  transversely isotropic, 105
Filament winding, 152
First ply failure, 207
Floquet theory, 233, 254
Flow rule, in plasticity, 23, 285, 287
Fluidized bed, stability of, 266
Flywheels, 213
Fourier law of heat conduction, 314
Fourier synthesis, 245, 251, 257
Frequency spectrum, 241

Gibbs free energy, 315, 326

Halpin-Tsai equation, 150
Hardening rule, 22. *See also* Work hardening
Head of the pulse method, 251
Heat conduction equation, 314, 316
Heat diffusion, 337
Heat flux vector, 312
Heat supply function, 312
Helmholtz decomposition, 236
Helmholtz free energy, 312, 326
Heterogeneous media, types of, 35
Hexagonal packing, 71, 150
Homogeneity, 33
Hybrid (fiber) systems, 213

Impedance, 231
Inclusion, spherical, cylindrical, lamellar, 36, 73
Initial value problem, 248
Interaction energy, 27
Internal energy, 312
Interpenetrating network, 36, 137
Invariants, 158, 218, 281
  of deviatoric stress tensor, 21
Isotropy, 6, 7
  single lamina, 165

Kinematics, nonlinear, 2, 301
Kirchhoff-Love hypothesis, 161

Lamé constants, 7, 36
Lamina, 152, 153, 189
Laminates, 152
  angle ply, 166, 186
  asymmetric, 166
  cross-ply, 166, 186
  symmetric, 166
Loading:
  in plasticity, 20
  surface, 22
Loading function, 19
Loss angle, 18
Loss tangent, 18, 294
Lubrication approximation, 63

Macroscopic homogeneity, *see* Equivalent homogeneity
Material derivative, 10
Memory, viscoelastic, 11

Micro-mechanic, 146
Mises criterion, 21, 275, 281, 282
Mixture theory, 262, 265
Moisture effects, *see* Environmental effects

Navier-Stokes equations, 10, 300
Neutral loading, in plasticity, 20
Newtonian viscous fluid, 10, 16, 46, 296
Nonlinear behavior, in plasticity, 18
Non-Newtonian effects, 69, 70, 298
Normal lines, 302
Normal stress effects, in nonlinear fluids, 69, 300
Notation, direct, 11
n point correlation functions, 337

Orthotropy, 6, 153, 190
  cylindrical, 214
  single lamina, 166

Periodic media:
  layered, 225, 228
  three-dimensional, 253
Phase angle, viscoelastic, 15
Phase inversion, 145, 150
Phase velocity, 227
Plane strain conditions, 167, 189, 190
Plane stress conditions, 129, 155, 189, 190, 214
Plasticity theory, inviscid, 18, 20, 23
Platelet(s), 73
  randomly oriented, 121, 137
  system, 100
Poisson's ratio(s), 7, 76, 83, 155
  bounds on, 79
Polarization stress, 109
Polycrystalline aggregate, 59
Porosity, 276
Prager's rule, in work hardening, 22
Prandtl-Reuss relations, 21
Pressure vessels, 208
  cylindrical, 210
  spherical, 212

Quasi-isotropic properties, 160
  in pressure vessel construction, 212

Rate of deformation tensor, 16, 312
Rate type theory, in plasticity, 18
Rayleigh scattering, 265
Relaxation functions, 11, 13
Representative volume element, 34, 35
Rule of mixtures, 58, 206

Scattered waves, 258
Self-consistent scheme, 59, 147, 337
Serrat-Frenet formulas, 303
Shear deformation theory, 174, 186
Shear lag analysis, 99
Shear modulus, 7, 76
  effective, 41, 51
  in fiber direction, 83
  transverse, 84, 122
Shear thinning, 70
Shrinkage effects, 337
Specially orthotropic, single lamina, 166
Specific heat(s), 325
  at constant deformation, 314, 336
  at constant stress, 316
  effective, 333
Stability parameter, in fluidized beds, 270
Statistical homogeneity, *see* Equivalent homogeneity
Stiffness tensor, 3
Stop bands, 241
Strain, 3
  average, 34
  concentration factor, 223
  infinitesimal, 3
  plastic component of, 19
Strain energy, 4
Strength, 149, 186, 199
Stress, 3
  average, 34
  concentration, 223
  interlaminar, 180
Stress-strain relations:
  isotropic, 7
  linear elastic, 3
  linear viscoelastic, 11
Suspension:
  dilute, 37, 41, 47, 69
  fluid, 265, 295
Swelling effect, 337

Temperature, 312

Tensor transformation law, 156
for strains, 125
for stresses, 127
Theorem of minimum complementary energy, 8, 9, 108, 116
Theorem of minimum potential energy, 8, 107, 109
Thermal behavior, 150
Thermal conductivity:
effective, 317, 320
tensor, 314
Thermal expansion:
effective, 321, 325
tensor, 314
Thermodynamics, 4, 298
Thermoelastic:
complementary energy, 326
potential energy, 326
Three phase model, 52, 60, 85
Toughness, 208, 292
Transient problem (dynamic), 231, 243
Transverse isotropy, 6, 75, 118, 154, 191
Tresca criterion, 21, 275

Uniaxial modulus, 7, 76, 81, 122
transverse, 122
Unloading, in plasticity, 20

Variational theorem, potential energy, 175
Velocity vector, 10
Viscoelastic behavior, 2
Viscosity, zero shear rate effective, 69
Viscous fluid, 9

Warpage, due to bending, 172, 174
Wave length:
long, 226
zero and infinite, 270
Wave propagation, viscoelastic, 17, 295
Wave(s):
dilatational, 238
longitudinal, 227, 238
number, 232
plane harmonic, 226
reflections, 230
shear, 238
solenoidal, 238
speeds, 228
transverse, 227, 271
Wave speed, viscoelastic, 18
Work hardening, 21, 22, 281
isotropic, 22
kinematic, 22, 23, 284
parameter, 287

Yield function, 19

Zeigler's rule, in work hardening, 22, 23